PHYSICS

ARISTOTLE (384–322 BC), is one of the two greatest philosophers of antiquity—the other being Plato—and some would rank him as the greatest philosopher of all time. He wrote extensively on a wide variety of subjects, including logic, metaphysics, ethics, and politics. But over one third of all his writings fall under the general heading 'the Natural Sciences', and treat of such subjects as astronomy, fundamental physics, chemistry, zoology, and psychology. This present work, called the *Physics*, is intended as the first of all these writings on natural science. It begins by introducing us to the study of nature, and it goes on to deal with some general topics common to all the sciences. It has been studied with close interest in every century of its existence.

ROBIN WATERFIELD was born in 1952. After graduating from Manchester University, he went on to research ancient Greek philosophy at King's College, Cambridge. He has been a university lecturer (at Newcastle upon Tyne and St Andrews), and an editor and publisher. Currently, however, he is a self-employed consultant editor and writer, whose books range from philosophy to children's fiction. He has translated, in particular, a number of Plutarch's essays, Xenophon's Socratic works, Herodotus' *Histories* and several dialogues by Plato, including *Republic*.

DAVID BOSTOCK is a Fellow and Tutor in Philosophy, Merton College, Oxford. His other publications with Oxford University Press are *Logic and Arithmetic I* (1974), *Logic and Arithmetic II* (1979), *Plato's Phaedo* (1986), *Plato's Theaetetus* (1988), *Aristotle's Metaphysics, Books Z and H* (1994), and *Intermediate Logic* (Clarendon Press, 1997).

OXFORD WORLD'S CLASSICS

For over 100 years Oxford World's Classics have brought readers closer to the world's great literature. Now with over 700 titles—from the 4,000-year-old myths of Mesopotamia to the twentieth century's greatest novels—the series makes available lesser-known as well as celebrated writing.

The pocket-sized hardbacks of the early years contained introductions by Virginia Woolf, T. S. Eliot, Graham Greene, and other literary figures which enriched the experience of reading. Today the series is recognized for its fine scholarship and reliability in texts that span world literature, drama and poetry, religion, philosophy and politics. Each edition includes perceptive commentary and essential background information to meet the changing needs of readers.

OXFORD WORLD'S CLASSICS

ARISTOTLE

Physics

Translated by
ROBIN WATERFIELD

With an Introduction and Notes by
DAVID BOSTOCK

OXFORD
UNIVERSITY PRESS

OXFORD

UNIVERSITY PRESS

Great Clarendon Street, Oxford OX2 6DP

Oxford University Press is a department of the University of Oxford.
It furthers the University's objective of excellence in research, scholarship,
and education by publishing worldwide in

Oxford New York

Athens Auckland Bangkok Bogotá Buenos Aires Calcutta
Cape Town Chennai Dar es Salaam Delhi Florence Hong Kong Istanbul
Karachi Kuala Lumpur Madrid Melbourne Mexico City Mumbai
Nairobi Paris São Paulo Shanghai Singapore Taipei Tokyo Toronto Warsaw

with associated companies in Berlin Ibadan

Oxford is a registered trade mark of Oxford University Press
in the UK and in certain other countries

Published in the United States
by Oxford University Press Inc., New York

Translation and Textual Notes © Robin Waterfield 1996
Introduction, Bibliography, Explanatory Notes © David Bostock 1996

The moral rights of the author have been asserted

Database right Oxford University Press (maker)

First published as a World's Classics paperback 1996
Reissued as an Oxford World's Classics paperback 1999
Reissued 2008

British Library Cataloguing in Publication Data

Data available

Library of Congress Cataloging in Publication Data

Aristotle.
[Physics. English]
Physics / Aristotle; translated by Robin Waterfield; with an
introduction and notes by David Bostock.
1. Aristotle—Physics—English.   2. Physics—Early works to
1800.   I. Waterfield, Robin, 1952– .   II. Bostock, David.
III. Title.   IV. Series
Q151.A7413   1996   500—dc20   95–4560

ISBN 978-0-19-954028-0

7

Printed in Great Britain by
Clays Ltd, St Ives plc

CONTENTS

INTRODUCTION

General

Aristotle was born in 384 BC in Stagira, which is a small town in northern Greece, roughly 40 miles east of the modern Thessaloniki. At the age of 18 he went to Athens, to be educated at Plato's Academy. But he stayed in the Academy for nineteen years, soon becoming a teacher and independent researcher, and did not leave until Plato's death in 347 BC. For the next twelve years he lived in various places around the northern part of the Aegean Sea, and it was during this period that he was for three years tutor to the boy who would become Alexander the Great of Macedon. In 335 BC he returned to Athens, and founded his own rival to Plato's Academy, the Lyceum, where he taught for another twelve years. But on the death of Alexander the Great in 323 BC political events forced him to leave Athens, and he withdrew to Chalcis in Euboea. At that time he was 61 years old, and he died only a year later.

In the early part of his life he published several works, which have now been lost (apart from a few fragments). The writings that have survived (which are very extensive[1]) were not published by him. They are lectures which he would use in teaching. We do not know their order of composition, but it seems probable that some parts of the *Physics* are relatively early, and belong to his first period in Athens. However, other parts may have been added later, as his thoughts developed and his lecture-course expanded. We also do not know to what extent the present arrangement of his works is due to Aristotle himself, and to what extent it was imposed by later editors. The usual view is that the division into books[2] and chapters is the work of these editors. I would add that Aristotle's original will not have

---

[1] Together they occupy nearly 2,500 pages in *The Complete Works of Aristotle: The Revised Oxford Translation*, ed. J. Barnes, 2 vols. (Princeton and Oxford: Princeton University Press, 1984).

[2] In ancient writings, a 'book' is all on one papyrus roll. So a long work is always divided into several books.

contained any punctuation or paragraphing, and that the headings in this translation are supplied by me, and are no part of the original.

When reading any work of Aristotle's one should bear in mind that he did not himself prepare it for publication, and it may well be that he never thought of publishing it at all.

1. *The Scope and Nature of the* Physics

The word '*Physics*', which forms the title to this work, is a transliteration of Aristotle's Greek, and not a translation. The proper translation would be '*On Nature*'. Over one-third of all Aristotle's many writings fall under this general heading 'On Nature'. Some of these are indeed devoted to topics that we would count as physics, or at least as falling under the more general title 'the physical sciences'. But the majority are in fact devoted to biology, where Aristotle made a very significant contribution. Our present work, called *Physics*, was apparently designed as Aristotle's first course on nature, to serve as an introduction to all the rest.

The work is in eight books, and it is really only the first two that constitute a general introduction to the study of nature. In Book I Aristotle discusses what he calls the 'principles' of natural objects, and this soon becomes an analysis of the principles of change—not only 'natural' change, but all change whatever. In Book II he begins by making clear why this was relevant: in his view nature is to be understood as a source of change, but change of a special kind. The rest of the book concerns the various kinds of explanation that the study of nature should aim to provide. Especially in this book, but to some extent in the first as well, it is important to remember that Aristotle is introducing us not just to what we call physics, but to natural science in general, including biology.

The remainder of the work is mostly devoted to a series of topics which do fall into the category we call physics, but very theoretical physics. Thus in Books III and IV there are important discussions of infinity, place, and time, which are fundamental notions of physical theory. Similarly Books V and VI are devoted to change and continuity. The following Book VII is something of a miscellany (and might reasonably be omitted on a first reading), but Book VIII is then devoted to a theme which we would hardly classify as physics.[3] It aims to argue that the changes we can see in the universe demand a single, eternal, and unchanging 'first cause' of all change to explain them. This is God.

In the third part of this introduction I give a detailed account of the content of the various books. This is best read in conjunction with the books themselves. The rest of this part contains some general comments on the nature and value of the work as a whole.

All of Aristotle's work is amazingly original, and the *Physics* is no exception. While he does of course owe something to his predecessors (as the next section will show), it is fair to say that no one before him had attempted a thorough and systematic treatment of topics so central to physics as infinity, continuity, place, time, and motion. Again, no one had seen, as he does, that biology is impossible without the assumption that an animal's parts serve some purpose, and that the same applies to its natural behaviour. Consequently no one had paid serious attention to classifying the various different kinds of explanation that a scientist should seek for. Again, though others had indeed argued for a 'first cause' of all nature, the argument that Aristotle constructs is all his own.

Not only is he a very original thinker, but he is also a very powerful thinker. In the first place, his thought is careful, well organized, and systematic; it is very seldom that he simply overlooks a worthwhile possibility. In the second place, he does not just assert; he argues. In fact he loves argument, and the pages of the *Physics* are everywhere stuffed with arguments. Moreover, the general standard of these arguments is clearly good. Of course, there are lapses from time to time, where to our eyes a mistake seems rather clear. But many of his arguments are very persuasive, and even when we are convinced, from our modern perspective, that he must be wrong somewhere, still it is often not easy to pinpoint the error. In the third place, he can be very inventive in seeking solutions to a problem. A specially good

---

[3] It is quite possible that Aristotle himself did not regard Book VIII as belonging to the same course of lectures. This is strongly suggested by the way it refers back to earlier books at 251^a9, 257^a34, 263^a11–12, and 267^b21.

example of this is the answer to Zeno's best-known paradox on motion that he gives in Chapter 8 of Book VIII. As I shall explain (pp. lx–lxi), this is a really ingenious suggestion. But there are many other good suggestions, all through the *Physics*.

There are other places where Aristotle's views are still worth very serious consideration, even today. I mention in particular his discussion of chance in Chapters 4–5 of Book II, and his claim that time depends upon change in Chapter 11 of Book IV. In some other ways too he can strike one as being surprisingly up to date, for example in his firm grasp of the fact that points of space, or instants of time, cannot be next to one another. But in fact he was writing well over 2,000 years ago, and it is hardly surprising that this quite often shows. For instance, he naturally accepts the common belief of his day, that the earth is at the centre of things and that everything else goes round it. Several of his arguments depend upon this background assumption, and here we can only say that his presuppositions are not the same as ours. A more interesting example of such a discrepancy concerns the laws of motion. We are all brought up to believe in Newton's laws, and to suppose that force is needed to initiate a movement but not to sustain it. Aristotle, however, takes it to be perfectly obvious that force is also needed to keep a movement going, so that if the force is withdrawn, the movement will stop. (He is thinking primarily of moving things by pushing and pulling.) It is fascinating to see how this assumption affects his thinking on a number of topics, and frequently leads him into error. We must learn to put aside our own point of view in order to appreciate how, for him, it is deeply mysterious that a stone thrown upwards should continue to move even after it has left the hand.

Although the *Physics* is Aristotle's prologue to natural science, he is writing here as a philosopher, and not as we would expect a scientist to write. This is not because he had no experience, and no conception, of what we would call science. On the contrary, in his biological works he argues exactly as we now expect of a scientist: he has amassed a great deal of observational data about all kinds of animals, and he is trying to systematize it and explain it. But we do not find that sort of thing in the *Physics*. Except where he touches on biology in Book II, he very seldom cites the results of observation or experiment, and when he does, the alleged 'result' is quite likely to be mistaken. (Usually one suspects that this is not because he has done the experiment, and it has somehow gone wrong, but because it was only a thought-experiment anyway.) The explanation, of course, is that he did not see how observation and experiment could be brought to bear upon the very general questions that he is raising in this work, namely questions such as 'What is change?' or 'What is place?' or 'What is time?' Nor did any of his predecessors, nor most of his successors (for many centuries), and so it became a tradition that such topics as matter, space, time, and motion were suited only to philosophical discussion. It is fair to say that Aristotle's *Physics* initiates that tradition, and that its contribution remained a dominant influence until in the seventeenth century Galileo and others founded modern physics.

Some Background

2. *Aristotle's Predecessors*

For some three hundred years before Aristotle was writing there had been a succession of Greek thinkers interested in nature. A *very* potted history would go as follows.

The story begins at Miletus, on the Aegean coast of what is now Turkey, with Thales (*fl. c.*600 BC), Anaximander (*fl. c.*575 BC), and Anaximenes (*fl. c.*550 BC). Thales claimed that everything began from water, and (perhaps) that everything now is made of water; Anaximander claimed the same for a stuff that he called 'the unlimited',[4] and Anaximenes made both claims for air. Some years later at Ephesus, which is about 25 miles north of Miletus, Heraclitus (*fl. c.*500 BC) proposed fire as the basic ingredient of the universe, underlying all else. (Heraclitus is perhaps better known for his 'flux' doctrine, that everything is changing all the time.) This group of thinkers is known as 'the Ionians', since they lived in Ionia, and as 'the monists', since each espoused the claims of one single element as the basic one.

---

[4] By this he probably meant 'the indefinite', i.e. a stuff that had no definite characteristics.

In opposition to all such speculations there then comes the group of thinkers known as 'the Eleatics', after their home in Elea in southern Italy. First, in a very influential work, Parmenides (*fl. c.*475 BC) argued that change is impossible, and more generally that *every* kind of diversity is impossible, and hence that reality must be single and changeless. His follower Zeno (*fl. c.*450 BC) aimed to support this view indirectly, arguing that the assumption of plurality gives rise to contradictions, and so too does the assumption of change. Finally Melissus (*fl. c.*440 BC)—who came not from Elea but from Samos in the Aegean Sea—offered a somewhat different version of Parmenides' argument. His conclusions are not quite the same, since he holds that reality is infinite whereas Parmenides holds that it is finite, but in other respects they are in agreement. In particular, both firmly deny the existence of change.

After these come 'the pluralists', who are alike in positing more than one basic entity, and in attempting to combat Parmenides' argument by insisting that each basic entity is in itself changeless, though their combinations may alter. But in other ways they are very different from one another.[5] Thus Empedocles (*c.*493–433 BC) gave equal prominence to four different stuffs of which the world is made, namely fire, air, water, and earth; he supposed that the world was eternally oscillating between a state in which each of the four elements is entirely separated from all the others, and a state in which all four are combined to form a wholly homogeneous mixture. (Our present state is one in which the elements are mostly separated, but not completely.) Anaxagoras (*c.*500–428 BC) claimed that all the infinitely many different kinds of stuff are equally basic, and that each of them contains a portion of each of the others, so that there is 'a portion of everything in everything'. He posited a beginning of the universe, when the rotation started, and the various different elements, and their opposite properties, began to be separated out from an original mixture. Finally we come to the atomists, whose founder Leucippus is a rather dim figure about whom little is known, since he was clearly overshadowed

[5] The pluralists have no common geographical base. Empedocles was born in Acragas (modern Agrigento) in Sicily; Anaxagoras in Clazomenae (near modern İzmir) in Turkey; Democritus in Abdera on the coast of northern Greece.

by Democritus (*c*.460–370 BC). Their claim is that the world is made of many indivisible atoms, moving in an infinite and otherwise empty space. (Atoms and space they called, respectively, 'the full' and 'the void', and even 'being' and 'not-being'.) These thinkers also held views about the basic causes of change: Empedocles proposed two basic forces, Love and Hatred, responsible for combining and for separating respectively; Anaxagoras instead invoked a single Intelligence, which started the process of separation; and Democritus spoke more abstractly of 'Necessity' as the cause.[6]

Aristotle frequently criticizes these predecessors by name, paying more attention to Empedocles and Anaxagoras than to any others. But in fact his most significant rivals are the atomists, and he argues against them both in his discussion of place and void in Book IV and in his discussion of continuity in Book VI. Yet curiously, in those discussions he names Leucippus and Democritus only once (at 213^a34).

Contemporary with Democritus was Socrates (469–399 BC), who, however, was not much concerned with natural science, and who is in fact never mentioned in the *Physics*. Socrates, of course, influenced Plato (428–347 BC), who in his turn influenced Aristotle, but mainly in metaphysics. In the *Physics* we see very little of Plato's influence, because Plato too spent little time on natural science. The one dialogue of his that engages in scientific theory is the *Timaeus*, and Aristotle had clearly studied this with care. In the *Physics* he cites its view of time in order to register his disagreement with it (251^b17–19), and he similarly complains that it assigns to space the function actually fulfilled by matter (209^b11–13, 209^b33–210^a2), but the *Timaeus* is hardly a positive influence. One might do better to cite Plato's discussion of causes in the *Phaedo* (95e–102a), which surely did influence Aristotle's discussion of this topic in Book II. But here again it is important to remember that Plato's aim is to dismiss

---

[6] In this very brief history I have omitted figures who are not mentioned in the *Physics*, e.g. Pythagoras and Xenophanes. For a useful and simple introduction to the period, see E. Hussey, *The Presocratics* (London: Duckworth, 1972). A more thorough treatment may be found in G. S. Kirk, J. E. Raven, and M. Schofield, *The Presocratic Philosophers*, 2nd edn. (Cambridge: Cambridge University Press, 1983). For some lively discussion see also J. Barnes, *The Presocratic Philosophers*, 2nd edn. (London: Routledge & Kegan Paul, 1982).

empirical enquiry, whereas Aristotle is recommending it. On the topic of natural science the two do take completely opposed attitudes.[7]

3. Aristotle's Logic and Metaphysics

In his short work the *Categories* Aristotle claims that all things that exist are either complex or simple, and the simple ones can be classified into ten categories. The first and most basic category includes particular individuals, such as a particular man, or horse, or tree. These he calls 'primary substances'. But the same category also includes what is predicated of these individuals when we say *what* they are, by giving their species (e.g. 'a man', 'a horse') or their genus (e.g. 'an animal'). So in this category of substance we have both the primary substances and certain special universals, called 'secondary substances', that are predicated of them. The remaining categories are intended to include all the other (simple) universals, and Aristotle claims that all of them are predicated of primary substances. In the *Categories* he lists them as: quantities, qualities, relations, places, times, positions, states, actions, and affections (i.e. cases of being acted on). Elsewhere we find the list abbreviated to eight categories, by omitting 'positions' and 'states', presumably on the ground that they are already included in the other categories. More commonly, the list is cited in an open-ended form, such as 'substance, quality, quantity, and the rest'. In any case, it is not very important to decide just how Aristotle's list should be composed. The important claim is the principle that underlies the list, namely the claim that every (simple) thing that exists either is a primary substance or is predicated of primary substances.

Aristotle never argues for this claim. It is presented in the *Categories* as if it simply states what is obvious, and it appears in many other places in his writings as something already established. In particular, it appears in this way at several places in the *Physics*, as we shall see. But it certainly is not true, as the *Physics* also reveals.

[7] It is often said that Aristotle's *Physics* has benefited from the puzzles elaborated in the second part of Plato's *Parmenides*, but I am somewhat sceptical of this view.

4. *Aristotle's Cosmology*

Aristotle's picture of the universe is that as a whole it is finite and spherical. It is made in concentric layers of five different elements, four of which Aristotle accepted from the tradition—namely earth, water, air, and fire—and a fifth which he added himself. The earth is at the centre, and around it is an (incomplete) layer of water, then a layer of air, and then a layer of fire. Outside this there is a succession of spherical shells, made of the fifth element (called *aithēr*), and rotating in various ways round the centre. Their role is to carry the heavenly bodies. Roughly, the idea is that the outermost shell carries the fixed stars, and this rotates on a north–south axis once every 24 hours. Inside it are other shells, rotating on different axes and with different velocities, carrying the planets, then the sun, and finally the moon. It is the nature of *aithēr*, of which these shells are made, to move in circles round the centre, and the fifth element is not capable of any other change than this. But inside the rotating shells, in the 'sublunary' sphere below the moon, there are all kinds of change. What happens here is of more relevance to the arguments of the *Physics*.

Whereas the celestial spheres move round the centre, Aristotle takes it to be an obvious fact of observation that earth, when left unsupported, will of its own nature move downwards, that is, towards the centre. Complementary to this, fire will of its own nature move upwards, that is, away from the centre. We may add that water will move downwards through air, though not through earth, and that air will move upwards through water (but not through fire, as the theory demands). As a result, there are 'natural places' for each of the four sublunary elements to be, and there are equally 'natural motions' for each, namely the motions that bring each to its 'natural place'. But one way in which these four sublunary elements differ from the fifth is that they can also be moved, by force, against their nature.

Another difference, which Aristotle accepted from the tradition, is that these four so-called elements can change into one another. (No doubt this relies upon the 'observation' that water, when heated, will turn into air, and that air, when cooled, will condense into water droplets.) To account for these transformations

Aristotle supposed that all four elements are made of a common matter, and that this matter is earth when it is cold and dry, water when it is cold and wet, air when it is hot and wet, and fire when it is hot and dry. This theory, however, is not often required for the arguments of the *Physics*, whereas the theory of natural places is crucial.

The Argument of the Physics

5. The Principles of Change: Matter, Form, and Privation

Book I opens by proposing to enquire into the 'principles' of the subject, and it begins this project by considering some opinions of earlier thinkers. In broad outline, the upshot of this is that in Chapters 2–3 Aristotle rejects the view of Parmenides (and Melissus) that reality is single and changeless, and in Chapter 4 he goes on to reject the view of Anaxagoras that there are infinitely many principles. He does not criticize other thinkers but instead looks for what is common to them all, and what he finds is that all of them agree in making use of opposites. This is a fair generalization, though one should note here an oddity in his method of argument. The views cited from earlier thinkers seem to identify the 'principles' in question with the ultimate constituents of things. Thus on one kind of view everything is made of the same basic stuff (water, or air, or fire), and on another there are four equally basic stuffs (adding earth to the three just noted). Aristotle describes these as the view that there is just one principle, and the view that there are four. Now it is probably fair to say that thinkers who held these views *also* made use of opposites (condensation and rarefaction in the one case, love and hatred in the other) but they did not suppose that these opposites were themselves 'principles' in the sense of basic constituents.[8] Aristotle, however, at once infers that opposites must be 'principles'. What does he mean by this?

The situation becomes clearer in Chapter 5, where Aristotle offers his own argument to show that the opposites must indeed

---

[8] But Anaxagoras perhaps did, and this may explain Aristotle's wording at 187^{a}20–3.

be accepted as 'principles'. For his argument is that all *change* is 'between opposites', so the claim that he is making seems to be that opposites are, as we may say, *principles of change*. Unfortunately, this claim is mistaken (and it is a mistake that Aristotle often repeats elsewhere). What is characteristic of a traditional pair of opposites, such as 'dense and rare' or 'hot and cold' is that they represent a scale, an ordering from more to less, allowing of intermediate degrees. But it is not true that all changes take place along such a scale, and in fact Aristotle's own examples in Chapter 5 reveal this perfectly well. Often the property acquired in a change does not lie on any appropriate scale. For example, there is no organization of bricks which is 'opposite' to their being organized into a house, and there is no shape for bronze that is 'opposite' to its shape when it is made into a statue (188^b16-21). Even where there is an appropriate scale, still it need not be relevant. For example, we may count white and black as opposites, with the various shades of grey in between, but if a thing comes to be white it need not do so from some point on this scale. Before being white it may have been red, or it may have been colourless (as glass is).[9] All that is required is that, before being white, it was *not* white, and there are all kinds of ways of being not white (similarly: of being not a house, or not a statue).

When he restates his position in Chapter 7, Aristotle remedies this error, for there he no longer speaks of opposites but just of a form and its 'privation'. If the form is the property acquired in a change, then Aristotle will say that whatever does *not* have the form *does* have the corresponding privation, so long as it is at least the right *kind* of thing to have that form. And this proviso must be satisfied if later the thing does come to have the form. So it is perfectly correct to say that change is between a form and its privation, though it need not be between opposites in the traditional sense.

So far, then, the point is just this: every change will involve the acquisition of some property, so that the initial state is one where the property is not present, but its privation is, and the

---

[9] Aristotle apparently thinks that *all* colours lie on a scale from pale to dark (188^b23-5). One wonders whether he was wholly colour-blind. But in any case he should recognize that colourless things are not on this scale.

final state is one where the property is present. In Chapter 6 Aristotle proceeds to argue that a further 'principle' must be recognized, for there must be what he calls an 'underlying thing', characterized first by the privation and then by the form.[10] So his final account may be represented in this way: every change has an initial state and a final state, and to describe these we shall need three distinct concepts—the concept of an underlying thing, present in both states; the concept of a form, present only in the final state; and the concept of its privation, present only in the initial state. A change is what happens when the underlying thing first has the privation and later has the form.

Aristotle's main argument for this account is in Chapter 7. First he takes the simple example of a person becoming educated, and shows in some detail how the analysis applies to this case. That is straightforward. Then he argues that *all* changes conform to the same pattern ($190^a 31-^b 10$). Given his own background assumption, that all existent things fall under his list of categories, we have two kinds of case to consider. The first is where the form acquired is a property in some category other than substance, and such cases will all be analogous to the example already discussed: the underlying and persisting thing will be a substance, which at first lacks and later has the form in question. But the second kind of case is where the form acquired is a property in the category of substance—for example, the property of being an oak tree—and where the change is that there comes to be an oak tree where there was none before. Considering this kind of case, Aristotle claims that even here we do not ever get something *from nothing*, for there is always a starting-point for the change. For example, living things come from seed, houses come from bricks and timber, vinegar comes from wine. But he does not show us how, in such changes, there is always a *persisting* thing which at first lacks and then has the form acquired.

In the case of houses, it is simple enough: the bricks and timber persist all through, first lacking and later having the form of a house. In the case of vinegar, it is clear what Aristotle

[10] One wishes to add: *or* characterized first by the form and then by the privation. For the most part, Aristotle does not add this, but presumably he would not object to it.

would say: we have the same stuff, the same matter, all through, first in the form of wine and later in the form of vinegar. In the case of an oak tree coming from an acorn, perhaps the best thing to say is that there are two distinguishable changes involved. In the first, the lifeless acorn germinates, to become a living thing; here the same matter which at first constituted an acorn comes to constitute something else, a growing shoot. (No doubt water needs to be added too.) In the second, the shoot becomes a tree, and here it is actually the same tree that persists all through— first as an 'infant' and then as a 'mature adult'—but of course it is no longer the same matter. It is clear that here I am filling in things that Aristotle does not himself say. He leaves it to us to work out how—and whether—his general account of change applies to these and other cases. It is a real question whether it can be applied to all cases without exception.

I add as a final comment on Chapter 7 that its ending should be noted. Here Aristotle says, in his summing-up, 'whether the form or the underlying thing is substance is not yet clear' (191^a19–20). Now where a change involves only the acquisition of a property not in the category of substance, it is perfectly clear that Aristotle must hold that the underlying thing is a substance and the form is not. The difficulty, then, arises where the form acquired is in the category of substance, and what acquires it will be some kind of matter (e.g. some bronze, in the case of a bronze statue). Elsewhere, Aristotle usually *restricts* the notion of form to just such a case, that is, to what are called 'substantial forms', and instead of speaking of 'what underlies' a change he soon speaks simply of matter (e.g. in Chapter 9), since matter is what underlies in changes of this sort. So the question that he is raising here is whether matter should itself be counted as a kind of substance, a substance that 'underlies' the various things made of it, which are also substances. Scholars do not agree on what his eventual answer to this question either is or should be, but we cannot pursue that problem here. (See note on 191^a20.)

In the remaining chapters of Book I, Aristotle first shows how his account of change resolves problems felt by earlier thinkers. The key notion here is that what comes to be so-and-so comes from what is not so-and-so, but not from what is nothing at all. Then he goes on to claim that there is a sense in which matter itself cannot either come to be or cease to be. His argument here is sound only if what he is speaking of is the 'ultimate' matter, of which his four elements are made. The matter of such a thing as a statue (e.g. bronze), may perfectly well come to be, but that is because it is made of something further (namely copper and tin combined). Similarly, on Aristotle's view, even the so-called elements—earth, water, air, and fire—may come to be and cease to be, as one turns into another. But the 'ultimate' matter is not made of anything further, or in other words there is nothing that underlies it, and for that very reason Aristotle's analysis of change does imply that it cannot come to be or cease to be. He is right to say that on his account there must be conservation of matter.

I return, briefly, to the question of what a 'principle' is supposed to be. Aristotle claims that matter, form, and privation are the three 'principles' of the study of nature, and what he has in mind is that all natural objects are subject to change, and that all change whatever is to be analysed in terms of these three. But he infers from this that whatever is subject to change is itself a compound, consisting of such-and-such a form in such-and-such matter ($190^{b}10$–11, 17–23), so he has not altogether lost sight of the view that the 'principles' of things are also the 'ingredients' of things. The chief distinction that he has introduced is that one such 'ingredient' will underlie, as matter does, while the other will characterize what underlies, as form does. This, one might say, is a 'logical' (or 'metaphysical') distinction, of which earlier thinkers were innocent. (But perhaps it should be granted that Plato is an exception.[11])

6. *What Nature Is*

Our subject is the study of nature, but in fact Book I merely discussed change, and said nothing about what nature is. Book II aims to fill this gap.

It opens with the thought that a natural object, as opposed to one made by man, has within itself its own source of change. I do not imagine that Aristotle would have been quite so happy

---

[11] In the *Timaeus* Plato proposes an analysis of things as constituted by *space* (as underlying thing?) taking on this or that form. His later 'unwritten doctrines' perhaps assigned a similar role (as underlying thing) to 'the great and small'. At any rate, that is how Aristotle interprets him at $192^{a}6$–14.

with this definition if he had been as familiar as we are with all kinds of automatic machines that contain within themselves a programme which guides their changes. No doubt he could claim, as at 192^{b}28–9, that these do not have in themselves 'the source of their production', but the point hardly appears relevant. For it seems equally true that a natural object—an animal, a plant, an element such as earth—does not have in itself the source of *its own* production. (No doubt many natural objects can reproduce their kind, and we do not yet have any machines that do quite this, but one feels that the day may not be too far distant.) Setting this suggestion aside, then, we are here faced with what is actually a large and interesting question: is there really, in principle, any distinction between natural objects and others? But I must leave that question unexplored.

Still, one may say, it is at least true that natural objects do contain within themselves a source of their typical behaviour, even if perhaps some artefacts may do so too. But even this requires some qualification. As Aristotle himself recognizes elsewhere (Book VIII, 253^{a}7–21 and 259^{b}7–16), one may perfectly well say that natural objects exhibit change only in reaction to some external stimulus. So perhaps it would be better to say that what explains their behaviour is *partly* the internal source that Aristotle is here thinking of, but partly also the external situation which stimulates that inner source. I think it reasonable, however, to say that we need not fuss too much about this point, and we may accept Aristotle's position that it would be absurd to try to *prove* the existence of these internal sources.

Aristotle next asks whether such an internal source should be taken to be the thing's matter or its form. (As he explains at 193^{a}30–1, he means 'the form which enables us to define what an object is', i.e. its 'substantial form'.) His answer is that both contribute to the internal source, and perhaps we should understand him in this sort of way. If one asks why a tree floats on water, the answer is that this is because of its matter: its matter is mainly wood, and all wood floats on water. But if one asks why a tree has roots, the answer is that this is because of its form, that is, because of what it is to be a tree: for (*a*) trees stand upright, and are not easily blown over, and (*b*) they are living things, and so need nourishment, which they get from the soil.

For both these purposes, a tree must have roots. But here one should notice that I am proposing these reasons on Aristotle's behalf; the reasons that he gives himself (at 193^{a}9–28 for matter, and at 193^{a}28–^{b}18 for form) seem just to side-step the issue, since they are not easily seen as citing any examples of a thing's behaviour being caused either by its matter or by its form. In any case, I think that the main question that should be raised at this point is a different one: why should one assume that *all* of a thing's natural behaviour must be due *either* to its matter *or* to its form? It is clear that Aristotle here gives no justification for this assumption. When one goes on to read his more detailed works on nature, and particularly the biological works, it is also clear that he does not abandon the assumption. As a result, the notion of a thing's form becomes stretched beyond recognition, as it is asked to explain so much, namely everything that is not explained by its matter. We shall meet an example later in this book. (See pp. xxviii–xxx.)

To conclude, Aristotle thinks that natural objects are distinguished by having within themselves a source of their own behaviour, and he calls this source the 'nature' of the object. This is the only kind of 'nature' that he believes in—that is, the nature of this or that particular object. (For example, he does not believe that nature can be thought of as a kind of impersonal force, arranging the smooth running of the world as a whole.) He claims that an object's matter, and its form, both contribute to its nature; and apparently he will not allow that anything else contributes to it. So we should not be surprised that in Chapter 2 he goes on to say that the natural scientist must study both the matter and the form of things, and he does *not* suggest anything else that the natural scientist should study. For example he does not say that the scientist should study the general laws of motion, or of heat, or of light, or anything of that kind. His interest is focused entirely on the *objects* that exist by nature.

7. *Kinds of Explanation*

Chapters 3–9 of Book II are all concerned, in one way or another, with the kinds of explanation that are required in the study of nature. In Chapter 3 Aristotle introduces his doctrine that there are four basic kinds of 'cause'; in Chapters 4–6 he breaks off to argue that chance is not a fifth kind of cause; in Chapter 7 he resumes his account of the four, and this leads into the main claim that he argues in Chapters 8 and 9, namely that teleological explanations have a fundamental role to play in natural science.

I begin with a few remarks on the treatment of chance in Chapters 4–6.

(i) *Chance*. Aristotle focuses on this problem. On the one hand when we say that something happened 'by chance' we appear to be denying that it had a cause, or perhaps citing chance itself as a cause. (Aristotle mostly considers the second of these suggestions.) On the other hand we do say that things happened 'by chance' even when we know perfectly well how they were caused, and do not include chance itself as one of their causes. Aristotle's example is this. Suppose that I owe you money, and that you happen to meet me in the city square, just as I am recovering a debt from someone else, and so have the funds to repay you. Then it was by chance that you met me at a time when I could hardly avoid repaying you. But this is not to say that there was no explanation for our meeting. You, no doubt, were pursuing a perfectly good plan of your own that brought you to that place at that time, and so was I. It is just that your plans did not include meeting me, and mine did not include meeting you.

Aristotle's diagnosis is, first, that we speak of chance only when we are concerned with events that do serve a purpose but which are not explained by the purpose that they serve. Thus your meeting me at that point did serve a purpose—it enabled you to collect what was due to you—and if you had known, you might deliberately have set out to meet me there, just for that purpose. But as things were you did not act for that reason, and nor did I. Second, Aristotle claims that the meeting must not be explained as a result of what happens always or usually. His thought is that if you standardly go to that place at that time, and so do I, then no one would say that we met 'by chance'. But I think he has not drawn quite the right moral from this thought. For whatever your reason was for going to that place at that time, we may surely suppose that, *whenever* you have such a reason, you act on it. Similarly in my case too. So it may well be that there are regularities which explain our meeting, but the point is that these regularities would not allow either of us to *predict* the meeting. It seems to be an important part of what we mean by a 'chance' outcome that it is not predictable. Aristotle's third point is that chance may be reckoned to be, in a sense, a cause, but only a 'coincidental cause'. This, however, seems to be a mistake on his part. On his account, *A* is a 'coincidental cause' of *B* if there is something *C* which is in its own right the cause of *B*, and by a coincidence *A* and *C* are the same thing. But his example of a chance meeting illustrates not this idea but the converse idea of a 'coincidental effect'. For by hypothesis there is a straightforward cause of your being at that place at that time, and the coincidence is that being at that place at that time is *also* being at a place where I am when I am in funds.

In Chapter 6 Aristotle distinguishes between chance outcomes, which are outcomes that might have been chosen, but were not, and spontaneous outcomes, which include also outcomes that might have been the purposes of nature, but were not.[12] This, of course, raises the question of how we are to understand these supposed 'purposes of nature', but I shall come to that later, when considering Aristotle's teleology. The more important question to raise about this whole discussion of chance and spontaneity is whether it is sufficiently comprehensive. There certainly are cases where we say both that something happened 'by chance' and that there is a perfectly good explanation of why it happened. I think myself that the important feature of these cases is that the explanation did not allow us to predict what happened, and that it does not really matter whether what happened did serve a purpose. But that is a point of detail. It is more significant that there *also* appear to be cases when we say that something happened 'by chance' and mean by this that there is *no* explanation for what happened. This might have been the attitude of some who said that the rotation of the world began spontaneously (196ª24–8). To update the example, it *is* the attitude of those who say the same thing of what we believe to be

---

[12] The discussion of spontaneity is, however, not very satisfactory. See note on 197ᵇ33.

the 'big bang' with which the universe began. But Aristotle never considers whether there can be chance events in this sense.

(ii) *The Four Causes.* Aristotle's four 'causes' are traditionally called the material cause, the formal cause, the efficient cause (which Aristotle himself describes as 'what started it'), and the final cause (i.e. the purpose or goal). But since it is only the efficient cause that is at all close to our way of thinking about causes, it is better to say that his theory is that there are four basic kinds of *explanation*. This is fair, for Aristotle himself insists that a 'cause' is always the answer to a question 'Why?', and that seems a good description of what an explanation is (194^b17–20, 198^a14–20). It is also well suited to some of his examples, where we are explicitly given a why-question and its answer, as in: 'Why is he walking? To get healthy' (194^b32–5), and in: 'Why did they go to war? Because they had been raided' (which gives the efficient cause), or 'To gain control' (which gives the final cause) (198^a19–20). But the puzzling feature of Aristotle's discussion is that, for the most part, he *fails* to specify any why-questions to which his 'causes' provide the answers.

This is especially notable with his examples of so-called 'material causes'. These are (194^b25, 195^a16–19):

The bronze is the material cause of a statue.
The silver is the material cause of a bowl.
The letters are the material cause of a syllable.
The elements are the material cause of a material body.
The parts are the material cause of the whole.
The premisses are the material cause of the conclusion.

The last of these is quite unexpected, and is probably best set on one side as an aberration,[13] but in all other cases there is no suggestion of a why-question to which the stated cause provides an answer; rather, the cause answers the different question 'What is it made of?' Now presumably Aristotle does think that some why-questions are answered by citing a thing's matter, for as we have seen he does allow that there is some truth in the suggestion that

---

[13] At 94^a20–4 of the *Posterior Analytics* we find a list of four causes in which the way that premisses 'cause' a conclusion is given *instead of* the material cause. That may help to explain why it is listed here as one kind of material cause. But the truth is that in *every* kind of explanation the explanans may be regarded as providing premisses from which the explanandum follows as a conclusion.

the matter of a thing is its 'nature', in the sense of its 'inner cause' of change or stability, and a cause of change must surely tell us *why* the change occurred. But he gave no helpful examples then, and he gives none now. So we are still rather at a loss to say just what is supposed to be explained by a thing's matter.

Just as the material cause seems to be introduced here not as answering 'Why?' but as the response to a different question, so too the formal cause is presented as answering the different question: 'What is it?'[14] So to ask for the formal cause is, as he says, to ask for a definition (194^b26-9, 198^a16-18). But again, we know that a thing's form is the major part of its 'nature', so Aristotle must hold that the account of what a thing is will also be an important part of the explanation of why it changes as it does. We begin to get some indication of how this might be when we find Aristotle claiming that a thing's form and its purpose are the same (198^a24-7).[15] At first this is just mystifying, for we would not know what to count as the 'purpose' of such a thing as a man, or a horse, or a tree. But we can obtain some illumination if we take into account Aristotle's views on teleology in nature.

(iii) *Teleology.* In Chapter 8 Aristotle argues that we must recognize that in nature things occur because they serve a purpose. The opposing view he first explains as the view that whatever happens in nature does so 'of necessity', and that it is merely an accident when what happens of necessity also turns out to be useful (198^b10-29). This theory one might fairly credit to Democritus. But then he also adds a brief reference to the evolutionary theory introduced by Empedocles, who had supposed that at first all kinds of combinations of living parts sprang up, brought together haphazardly by the operation of Love, but thereafter only the useful ones survived (198^b29-32). This is evidently a rather different theory.

The first objection that he raises, at $198^b34-199^a8$, has no force

---

[14] At 90^a14-34 of the *Posterior Analytics* Aristotle explicitly argues that the question 'What is it?' *is* a why-question, giving the example that 'What is an eclipse?' is the same question as 'Why is the moon eclipsed?' Compare also *Metaphysics* Z, 1041^a32-^b9.

[15] This was also the thought behind the earlier, but very obscure, argument at 193^b12-18.

against Democritus. He argues that in nature we find regularities, that what is regular cannot be due to chance, and that
what is not due to chance must be for some purpose.[16] But to
this Democritus will certainly reply that on his account the
regularities in nature are due neither to chance nor to purpose
but to necessity. Empedocles, however, needs to maintain that a
regularity can be initiated by chance but then perpetuated in
some different way. This is a more difficult position.

For the remainder of the chapter (from 199ª8) Aristotle mainly
develops an analogy between the working of nature and the
working of human design. Put succinctly, his claim is that almost all of the features of animals and plants are *as if* they were
designed to serve a purpose, namely the purpose of enabling that
animal or plant to live.[17] This is a perfectly fair claim, and it is
enough to destroy Democritus' theory. For, if everything were
due to a necessity which pays no attention to what is useful and
what is not, one would expect animals to show a roughly equal
balance of useful and useless features. But that is certainly not
what we actually find. Aristotle also pays further attention to
Empedocles' theory, and seeks to impale him on the horns of a
dilemma: either animals and plants breed true, or they do not.
Aristotle insists that the correct answer is that they do (199ᵇ13–
26), but in that case Empedocles cannot suppose that there ever
was a stage of haphazard combinations brought together by
Love; for such combinations could only come from the seed of
parents of a like kind. If, on the other hand, Empedocles does
insist that animals were first created in this way, then he is
without any explanation of why they have bred true thereafter.
Certainly, Empedocles failed to appreciate the peculiar combination of chance and necessity that provides the mechanism for
our evolutionary theory, and it is not surprising that Aristotle
does not see it either.

Aristotle is aware, then, of the first attempt at an evolutionary

[16] Presumably what Aristotle means to say is that a regularity which *seems* to
serve a purpose, if it is not due to chance, must *actually* serve a purpose. But even
when thus corrected his thought is not particularly convincing.

[17] From our evolutionary perspective, the 'purpose served' is not so much the
survival of the individual as the survival of the species, or—what comes to the
same thing—the successful *reproduction* of the individual.

explanation of why the parts of animals, and their instinctive behaviour, are so well adapted to their environment. But he does not see how it could be made to work. Apparently he never seriously considers Plato's theory (in the *Timaeus*) that things appear to be designed because they *were* designed, by some divine artificer. I guess that his main reason for rejecting such a suggestion was that he thought that it would have to involve a time at which the world was created—or anyway, a time at which it was organized into its present form[18]—and he believed this to be impossible. (We shall see his reasons in Book VIII.) In any case, he rejects both of these explanations, and what he says instead is that the parts of animals, and their instinctive behaviour, are to be explained in the same way as human artefacts are explained: each serves a purpose, and that is why it exists. But in nature's case the purpose in question is not the *conscious* purpose of anything, and so from our point of view this seems a very doubtful metaphor. If we seek to cash the metaphor, then all that Aristotle tells us is this: human purposes are always connected with goodness, namely with what seems good to the human agent in question, and a 'natural purpose' is equally connected with goodness, namely with what is good for the animal or plant concerned. If we dispense with the metaphor, then, what we can say by way of explanation is just this: the animal has such-and-such a feature (e.g. sharp front teeth) just because that feature is *good for it*. The goodness of something is in this way an ultimate and basic explanation of why it exists.

It must be admitted that Aristotle seldom does dispense with his metaphor. Indeed, when he is putting his principles into practice in the biological works, he standardly speaks of two kinds of cause: first 'Necessity' supplies matter of a distinctive kind—for example tough, earthy matter in the region of the head; and then what appears to be a personified 'Nature' fashions this matter into whatever is best for the animal in question—for example horns, or tusks, or a thick, hairy mane, or whatever it may be.[19] In his practice, then, Aristotle does admit 'laws of matter' taking this sort of form: when such-and-such

---

[18] Cf. iii, 203ª28–33, on Anaxagoras.
[19] Plato's *Timaeus* similarly contrasts 'Necessity' and 'Intelligence'.

matter is present, in such-and-such conditions, then such-and-such a further kind of matter will inevitably be produced. And in practice he pictures Nature as a beneficent craftsman, fashioning the given matter into suitable shapes and sizes. When one thinks of the state of chemistry in his day—and, indeed, for very many centuries afterwards—it is hardly surprising that he did not see how there could be 'laws of matter' which would determine such things as the shape of the teeth, or the thickness of the hair. But he is aware that his picture of Nature as a craftsman is only a metaphor, and he supposes that what he is really talking of is the 'nature' of the animal or plant concerned. As we have seen, he has assumed that this must be either its matter or its form, so in the present case it must be the form, which he has already identified both with the definition of the thing and with its final cause—that is, what it is for.

His theory, then, is this. The various parts of an animal are each for something, namely their contribution to the efficient functioning of the body as a whole. And the body too is for something, namely for living the kind of life appropriate to the animal in question (*De Partibus Animalium* i, 645^b15–20). It is this life, then, that is the ultimate goal, and other features of the animal are to be explained by showing how they contribute to this. So it is this life too that is the animal's form, and this that one specifies when defining what it is to be that kind of animal. So it turns out that what is metaphorically explained by invoking a benevolent Nature is, on Aristotle's account, more soberly explained simply as required by a definition, a definition that specifies a kind of life.

This is a mistake, for the explanation in terms of goodness achieves something which an explanation based simply on a definition cannot. To continue with Aristotle's own example (198^b23–7), we can appreciate that the way human teeth are arranged is more useful to us than other conceivable arrangements would be, and we gain at least something by being told that this is not an accident, for the teeth come up in that arrangement just *because* it is the most useful one. But we gain nothing if we are told that they come up in that arrangement because that is what is specified in the definition of 'human being'. Besides, a specification of the kind of life that humans lead will not actually entail anything about the arrangement of their teeth, and it is surely stretching the notion of a definition beyond any reasonable limit to suppose that it should include a specification of *every* feature of an animal that is useful to it. Yet Aristotle does actually accept this consequence (*De Generatione Animalium* v, 778^{b}12–14). Clearly the cause of the trouble is that the notion of form is being asked to do too much.

(iv) *Necessity*. In the light of what I have just been saying about Aristotle's practice in biology, you might expect him to say that there are *two* kinds of necessity to be found in natural phenomena. On the one hand there are what one might call the necessary laws of matter. One example would be (as at 198^{b}19–20) that when air rises it cools, becomes water, and so falls as rain; another would be (as I was suggesting above) that in such-and-such conditions earthy matter is necessarily produced; another would be (as indicated at 200^{a}11–13) that only iron will make the strong kind of teeth that a saw requires. On the other hand there is also what Aristotle here calls 'conditional necessity', which requires merely that *if* such-and-such an end is to be realized *then* such-and-such a kind of matter must be present. It appears that his view should be that the natural scientist will need to invoke both of these kinds of necessity.

It is possible that this is indeed the view that he means to be propounding in Chapter 9, but on the face of it what he says is that we find only *one* kind of necessity in natural phenomena, namely 'conditional necessity'. At the very beginning of the chapter he puts his question in this way: 'Is necessity present in nature conditionally, or *also* unconditionally?' This evidently suggests that the two alternative positions that he wishes to consider are (*a*) that *only* conditional necessity is present, and (*b*) that both kinds are present. But if these are indeed the alternatives, then it seems clear that what he is arguing for is (*a*). For at 200^{a}5–15 he is certainly claiming that the necessity there being considered is only a conditional necessity: *if* there is to be a wall *then* there must be stones for the foundation, then earth (i.e. mud-bricks), and then timber; *if* there is to be a saw *then* there must be iron for its teeth. Similarly at 200^{a}15–30 he surely is not saying that whereas there is only one kind of necessity in mathematics there are two different kinds in natural phenomena.

On the contrary, he appears to be saying that in each case there is only one, and they are the reverse of one another.

In that case, his position in this chapter must be either that there are no necessary laws of matter, or that there are such laws but they are necessitated by something more fundamental, namely the needs of the living plants and animals. Scholars dispute whether this is the right interpretation of the chapter, and whether—if so—the position that Aristotle seeks to occupy is coherent.[20]

8. The Definition of Change

Near the beginning of the *Physics* Aristotle says that the natural scientist does not need to *prove* the existence of change, for that is the foundation of his whole subject ($185^{a}12$–20). I think he might usefully have added that there is equally no need of a *definition* of change, for essentially the same reason: change is one of the basic and fundamental concepts in natural science, and cannot be defined in terms of anything more fundamental.[21] It is useful to offer an analysis of the basic factors involved in change, as Book I does; and it is useful too to determine how many different kinds of change there are, and to raise other general questions about change, as Book V does. But Book III opens with an attempt at a general *definition* of change, and this was a mistake, for Aristotle's proposed definition achieves nothing.

The definition is:

Change is the actuality of that which exists potentially, in so far as it is potentially this actuality. For example, the actuality of a thing's capacity for alteration, in so far as it is a capacity for alteration, is alteration. ($201^{a}10$–12)

It is a perfectly general truth that the actuality of the potentiality for X is X. So certainly the actuality of the potentiality for alteration is alteration, and in general the actuality of the potentiality for change is change. But this tells us nothing about what change

---

[20] A useful parallel passage is *De Partibus Animalium* i. 2–3. (Cf. also *De Generatione et Corruptione* ii. 11.)

[21] We might say that change may usefully be defined in terms of time. But Aristotle wishes to define time in terms of change.

is, since it is also true that the actuality of the potentiality for rest is rest, the actuality of the potentiality for being a statue is being a statue, and so on (201^a29-34). It appears that Aristotle hopes to narrow down the potentialities that his definition applies to, so that only potentialities for change are included, by insisting that he means 'the actuality of what is potential, *as potential*' (201^b4-5). But it seems to me that this is merely playing with words.

Even if we try to look more sympathetically at what Aristotle must have had in mind, still I think we have to conclude that the basic idea is mistaken. For the thought appears to be this. This lump of bronze, for example, is not now a statue, but it has the potentiality of being a statue (which just means that it could be a statue). Suppose, then, that this potentiality is realized, so that at some future time it is a statue. Then there is bound to be a process which takes it from its present state of not being a statue to its future state of being one, that is, a process by which what it was potentially becomes what it is actually. And the thought is that while this process is going on, the potentiality, instead of being 'latent', is 'active' or 'alive', so that we can define a change as what happens when a potentiality is 'alive'. But still this does not work, for the same reason as before. For the potentiality that is 'alive' during this process is the potentiality for *becoming* a statue, that is, a certain potentiality *for change*. By contrast, the potentiality for *being* a statue is 'alive' only when the change is completed, and so no longer exists. Thus the only relevant potentialities are the potentialities for change, but it is evidently circular to define 'change' in terms of 'potentiality for change'.

Towards the end of Chapter 2 Aristotle makes some further claims about changes. In particular, he implies (i) that *every* change is due to an agent acting on a patient, (ii) that *always* the agent acts by contact, and (iii) that *always* the agent brings with it a form which it imparts to the patient (202^a3-12). In Chapter 3 he goes on to argue (iv) that the change is *always* located in the patient and not the agent. In all cases the claim appears to be exaggerated, but I merely observe here that (i) will be argued for in Chapter 4 of Book VIII, (ii) will be argued for in Chapter 2 of Book VII, and (iii) is perhaps qualified at 257^b11-12 in Book VIII.

9. *Infinity*

In the second part of Book III Aristotle discusses infinity. In Chapter 4 he summarizes some of his predecessors' views on the topic, and in Chapter 5 he begins his own account with the claim that infinity cannot be construed as a substance in its own right. If it exists at all, it can only be as an attribute of something else (204^a8–34). This claim may surely be granted.[22] The bulk of Chapter 5 (from 204^b1) is then devoted to showing that there cannot be any material body that is infinitely large. This is an important part of Aristotle's theory of the universe, for he holds that the universe as a whole is finite, and so of course any material body within it must be finite too. But his arguments here are disappointing.[23]

His first point is, as he says himself, made 'at an abstract level' (204^b4–7), and it is clear that no opponent need be much disturbed by it. Passing on to arguments 'more in keeping with natural science', at 202^b10–205^a7 he considers first the suggestion that an infinite body might be composed of several elements, such as his own elements earth, air, fire, and water, and then the suggestion that it might be simple and uncompounded. Against the first he argues (*a*) that if just one of these elements was infinite it would 'destroy' the others, and (*b*) that there could not be more than one that was infinite, since even one infinite element would leave no room for anything else. Clearly one need not agree with either of these claims. Against the second he argues that we cannot suppose that there is some one infinite stuff, more basic than the familiar four elements and underlying them all, because if there were we should perceive it, which we do not. It is evident that this sort of argument will cut no ice against an opponent such as Democritus, who believes in an infinity of imperceptible atoms spread through an infinite space.[24] The remaining arguments (from 205^a8) all rely in one

[22] Aristotle says that it was denied by the Pythagoreans and by Plato, but this seems to be a misunderstanding on his part. See note on 203^a4.

[23] Aristotle also argues for the finitude of the universe elsewhere, most notably at *De Caelo* i. 5–9. See also n. 25.

[24] I remark also that Aristotle should have seen how this argument would endanger his *own* conception of the common matter of which the four elements are made. He thinks of this as being 'imperceptible' (*De Generatione et Corruptione* 332^a27, $^b1$).

way or another on Aristotle's own theory of how things have a 'natural place' in the universe, and a 'natural motion' to that place. But here an opponent such as Democritus will simply deny the theory that Aristotle is presupposing. Apparently Aristotle fails to see that this theory is itself one of the points at issue.[25]

However, the main interest of Aristotle's discussion of infinity is not his denial of the infinitely large—a denial which happens to be endorsed by modern physics—but the more positive proposals that he goes on to make in Chapter 6. Here he is trying to accommodate the very natural views (*a*) that time is infinite, (*b*) that magnitudes are infinitely divisible, and (*c*) that number is infinite (206ª9–12). His opening remarks are confusing, for he appears to equate being infinitely large with being actually infinite, and being infinitely divisible with being potentially infinite, and to be saying that there is only a potential infinity and never an actual infinity (206ª12–18). This is scarcely comprehensible, since it seems obvious that x is potentially ϕ only if it is possible that x should be actually ϕ, and yet Aristotle seems to be denying this for 'ϕ' as 'infinite' (206ª18–21). But the truth is that he is not denying it, and the apparent equations in his opening remarks are not what he intends. His real doctrine begins to emerge as soon as he begins to talk about the sense of 'to be' in which the infinite can be said to 'be' (206ª21 ff.).

His leading idea is that there are infinite *processes*, but that a process is the only sort of thing that is, or could be, infinite. As he believes, time is an example of an infinite process; it will continue without end. In just the same sense, the human race can be said to be infinite, for it too will continue without end (or so Aristotle believes, 206ª25–6). These are processes that are 'actually' infinite, which is just to say that they really are endless. By contrast, the continued division of a magnitude is only 'potentially' infinite, because although it is theoretically possible for such a division to go on endlessly, this never happens in fact. Aristotle does not mean by this to point to some practical problem in continuing the division beyond a certain stage (for example,

---

[25] He does more to argue for his position in his discussion of 'the void' in Chs. 6–9 of Book IV.

when the magnitude to be divided has become too small to be seen with the naked eye). On the contrary, he pays no attention to such practical problems, and he insists that however far a division has gone it could always be taken further. So I think he would have no objection to the suggestion that there *might* be an 'actually infinite' division of, say, a length. This would be an infinite *process* of dividing the length, for example a process which made one further division in each further year, for ever and ever. But what he wishes to insist upon is that there cannot be any time at which such an infinite process is *completed*, that is, a time at which infinitely many divisions have been made. For, if there were such a time, it would be a time at which there existed an infinite totality of different objects (namely the different parts of the line, which have been distinguished from one another by the division), and his claim is that there cannot be any infinite totality in this sense. On the contrary, the *only* kind of infinity there can be is an infinite *process*.

Aristotle does not explain why he denies the existence of infinite totalities. For example, one might very well hold that the parts of a finite line do constitute an infinite totality, whether or not there has been any process of dividing the line into those parts. A possible suggestion is that he rejects such totalities because of what one might call their 'absurd arithmetic'. For example, if we take the positive integers as an infinite totality, then we can divide them into two parts—for example, the even integers and the odd integers—which *each* form another infinite totality, similar to the first. Thus infinity plus infinity equals infinity, and infinity minus infinity *may* equal infinity again, or it *may* equal any finite number you desire, including zero. There are a few small hints that Aristotle saw such consequences as these, and found them objectionable (e.g. at 204^{a}25, 204^{b}19), but they do not seem to me to give much probability to this interpretation. My own view is that Aristotle was led to deny the existence of infinite totalities by his reflection on Zeno's paradoxes. But we shall come to that in section 11.

I end this section with a brief remark on how Aristotle's denial of the infinite affects mathematics. First, concerning the numbers, his position is that a number exists only if it is the number of some group of independently existing objects. Consequently, since he denies the existence of infinite totalities of objects, he also denies that the numbers themselves form an infinite totality. Instead, the numbers are to be construed as participating in an infinite process. Since the process of halving a given magnitude, and then halving what remains, and so on, is a potentially infinite process, the number of the halves resulting is, as one might say, a 'potentially infinite' number. But what this means is that for any number, however large, there *can* be a larger number; it is not always true that there actually *is* a larger number (207^b10–15). So apparently we are to think of the number of numbers as increasing over time, but it will always be a finite number. (How many exist today, one wonders.) What is nowadays called 'classical mathematics' certainly cannot accept such a view, since it is fully committed to infinite totalities, but perhaps the mathematicians of Aristotle's day would see less reason to object. At any rate, Aristotle does not anticipate any objection on this score, though he does anticipate that the geometers will object to his denial of the infinitely large. Here he is quite right to reply (207^b27–34) that geometry does not need to assume the existence of lines that are infinitely long, or anything of that sort. For of course there is such a thing as the geometry of a finite space, and Aristotle almost puts his finger on the crucial point: the geometry will be Euclidean provided that there are similar figures of different sizes.

10. *Place*

Aristotle discusses place in Chapters 1–5 of Book IV. Usually he begins such a discussion with a résumé of the views of his predecessors, but in this case he does not, telling us instead that his predecessors had nothing helpful to say on the topic (208^a34–$^b1$). This perhaps helps to explain why his own discussion fails to come properly to grips with what seems to us to be the main issue concerning places, namely how to identify the same place over time. It is not that Aristotle ignores the question altogether, but that he has not appreciated the problem that it creates for what he does say.

His discussion begins very appropriately, with the observation that we must accept that there is such a thing as place because we see that different objects may succeed one another in the same place (208ᵇ1–8). As we soon find, this observation gives rise to the question on which all his discussion is concentrated, namely this: 'Every object is in some place. What, then, is the place that the object is in?' Perhaps this was not the best question to choose; it leads him to pay little attention to the suggestion that there might be places occupied by no objects.

In Chapter 1 he develops some further prima-facie arguments for saying that there is such a thing as place (208ᵇ8–209ᵃ2), and then some problems for this idea (209ᵃ2–30). Most of Chapter 2 is then devoted first to putting forward, and next to rebutting, the two suggestions that the place of a thing is either its form or its matter. The main objection to these suggestions is straightforward, namely that a thing may change its place without changing either its form or its matter. Chapter 3 then begins by distinguishing various senses in which one thing may be '*in*' another (210ᵃ14–24). It goes on to argue that, when we are speaking strictly, nothing can ever be said to be 'in itself' (210ᵃ25–ᵇ2), and finally it applies these points to a problem raised by Zeno (210ᵇ22–7, cf. 209ᵃ23–5). It is not until Chapter 4 that Aristotle sets himself to answer his own question.

He accepts from Plato[26] the common Greek idea that to be somewhere is to be *in* something, and hence that the place of an object *contains* it. But he adds that the (immediate) place of an object must in addition be the same size as it, and that it can be vacated by the object (210ᵇ34–211ᵃ3). As a further axiom he also adds that the account must leave room for the doctrine of natural places (210ᵃ3–6), but this plays no role in his explanation of what an object's place is, which is based on the first three axioms. For he says that they leave us only four candidates to consider, namely (i) the limits of the object (i.e. its shape, i.e. its form), (ii) what is between those limits (i.e. its matter), (iii) the limits of what contains it, and (iv) what is between those limits, construed now as an immaterial extension which remains the same whatever body may occupy it (211ᵇ5–9). But of these the first two have already been rejected for good reasons, and Aristotle now rejects the fourth as well, so that he ends by endorsing (iii): the place of an object is the limit of what contains it.

---

[26] See e.g. Plato, *Parmenides* 145d7–e6.

His argument against (iv) is condensed and obscure, but I think it is this. Imagine that we begin with an ordinary jug of water, so that on Aristotle's own theory the place of the water is the inside surface of the jug, and the surface of the air, where each is in contact with the water. Then (*a*) suppose that the water is poured or siphoned out of the jug. This means that different parts of the inner surface of the jug come to be in contact with the water, and the air–water boundary shifts also. So during this time the water that is left in the jug is coming to occupy all the time a smaller and smaller place, until eventually there is no water left in the jug at all. Now Aristotle appears to suppose that if we construe a place as an immaterial extension which remains the same while its contents are changed, then we must say that all of these infinitely many smaller and smaller extensions will continue to exist even when their boundaries have ceased to exist, since the water has now gone altogether. So the one jug will contain infinitely many of these places, a result which he evidently treats as objectionable ($211^b19–22$). He also supposes (*b*) that on this theory, when the jug itself is moved, the immaterial extension between its inner surfaces *both* persists as the same extension (i.e. the same place) *and* moves to another place. So on this account a place may change its place, which again he finds objectionable ($211^b22–5$). His own position, it appears, is that when the jug is moved as a whole, its contents maintain the *same* place throughout ($211^b25–6$; but he will modify this position later, at $212^a14–21$).

The first of these two arguments very strongly suggests that the theory it is attacking is a rival theory which has more to it than Aristotle himself tells us. In fact it reminds us of the view that a place is just a part of space, capable of being occupied by a body, but not needing to be so occupied in order to exist. We do not find it at all absurd that, on this view, there are always infinitely many different places contained within the one jug. However, we would not want to say that when the jug moves then these infinitely many places move with it, so that a place may change its place. One wonders whether the theory that Aristotle is criticizing did really take that view. Or is it, perhaps, that Aristotle is here bringing in his *own* view that when the jug moves then so does the place it contains, and *combining* this with the rival view that places are just immovable parts of an immovable space, so that whatever moves must change its place? Unfortunately one cannot be sure, since the theory that he is criticizing is not known to us from other sources. But I remark that it may perhaps have been the atomists' theory; at any rate, in Chapters 6–9 Aristotle constantly associates it with their theory of the void. (See pp. xlii–xliii.)

In any case, Aristotle does eventually come to recognize that his own theory needs to take account of the fact that a container may itself move. This leads him to introduce a qualification, and to say that a thing's place is the containing limit of its nearest *unmoving* container (212^a14–21). But it is clear that this will not do. According to his own illustration, if I am sitting in a boat which is floating down the river, then the nearest *unmoving* container is the river as a whole.[27] But the river is in contact only with the boat, and not with me, so the boundaries of the river cannot provide a place for me that is no larger than I am, contrary to one of Aristotle's axioms for places.

There are other internal incoherences in his account, notably concerning the doctrine of 'natural places', and concerning the movement of the fixed stars. On the first, Aristotle standardly says that the natural place for earth is 'at the centre' of the universe, and the natural place for fire is 'at the periphery' (sc. of the sublunary sphere). But it is impossible to see how 'at the centre' can be taken as designating what he calls a place, for 'the centre' is not the inner limit of any containing body. Similarly 'the periphery' is the inner limit of a body that contains *all* the sublunary world, and not just fire. Here Aristotle is in fact specifying places that are not containers. As for the fixed stars, he is committed to saying that the outermost shell of the universe, which carries these stars, both has no place (since nothing surrounds it) and rotates once every 24 hours. But rotation is surely some kind of change of place. He attempts to evade this objection by saying that the spherical shell does not change its place, since it has no place, but its *parts* do (212^a31–$^b2$, $^b8$–22). We may reply, however, (*a*) that an outermost part still is not *surrounded*

---

[27] We might prefer to say not 'the river as a whole' but rather the river-bed (or the banks). This makes the objection even more telling. But for the sake of argument I allow Aristotle his own description of the example.

by anything, (*b*) that while it has other parts as neighbours nevertheless there is no genuine boundary that separates it from its neighbours (212^{b}4–5), and anyway (*c*) these other parts are also moving, and so will not provide a place for it according to Aristotle's revised criterion. (On the original criterion, the neighbouring parts may perhaps be reckoned as providing a place for our initial part, but if so then it is always the same place, so again we still cannot explain how the part can be rotating.) The moral of these objections, and others that one may propose, is that, although indeed we do speak of an object as being '*in*' a place, still it is a mistake to think of a place as a container.

It is hardly surprising that Aristotle should have spent no time in the consideration of what has become *for us* the crucial question, namely: is a thing's place 'absolute', or merely 'relative' to other things? Since his belief is that the earth is, by its nature, at rest at the centre, he would evidently take the state of the earth to be one of 'absolute rest'. But then the logical position to take is that the rest or motion of other things is to be determined by their rest or motion relative to the earth, and hence that the place of a thing is equally to be determined by its place relative to the earth. So the fundamental notion here is the notion of how a thing may be spatially related to other things. *Sometimes* the relevant relation is one of containment, but often it is not. To put this in simple terms, the question 'Where is it?' *may* be answered by specifying something that contains it (e.g. 'In the box' or 'In the kitchen'), but equally it *may* be answered by specifying the thing's distance and/or direction from a given thing (e.g. 'It's behind you'). Aristotle fails to pay any attention to the second kind of answer, but from his own perspective on the general organization of the universe it must surely be this second kind of answer that is the more fundamental.

11. *Void*

In Chapters 6–9 of Book IV Aristotle argues that there is no 'void'—that is, empty space. He begins in Chapter 6 by setting out the existing arguments both for and against void. As he rightly comments, the arguments against are scarcely relevant to the point that is actually at issue, and as he goes on to show in Chapter 7, the arguments for are inconclusive. (If we begin from the idea that all matter exists in the form of solid particles, not divisible and not compressible, then indeed a void will be necessary to explain the phenomena of motion and compression. But if we suppose, as Aristotle does, that matter may also exist as a continuous fluid, and that it is itself capable of contracting and expanding, then there is no such necessity.) Then in Chapter 8 he sets himself to argue directly that there is no void.

His main target in this chapter is the doctrine of the atomists, that the universe as a whole consists of infinitely many atoms moving in an otherwise empty but infinite space. It is fair to say that on this picture of the universe there is no room for anything like Aristotle's own doctrine of natural places and natural motions (though this is more because the space is infinite than because it is mainly empty), and it is fair to say too that the theory he is attacking did not have much to offer by way of an alternative explanation of gravitational phenomena. The best way to make sense of Democritus' position would seem to be to suppose (*a*) that he assumed a modern view of inertial motion, that is, that an atom moving in any direction through the void needs no force to keep it going at the same speed in the same direction, and (*b*) that we may assume an initial state of the universe in which atoms are moving at random in all directions, and which then develops into the present state as a result of the chance collisions of these atoms, which may result in some of them being hooked onto others. Admittedly this does not actually explain why it is that earth falls downwards whereas hot air rises upwards, but Aristotle's objection goes back to an earlier stage. Since he does not believe in inertial motion (as he very clearly says at 215^a19–22), the first phase of his attack, at 214^b12–215^a24, is to press the objection that Democritus needs to assign a cause for any motion through the void, but cannot do so. (He overstates his objection, however, when he implies that on this theory the void itself was supposed to be the 'cause' of motion, 214^b13–17.)

The second phase of the attack, at 215^a24–216^a11, contains Aristotle's best-known argument against motion in a void. He correctly observes that the same object falls more slowly through a dense medium such as water, and more quickly through a rarer medium such as air. He also 'observes' that in the same medium, heavier objects fall faster than lighter ones. (Here he is no doubt generalizing from the behaviour in air of such things as a leaf and a stone.) He at once concludes that the three quantities in question satisfy the simplest mathematical relation that yields these results, namely (for suitably chosen units)

$$\text{Velocity} = \frac{\text{weight}}{\text{density of medium}}.$$

He then argues at length that, since a void has zero density, this implies that velocity in a void must be infinite, which he reasonably rejects as impossible. But then, to one's surprise, at 216^a11–21 he comments that the reason why heavier objects fall faster in the same medium is that they more easily overcome its resistance, and so he reaches what is actually the correct result, that in a void, where there is no resistance to be overcome, all objects will fall at the same speed. But he fails to see that this reasoning upsets the equation he has been relying on, and instead supposes that its result is itself absurd. It should surely have occurred to him that his equation might itself be oversimple. (The chapter concludes with two further arguments, of which the first seems to us very naïve (216^a26–$^b12$), and the second very obscure (216^b12–16), so I here pass over them.)

The arguments of Chapter 8 have concerned an otherwise empty space in which objects are supposed to be moving. Aristotle calls this a 'separated void' (214^b12, 216^a24, 216^b20), and he several times says that to suppose the existence of such a void is to suppose that places exist independently of the bodies occupying them, a theory that he has already refuted (213^a12–19, 214^a16–24, 216^a24–6, 216^b31–2). Now these two theories are certainly not equivalent, for one could perfectly well hold that places exist independently without supposing that there is any void, but Aristotle would seem to be right to say that one cannot both hold that the void exists and accept his own theory of place. This is because the place of a body might then be the boundary of a merely empty space that contains it, but (*a*) such a boundary seems to be too dependent upon the body enclosed for us to say that that *same* boundary may enclose now one body and now another (e.g. as air and water change places with one another), and (*b*) there is surely no sense in stipulating that the *void* that surrounds a body must be 'motionless' if it is to provide that body's place. For motion is ascribed not to the void but to the bodies in it. The theory of void does clash, therefore, with Aristotle's theory of place, but it is clear that it is the question of void that is the prior question.

In Chapter 9 Aristotle passes to the question of whether there can be a void within bodies, as the phenomenon of compression seems to indicate. He remarks that if this means that an apparently solid body may contain many pockets of 'separate' void, then this view has already been refuted, and so he instead considers the theory that there is in bodies a void which is 'not separate' (216^b30–4). It is not easy to see what this theory is supposed to be (and I do not imagine that anyone ever held it). I guess that the idea is that we may consider void as an *ingredient* in the mixture of stuffs that constitutes a body, but that this is a mixture that is homogeneously mixed, so that *every* part of the body, however small, contains *all* the ingredients (in the same proportion). Aristotle also characterizes this theory as the theory that there is void *potentially* (217^b20–1), perhaps on the ground that such a void would come to exist as an actual void only if the mixture were separated into its various ingredients. It hardly seems satisfactory to us to treat a void as just like a material ingredient in things, except that it is immaterial, and we may perhaps accept Aristotle's main counter-argument, that there is no need for such a theory. For in his rival view the same matter can itself both contract and expand, while all the time filling space uniformly, and this appears to get just the same effect in a more straightforward way. In any case, it was the 'separate' void that was the real issue.

12. Time

Of all the discussions in the *Physics*, the treatment of time in Chapters 10–14 of Book IV is the least well organized. It begins appropriately enough, as Chapter 10 first presents some problems about time, and then discusses some previous views on what time is. This leads naturally into Aristotle's own account in Chapter 11 of what time is. But then in Chapter 12 we find first a miscellaneous series of notes, apparently arising from the account just given (220^a27–$^b32$), and after that a lengthy treatment of what it is to be 'in time' (220^{b}32–222^{a}9). Chapter 13 is then concerned mainly with the definition of various temporal terms (such as 'now', 'recently', 'long ago'), but it twice digresses from this theme (222^{a}30–^{b}7, ^{b}16–27). Its final sentence then briefly summarizes Chapters 11 and 13, passing over Chapter 12, and apparently it concludes the discussion. But the following Chapter 14 begins with a further treatment of being in time (222^{b}30–223^{a}15), goes on to raise a new question about whether time would exist if no one were conscious of it (223^{a}16–29), and then continues with further treatments of several topics already covered (223^{a}29–224^{a}17). When considering this chapter one should bear in mind that it is quite possible that it contains some passages that Aristotle rejected from earlier versions when he put together our present Chapters 10–13, and not only his later reflections on the topic. Indeed one cannot be sure how much of Chapters 12–14 would have survived in its present form if Aristotle had ever written up a 'final version' of his thoughts on time. In what follows I shall pay little attention to these chapters, concentrating instead upon the main theory presented in Chapter 11. For this is certainly Aristotle's most interesting contribution to the topic.

Chapter 10 has concluded by pointing out that all those of Aristotle's predecessors who had something to say on what time is connected it very closely with change (218^{a}30–^{b}20). Aristotle has replied that time cannot be identified with any particular change (e.g. because a particular change is 'in' the thing changed, and nothing else, whereas time is everywhere at once), but the connection between time and change is the basis of his own account. In Chapter 11 he begins by claiming that time is 'not without change', offering as an argument that we notice that time has passed when and only when we notice that there has been a change. From this he infers that since time cannot be identified with change it must be 'some aspect' of change (218^{b}20–219^{a}10). I observe here that the premiss is mistaken, and the conclusion does not follow. The conclusion does not follow because we cannot rule out the possibility of time passing without out anyone being aware of it. (Admittedly, Aristotle himself does try to rule out this possibility at 223^{a}16–29.) The premiss is mistaken because we *also* notice that time has passed when we notice that *no* (perceptible) change has occurred, for example between two ticks of a clock. Time, therefore, is just as much 'an aspect' of rest as of change, as Aristotle himself recognizes at 221^{b}7–23. But he pays no attention to this point in his main discussion, where he attempts to deduce the characteristics of time from those of change, and not from what is common to both change and rest.

In fact Aristotle focuses upon a particular kind of change, namely a change from one place to another, and he argues that, since this change takes place over a continuous magnitude (i.e. a line), the change itself is also continuous, and so too is the time of the change. In addition, one point on this line is 'before' another and so equally one instantaneous state of the change is 'before' another—that is, the states of the moving objects being first at this point and then at that—and so again one 'now' of the time taken is 'before' another. Thus the distance corresponds to the change, and the change to the time. And as distance is what is between two points, and similarly change is what is between two instantaneous states, so time too is what is between two 'nows'. Here Aristotle fails to explain how 'a now' differs from what he has called 'the before and after in change', which I have been taking to be the instantaneous states through which the change passes. But I think that the explanation can easily be drawn from 218^{b}10–13: the instantaneous states of a given change are special to it, whereas a now belongs equally to all instantaneous states that are simultaneous with one another. Given this explanation of a now, we may then understand a time simply as what is between two nows (219^{a}10–30).

The picture so far is reasonably straightforward, but Aristotle at once introduces a more puzzling suggestion when he says that time is 'a number' of change, and in particular a number 'in respect of before and after' (219^{b}1–2). It is natural to suppose that in this context Aristotle must mean by 'a number' something like what we would call 'a quantity', and that the phrase 'in respect of before and after' is intended to pick out *which* quantity, that is, it is intended to pick out the time of the change, rather than the distance covered, or the work done, or any other quantifiable aspect of the change. Whether the phrase can do this without introducing some circularity may be debated. More seriously, it can also be debated (*a*) how Aristotle conceives this quantity (i.e. how he would understand the expressions 'the same time' and 'the same quantity') and (*b*) whether he does really mean us to understand this odd use of the phrase 'a number' in *any* quantitative way at all. Concerning (*b*), one should notice that he *also* calls the now 'a number' (219^b22–8, 220^a1–4, 220^a21–4), but it is quite unclear how a single now could be any sort of quantitative measure, and from 220^a21–4 it appears that all that Aristotle means to suggest by this terminology is the simple point that a now is a universal, shared by all instantaneous states that are simultaneous with one another. So it *may* be that when he says that a time is 'a number', again he means to suggest no more than that a time is in a similar way a universal. But I confess that I do not think this very likely, for at several places Aristotle emphasizes the idea that time is a *measure* of change.

Concerning (*a*), one needs to notice first that when Aristotle speaks of 'a time' he always means a stretch of time, and never a date. So he is thinking of stretches of time when he says that time is measured by motion, and in particular by the revolution of the sphere of the fixed stars, since this alone is a change that is never-failing and always uniform (220^b14–24, 223^b12–224^a2). Thus one revolution of the fixed stars, namely one (sidereal) day, provides a unit of time, by reference to which other times can be assigned numbers, that is, as fractions or multiples of one day, and these same numbers may then be used to measure the times of other changes. But this does not determine what is to count as 'the same time'. For example, one may say that one change took 'the same time' as another, because each took one whole day, though a different day in each case; or one may insist that 'the same time' implies that it is the same day in each case. For the most part Aristotle seems to be thinking in the second way, but he is not entirely consistent. (For example, 219^b5–12 and 12–14 apparently contradict one another on this point.)

As these last remarks indicate, the details of Aristotle's account are often left obscure, and sometimes inconsistent. But the main thrust is clear. Just as he had argued that place has no existence apart from bodies, so here too he wishes to claim that time has no existence apart from changes, and that changes in turn are always changes of bodies. We are not likely to feel much sympathy with his claim that 'empty space' is an impossibility, since on this topic we have all been brought up to be familiar with Newtonian ways of thinking, but many philosophers *are* sympathetic to the view that there cannot be such a thing as 'empty time'.

13. *The Varieties of Change*

In Book V Aristotle returns to the topic of change, and first he tries to say how many kinds of change there are. After some preliminary distinctions (224^a21-^b35), his classification begins in Chapter 1 with the claim that there are three basic kinds of change: (*a*) when the subject of the change comes into being (generation), (*b*) when the subject ceases to be (destruction), and (*c*) when the subject continues throughout but some attribute is varied (variation) ($224^b35-225^b5$). One may observe that there appears to be some tension between the doctrine of this passage and the doctrine of Book I, Chapter 7, which claims that in *all* changes, including generations and destructions, there is a subject which persists throughout. But we can offer this reconciliation: in Book I Aristotle wishes to stress that when a statue is generated there is *something*, namely the bronze, which exists beforehand; but here he wishes to stress that *the statue* does not exist beforehand (225^a20-32). That is why its generation cannot be regarded as a variation in it. (But presumably it can be regarded as a variation in the bronze.[28])

One should notice Aristotle's unargued assumption that for any change there is always some one thing which is the subject of the change. This comes from his doctrine of the categories (§3 above), for a change is not itself a substance, so it must be predicated of a substance. The same doctrine of the categories also underlies his further subdivision of variation in Chapter 2. For he assumes that any change must take place with respect to one of his categories of predication, which he proceeds to list at 225^b5-9. (This list is the familiar list of eight, but with time omitted. Perhaps this omission is just a slip. More probably Aristotle would be ready to argue that since every change takes place *in* time there is no such thing as a change which is merely

---

[28] Cf. vii. 3, 246^a4-9, 246^b14-17, 247^a14-19.

a change *with respect to* time.²⁹) Then in Chapter 2 he argues
that there is no variation in respect of substance, relation, or
action and affection, and so he concludes that there are just
three kinds of variation, namely: in respect of quality (i.e. alteration), of quantity (i.e. increase or decrease), and of place (i.e.
movement).

The argument against there being a variation of substance is
brief and unconvincing. Aristotle says simply that one kind of
substance is not opposite to another (225^b10-11). Here he is
relying on the claim that variation is always between opposites,
or intermediates, but I have already observed that this claim is
mistaken, unless mere privations are permitted as opposites
(p. xvii.). But then there is nothing to stop one kind of substance
implying the privation of another. It appears that what Aristotle
should have said is that the same thing cannot be first one kind
of substance and then another, since this would rather be a case
where the first substance was destroyed, and the second generated from it (as, for example, when wine turns to vinegar). But
such an argument would again create a tension with Book I,
which appears to allow that the same *matter* is first one kind of
substance and then another. The argument against there being a
variation of relation is equally brief, and apparently it claims
that a change in the relation between two things can always be
analysed as a more fundamental change that is in only one of the
two, or perhaps as two different more fundamental changes, one
in one of the two and one in the other (225^b11-13). No doubt
in some cases this is so. As Plato observed,³⁰ Socrates may become shorter than Theaetetus just because Theaetetus grows while
Socrates remains the same size, and presumably this is the kind
of thing that Aristotle is thinking of. But not all changes in
relation are like this. For example, if two people marry, or di-
vorce, this is a change in their relation which is not just the
logical consequence of changes in each individually. Finally,
Aristotle spends almost all of Chapter 2 arguing against the
suggestion that there could be a variation in action or affection,

---

²⁹ Aristotle thus overlooks or denies a kind of change that interested Plato,
namely getting older, construed merely as having existed for a longer time. (See e.g.
Timaeus 38a; *Parmenides* 140e–141d, 151e–155c.)
³⁰ *Theaetetus* 155a–c.

that is, a variation in what one is doing or what is being done to one. His argument is that this would be a variation in a change, and that there cannot be such a variation. From our point of view, this is an extraordinary claim.

Aristotle's main argument for it is merely a restatement of his doctrine of categories: changes are not themselves substances, but only substances can be proper subjects of predication, and therefore nothing at all can properly be predicated of a change (225^b16–21). Consider the implications of this position. You cannot say that a change is quick or slow, you cannot assign it a date or a time, you cannot say that it is continuous, or uniform, or interrupted, and so on. But all of these things Aristotle himself wants to say of changes. Indeed he wants to say that some changes begin slowly but get quicker as they go on—that is, they accelerate. This seems a perfectly clear case of a change undergoing variation. But all this is denied by the argument that Aristotle gives here, which is that since changes are not substances they have no properties, and so of course their properties cannot vary. It is to our minds an extraordinary thing that he was prepared to rely just on the unargued doctrine of categories for such an outrageous claim.

Having thus dismissed the obvious way in which the same change may vary as it goes on, Aristotle next goes on to consider whether perhaps one change may turn into a different change, but his argument on this point clearly begs the question (225^b21–33). Finally, he broadens his claim by arguing in addition that there cannot be either a generation of a change or a destruction of a change (225^b33–226^a18). To understand his position here one must appreciate that in his view a generation or destruction is itself a process which, like any other change, is not instantaneous but occupies a whole period of time. (He will argue for this in Book VI.) If we grant this then his position does become more comprehensible, but even so the main argument appears to be oversimple. For it assumes that if in even one case there is a period during which a change is being generated, but has not yet come into existence, then the same must apply to all changes whatever, including any change which is itself the generation of a further change. Aristotle is certainly right to say that *this* would give rise to an infinite regress, but whether the regress would be vicious is a difficult question, which I leave undiscussed.[31] The more pertinent observation is that no regress arises if we suppose merely that *sometimes* a change has a preceding period of gestation, existing before that change itself exists, and bringing that change into being.

The remaining chapters of Book V are of less interest. Chapter 3 gives definitions of various terms, including a definition of 'continuous', which I shall comment on in connection with Book VI. Chapter 4 asks what is to count as a *single* change. Under this heading are comprised the two questions: (i) when is change A the same change as change B, and (ii) when are changes A and B different parts of the same one change that includes them both? Aristotle has the first question mainly in mind in 227^b3–228^a3, which distinguishes between changes that are the same in genus, in species, and in number. He slips into the second at 228^a3–19, where he first asks whether two repetitions of specifically the same change might be counted as numerically the same, but almost at once reconstrues this as the (more plausible) question of whether they might be counted as different parts of some one continuing change. (His unspoken answer to this latter question is 'No', since the repetitions he is envisaging occupy non-consecutive periods of time.) It is then the second question that occupies him for the rest of the chapter, but now applied to changes that do occupy consecutive periods of time. Chapter 5 concerns the notion of one change (i.e. variation) being opposite to another, and Chapter 6 extends this to the different idea that change is opposite not to change but to rest. Finally, the second part of Chapter 6 (from 230^a18) raises a question about the idea that changes may be either natural or unnatural (230^a18–$^b10$), and then it notes how this distinction between changes is related to some earlier proposals (230^b10–21). The chapter concludes with two rather unexpected footnotes on points already treated, which one should think of as afterthoughts (230^b21–231^a2).

---

[31] e.g. suppose that for every change A occupying x minutes there is a previous change B, which is the generation of A, and occupies $x/10$ minutes. Then before any given change there were an infinite number of previous changes, all occurring in a finite time. But whether we say that this is impossible will depend upon our reaction to Zeno's paradoxes, which I will discuss in §15.

14. *Continuity and the Attack on Atomism*

In Book VI as a whole Aristotle is arguing for continuity and against atomism. The argument concerns the nature of space, of time, of motion, and of bodies. The theory that bodies are made up of atomic (i.e. indivisible) particles is familiar; it was introduced at least two generations before Aristotle, by Leucippus and Democritus, and Plato had endorsed a version of it. But this book of Aristotle's *Physics* is our first evidence for the idea that space and time themselves might have an atomic structure.

It is a probable speculation that this idea arose from reflection on a development of Zeno's paradox of division, which we may briefly summarize in this way. Suppose that a finite line is everywhere divisible. Then let the line be everywhere divided, and consider what results. We see that no resulting part can have a positive magnitude, for any such part could be divided further, and we were supposed to have made all the possible divisions. It follows that the resulting parts must each have zero magnitude. But that too is impossible. For $0 + 0 = 0$, and so there is no way in which a positive magnitude can be reassembled from a collection of parts which each have zero magnitude. The moral must apparently be that it is not true to say that a finite line is everywhere divisible, and the only alternative seems to be that any process of division must eventually reach some very small parts of the line which are not further divisible, that is, which are atomic. And since this argument evidently applies to any magnitude whatever, we may conclude that space itself must be composed of indivisible minima, and presumably time as well.

In the *Physics* Aristotle makes no direct reply to this argument,[32] but it is perfectly clear that he does not believe its conclusion. His own view is that what is genuinely indivisible must have zero magnitude, like a point of space or an instant of time. But at the same time he means his arguments in this chapter to be a response to atomist theories. This is particularly clear in Chapter 10, where he argues about things that are not divisible,

---

[32] Aristotle cites the argument, and attempts to reply to it, in *De Generatione et Corruptione* i. 2. So far as the *Physics* is concerned, one might say that the doctrine of infinity in Book III yields a solution, for it claims that no infinite division could ever be completed. But it is interesting to notice that Book VI of the *Physics* never relies upon what Book III has said about infinity.

and without parts, but nevertheless treats such things as having magnitude.[33] But all through the book it is important to bear in mind that in the background there is a rival atomic theory to which he is responding.

We may illustrate this from the very first argument in the book, which aims to show that what is continuous cannot be made up of indivisible things, such as points ($231^a21–^b18$). From a modern perspective, a continuous line *is* regarded as made up of indivisible points, so *we* look at the argument with that as our focus of interest. Considered in this way, the main thing that one notices is that Aristotle has omitted to state the definition of 'continuous' on which his argument depends. Certainly in Chapter 3 of Book V he has said that x is continuous with y if and only if x and y share a limit, and he begins by reminding us of that ($231^a21–3$). But this does not yet tell us what it is for a single thing, for example a line, to be a continuous thing. One suggestion, which seems to be in line with his own discussion, is that x is continuous if and only if (i) x has at least two parts, and (ii) any division of x into *two* parts must divide it into parts that share limits. It will then follow that a division of x into any *finite* number of parts must yield parts which share limits. Construing the definition in this way, a finite line will indeed count as a continuous thing, but it will not follow that the line cannot be made up out of points. An alternative suggestion is that clause (ii) of the definition should rather be: 'Any division of x into *any* number of parts, including an infinite number, must yield parts which share limits with one another.' On this definition it will indeed follow that a continuous whole cannot be made up out of points, since no two distinct points can share a limit, but from our perspective the definition now does not define what it is meant to, since a finite line does not satisfy this condition.

All this, however, looks at Aristotle's argument from *our* perspective. But if we look at it from the perspective of his own time, the controversial claim must be that two distinct and indivisible

---

[33] See e.g. 240^b20, where we are asked to consider the hypothesis that an indivisible thing moves 'from *AB* to *BC*'. The choice of symbols clearly suggests that our indivisible thing begins by occupying the *stretch AB*, and then moves so as to occupy the adjacent stretch *BC*.

things cannot share a limit. For the atomists would certainly have thought otherwise. This disagreement, however, stems from different ways of construing the word 'indivisible'. Aristotle insists that anything extended must be divisible, and what he has in mind is that it will certainly be divisible *in thought*.[34] This may be granted. But when the atomists call something indivisible they mean that it is not divisible *in practice*. So far as the atoms of bodies are concerned, this means in the first place that there is no way of physically separating one part of an atom from another, and it is probably intended to imply further that one part cannot even be distinguished from another by their having different physical properties (e.g. that one part is hotter than another). Applying this view to atoms of space or time, where literal 'separation' is anyway impossible, it is the second point that is important: it is characteristic of a space-atom, or a time-atom, that no physical property could belong to one part of it but not another. For example, there could not be a situation in which a body occupies one half of a space-atom and not the other, or a situation in which a state of affairs continues through one half of a time-atom and not the other.

In the remainder of Chapter 1, and for all of Chapter 2, Aristotle seeks to show that there cannot be such atoms either of space or of time. His arguments, however, depend upon the premiss that *motion* is continuous, and while they do follow very nicely if that premiss is granted, still a determined opponent need not accept the premiss. Aristotle attempts to argue for it at $231^{b}18$–$232^{a}17$, but in the course of the argument he himself describes a perfectly coherent alternative, namely the suggestion that motion proceeds by a series of instantaneous 'leaps'. This means that the moving object stays at one position for a while, and then at once appears in a different position, without there having been any times at which it occupied intermediate positions. The atomist hypothesis is that *all* movement is really a series of very small leaps of this kind, though admittedly that is not how it looks to the naked eye. Aristotle's objection is that on this theory an object *has crossed* an (atomic) space, without

[34] Ironically, one may note that at the end of Book III Aristotle himself argues that it is irrelevant that numbers do not give out *in our thought* ($208^{a}14$–19), for his position is that in practice they *do* give out.

liii

there being any time when it *was crossing*, and it is fair to say that such a suggestion does offend common sense. However, the atomist may perfectly well reply that common sense is here mistaken, for common sense relies upon perception, but perception does not reveal either the atomic structure of objects or the atomic structure of motion. It is not easy to see how Aristotle could defend himself against this reply.

Moving ahead to Chapter 6, we may note that Aristotle there generalizes his principle. In Chapters 1 and 2 he had been considering movement in particular, and had claimed that every movement takes time, since if a thing has moved from A to B then there must have been a time when it was moving from A to B. In Chapter 6 (from 237^a17) he claims that the same applies to all changes without exception. His arguments, however, beg the question. At 237^a20-5 he supposes that if a thing changes from being in state A to being in state B, then there must be *both* a last instant of its being in state A *and* a first instant of its being in state B. If so, then these must be different instants, so there must be a stretch of time which separates them, and during this stretch the thing must be neither in state A nor in state B but changing from one to the other. However, one can see that there must be a fallacy in the reasoning, by taking the special case in which A is the negation of B, so that a thing is in state A at *every* time at which it is *not* in state B. Change that is in this way between contradictories cannot be reconciled with Aristotle's theses (despite what he says himself in Chapter 9 at 240^a19-29).[35] The error comes about because Aristotle is exaggerating his claims against the atomist: he thinks of the atomist as claiming that *all* changes are, at bottom, instantaneous; he therefore replies that *no* changes can be instantaneous; but he does not really need so strong a thesis. (Somewhat similar to this is his claim in Chapter 4, repeated in Chapter 10, that in *every* change the changing object must itself have parts. Evidently this claim overgeneralizes, and the arguments for it are clearly fallacious. Aristotle is opposing too strongly the atomist claim that

[35] In Book VIII he appears to recognize that some changes may be instantaneous (253^b14-26; cf. 186^a15-16), and he also gives a much more careful account of changes between contradictory states (263^b9-264^a6).

every change is, in the last analysis, a change of atoms which have no parts.)

Of the various further topics treated in Book VI, the most interesting is perhaps the discussion of Zeno's paradoxes on motion, which occupies most of Chapter 9. I devote my next section to this. Here I draw attention to just one further issue, since Aristotle's position on it is somewhat puzzling. He claims that one cannot speak of anything as either being in motion, or being at rest, at an instant ($234^a22–^b9$, $239^a10–22$; the point is generalized to apply to all changes at 237^a14). Now no doubt motion and rest apply in the first place to what is happening to an object during a stretch of time, and an object's state of motion at a particular instant cannot be defined without reference to its states in stretches of time that include that instant. But that is not a good reason for saying that there are no such instantaneous states, for they are frequently invoked in our ordinary ways of thinking, and in fact it is quite difficult to do without them.[36] The puzzling question, then, is why Aristotle should adopt this position, and whether he thought that it too was somehow required for his attack on atomism. For his arguments on the topic are clearly not compelling,[37] and they do threaten to introduce an inconsistency. This is because in Chapter 5 he claims that there is always a first instant when any change is completed, and that at that instant the object is in its final state ($235^b6–236^a7$). But in Chapter 8 he is forced to deny this for coming to be at rest, precisely because it would imply a first instant when the object is at rest ($239^a10–22$). To avoid a charge of inconsistency, then, he would have to claim that coming to rest is not to count as a 'change'. This would indeed be a consequence of the doctrine of Book V that there is no change of change, but it is hardly a satisfying position.

[36] They are naturally called upon when describing acceleration and deceleration, or simply when raising such a question as: 'How fast was he running when he broke the tape?' For an example where Aristotle himself must apparently invoke them, see my note on 236^b33.

[37] The argument on which he seems to put most weight is that the instant of change between a state of motion and a state of rest would have to be an instant of both motion and rest if it were an instant of either ($234^a31–4$). But later in Book VIII he seems quite ready to resolve such a question by stipulation ($263^b9–264^a6$). Another possibility is that he was influenced by the argument set out in my note on 232^a1.

15. *Zeno's Paradoxes on Motion*

Aristotle attributes to Zeno a set of four paradoxes on motion, designed to show that motion is impossible.

The first he calls 'the dichotomy' (239^b18–22), and he describes it simply as claiming that before one can reach the end of any movement one must first reach the half-way point (239^b11–14). It is clear that the argument functions by claiming that after reaching the first half-way point one must then reach the half-way point of what remains, and after that the half-way point of what still remains, and so on. But this gives an infinite series of half-way points, that is, a series which has no end, but which must all be passed before the goal is reached.

The second argument he calls 'the Achilles' (239^b14–18), since Achilles was famed as a very fast runner, and the tradition adds that the very slow runner against whom he is competing is a tortoise. The argument is that if Achilles and the tortoise are to have a race, and if the tortoise is given a start, then before Achilles can overtake it he must first reach the point p_1 where it started. But when he does reach p_1, the tortoise will still be ahead, say at p_2, so before he can catch it he must first reach p_2. But again, when he reaches p_2 the tortoise will still be ahead, say at p_3, and so on. Once more there will be an infinite series of points that Achilles must pass before he reaches the tortoise. As Aristotle rightly remarks, these two arguments give rise to exactly the same problem, and a solution to one will automatically be a solution to the other.

The third argument is the paradox of the arrow (239^b5–7), and it claims that the moving arrow must be at rest, since 'everything opposite to something equal to itself is at rest, and what is moving is always at a now'. Aristotle comments on this argument that it assumes a false premiss, namely that time is made up of nows (239^b8–9, $^b31$–3), but he does not spell out for us just how this premiss is involved.

The fourth argument is called 'the stadium' (i.e. 'the race-course') (239^b33–240^a18), and our interpretation of it must be somewhat tentative, since (*a*) Aristotle's text appears to be corrupt at this point, and some emendation is needed if it is to make any sense at all, and (*b*) in any case many commentators

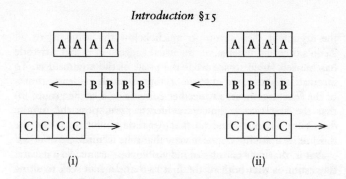

(i)   (ii)

have held that Aristotle has in fact missed the essential point of
the argument. The best reconstruction of Aristotle's own version
of the argument would seem to be this.[38] We have three rows of
bodies—the *A*s, the *B*s, and the *C*s—with each body the same
size as every other. (For definiteness, let us suppose that there
are four bodies in each row.) The row of *A*s is stationary, and
the row of *B*s and the row of *C*s each move past it, at the same
speed and in opposite directions.[39] We consider what happens
between positions (i) and (ii), which are as shown in the diagram.

The argument is that the leading *B* has passed only two *A*s,
whereas the leading *C* has passed all four *B*s. But since all the
bodies concerned are the same size, and the *B*s and the *C*s are
each moving at the same speed, it must take twice as long to
pass four bodies as to pass two. Hence the time taken between
(i) and (ii) must be twice as long as itself.

Now if this is indeed Zeno's argument, then Aristotle is clearly
right to reply as he does, that the fallacy is obvious: it takes
longer to pass a stationary body than it does to pass a body of
the same size moving in the opposite direction. But commentators have not been satisfied with this rather simple account, since

[38] Our text and interpretation follows Ross (1936).
[39] The *B*s move from the middle of the race-course, and the *C*s from the end.
A Greek race-course was shaped thus:

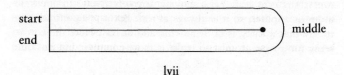

the argument then seems so much below the standard of all Zeno's other arguments. So the usual suggestion is that Aristotle has himself failed to see what the point of the argument is. To motivate this idea, we observe (*a*) that Aristotle evidently thinks of the four arguments as together comprising a single group, (*b*) that the first two arguments evidently presuppose the infinite divisibility of space, and (*c*) that Aristotle himself says that the third relies upon the opposite view that time is 'made up of nows' —that is, that time consists of indivisible time-atoms. So a natural suggestion is that both of the first two arguments seek to show that it is impossible for space and time to be infinitely divisible, while both of the second two seek to show that, on the other hand, space and time cannot be atomic. The argument of the stadium should therefore be directed against an atomic standpoint.

Given this idea, it is easy to see how the argument may be made more interesting. For suppose that each of our moving bodies occupies just one atom of space, and that the moving bodies are moving at a speed that covers one space-atom in one time-atom. Then in one time-atom the leading *B* moves from being opposite to one *A* to being opposite to the next, but it also moves from being opposite to no *C* to being opposite to *the second* C. Yet Zeno could clearly appeal to our intuitions to recognize that there must be a time and a place when the leading *B* is opposite to *the first* C. If so, then that time must divide a supposedly indivisible time-atom, and that place must divide a supposedly indivisible space-atom, so neither time nor space can be atomic after all. This argument is in fact very similar to one that Aristotle uses himself, at 233^b15–31. The correct reply for the atomist to make is, of course, that there is *no* time, and *no* place, when the leading *B* is opposite to the leading *C*, but it is true that common sense will not find that reply attractive.

If the paradox of the arrow is an attack on time-atoms, then presumably it goes like this. By hypothesis, nothing is the case in one part of a time-atom that is not also the case all through that time-atom. But that means that during any one time-atom, everything is at rest. Yet a moving arrow is always in one time-atom or another, so it is always at rest. But it is a contradiction to say that a thing is both moving and at rest. Here the correct reply for a time-atomist to make is more complex, but he could well begin by accepting Aristotle's own thesis that talk of motion and rest applies in the first place to stretches of time, containing many time-atoms. If he wished, he could go on to claim (like Aristotle) that such talk does not apply at all to a stretch that is only a single time-atom. Or he could be more sophisticated, and might offer to define velocity during a time-atom as a limit of velocities in periods including that atom, in much the same way as we would define velocity at an instant. But, in any case, even though he accepts that *in one sense* nothing moves during a time-atom, still he could suggest that in our ordinary ways of talking we use a *different* sense, which is equally defensible.

The last three paragraphs have presented what nowadays must count as the usual view of Zeno's third and fourth paradoxes. But it is very conjectural, and for my part I find the conjecture implausible. My main objection is that it is very unlikely that at the time when Zeno was writing anyone had thought out an atomic theory of either space or time. I observe further that if the fourth paradox really was directed at atomism you would certainly expect Aristotle to recognize that fact, since Aristotle was himself searching for arguments against atomism. And finally we should note that it is Aristotle who says that the third paradox was directed at time-atomism, but it is not at all clear that he is right about this. For, on his own account of what the argument is, it makes no mention of atoms, but simply starts from the premiss that anything is at rest which is 'opposite to something equal to itself'. (I take it that this in effect means 'when it occupies a space no bigger than itself'.) But this is true of every object 'in every now', whether 'a now' means here an instant of time or an atom of time, and I see no reason to suppose that Zeno was specifically thinking of the latter rather than the former. Presumably Aristotle thought that only a time-atomist would find Zeno's definition of 'at rest' at all plausible—for that is, on either account, the basic mistake in the argument —but we need not suppose that Zeno shared this opinion.

Let us now turn to the dichotomy and the Achilles, which evidently presuppose the infinite divisibility of space, and which are much more threatening both to Aristotle and to our own views. Aristotle offered an answer to the dichotomy earlier in Chapter 2, where he took it that Zeno's point was that it must be impossible to traverse an infinite number of distances in a finite time. He replied that there is no difficulty, for the time is just as divisible as the distance is, so we also have an infinity of different parts of the time, with one part for each distance to be traversed (233^a21–31). But later in Chapter 8 of Book VIII he returns to the same problem, and now he says that while his former reply would meet what Zeno himself had argued, still it does not get to the heart of the problem. For if we just consider the infinite divisibility of time, and forget about space altogether, still the real difficulty remains (263^a11–23). Aristotle does not spell out exactly what this difficulty is, but presumably it goes thus. We can divide a single minute into an infinity of ever-diminishing parts, so that in order to exist through a whole minute I must first exist through each member of this infinite series. But that appears to be impossible, for how can one reach the end of a series which has no end?

Aristotle's new reply to this problem relies upon the doctrine of infinity elaborated earlier in Book III, which claims that there are no actually infinite totalities. Applying this thought, Aristotle now says that the infinitely many parts of a stretch of space or time, and the infinitely many points that would mark them out, do not actually exist. They have a *potential* existence, and a process of actualizing them would indeed be an infinite process, but for that reason it could not be completed. He gives no general account of what it is to 'actualize' a point, but the basic idea seems to be that a point is not actualized unless something is done to it or at it which distinguishes it from its neighbours. Thus one would actualize a point by pausing at it, by changing direction at it, by counting it, or merely by thinking of that point in particular. But simply moving across a point of space, or existing through a point of time, does not actualize that point (263^a23–$^b9$; cf. 262^a19–28, $^b30$–2).

This is a suggestion of some interest, and it does respond in the right way to Zeno's problem. For let us now try to be clear about what that problem really is. First, there is no *logical* difficulty about completing an infinite series of tasks in a finite time. If there appears to be such a difficulty, then that is simply due to an equivocation on the phrase 'come to the end of'. It is true, of course, that one cannot reach 'the end' of an infinite series of tasks, if that means to perform the last member of the series, for an infinite series has no last member. But it in no way follows from this that one cannot 'complete' the series, in the sense of getting oneself into a position in which every one of those tasks has been performed. On the contrary, Achilles does complete such a series when he catches his tortoise. But second, although there is no *logical* impossibility in completing an infinite series of tasks, there are many, many examples where we do nevertheless feel that there is *some* kind of impossibility. (For example, Aristotle says that everyone agrees that it is not possible to be in a position where one has counted infinitely many numbers: 263^{a}9–11.) So the problem is to say which infinite series can be completed, and which cannot, and why.

Aristotle's answer is that Achilles can run smoothly until he catches his tortoise, despite the infinitely many points that he must pass. He can do this because his running in this way does not single out any point on the journey except the first and last. (We do not have to add, as Aristotle does, that points do not actually *exist* unless singled out; that is an optional extra.) But what Achilles cannot do is to single out infinitely many different points on his run. For example he cannot make it a staccato run, in which he pauses at infinitely many points along the way. (Since the time is infinitely divisible, we could specify the task so that there was time to fit in each of these infinitely many pauses; but still the task would not be a possible one on Aristotle's account.) Again, Achilles cannot stick in a flag at each of infinitely many points as he passes it, and he cannot mark them out merely in thought either. Common sense would, I think, agree with Aristotle on these examples of what Achilles can and cannot do. So, as I say, here is a suggestion that deserves to be taken seriously about which infinite series of tasks can be completed, and which cannot. I do not think that the suggestion is ultimately successful, but that is a large issue which I cannot pursue here.

16. *The Unchanging Cause of All Change*

In Book VIII Aristotle sets himself to argue that there is a single cause of all change, that it is itself unchanging, that it is not material, that it has infinite power, and that it is located at the circumference of the universe.

His argument begins in Chapter 1 with the claim that change is eternal, since there could not be either a first change or a last change. He offers two lines of argument. The first begins with the thought that materials capable of change must pre-exist any change (251^a8–16). If, then, there were a first change, say at a time t, then the materials for it must have existed at all times before t, but without producing any change. But this can only have been because the conditions required for the change were not in place before t, and this in turn implies that there was after all another change before t, which brought those conditions into existence at t. So the hypothesis of a first change yields a contradiction (251^a26–$^b10$). This is a powerful line of argument. In effect, it claims that there must be a reason why a thing happens at one time rather than another, but that there could not be such a reason for a supposed first event. I remark incidentally that the same argument would apply even if the first event was a generation of something from nothing (and it applies too to any event—whether or not it is the first event—which is preceded by a period during which nothing happens).

From our perspective, the argument may seem to be a deterministic one, relying on the thought that every event is caused by a preceding event. But we do not need to interpret it in this way, and it is better that we do not, since Aristotle was not a determinist. On the contrary, he believed that living things can *initiate* chains of causation, and that is why they play such a large role in his discussion. He draws attention to them in the very next chapter, noting that the fact that living things can initiate change where there was none before is an objection to his argument (252^b17–28). But he replies that living things initiate *motion*, as a result of their thoughts and desires, and that these in turn are always a response to some stimulus from the environment (253^a7–21). We do not need to construe this as claiming that the stimulus *determines* the response; the stimulus may explain why the response came when it did—which is all that the argument requires of it—while nevertheless it was the agent who determined what the response was to be.[40]

---

[40] Aristotle's initial statement of his position, at 253^a7–21, is clearly not committed to determinism. But I must admit that the recapitulation at 259^b1–20 does sound rather deterministic.

This first argument against the possibility of a first change carries some weight. But Aristotle supposes that there is also a symmetrical argument against a last change (251^b28–252^a5), and here he is very much less convincing. In between, he has offered a second argument from the nature of time, and this genuinely is symmetrical between past and future time. It claims that there cannot be a first or a last *time*, and infers from this that the same must follow for change, since it has already been argued (in Book IV) that time is 'not without change' (251^b10–28). But while it is fair to say that people do find it difficult to envisage time itself beginning or ending, that is because they do not connect time so closely with change as Aristotle does. Besides, the argument that Aristotle himself supplies is quite unpersuasive, for we surely need not agree that every instant of time *must* limit *both* a past *and* a future stretch of time.

Anyway, let us grant for the sake of argument that there always has been change and always will be. The following Chapter 3 is then rather programmatic, and I here pass over it.[41] The main argument resumes in Chapter 4, which sets itself to prove that everything which changes is changed by something. It discusses three cases: (*a*) changes in a thing that are contrary to its own nature must be caused by something other than it (254^b24–7); (*b*) changes in a living thing that are in accordance with its own nature are caused by itself, which means that one part of it causes the other part to change (254^b27–33); (*c*) changes in a non-living thing that are in accordance with its own nature are problematic. Aristotle has in mind the downwards motion of earth or water, and the upwards motion of air or fire (254^b33–255^a5). He says that these things do not move themselves in the way that living things do, (i) because there is only one motion that is natural to them, and they cannot stop themselves moving in this way, and (ii) because they do not consist of two parts of a different kind, one of which might move the other (255^a5–18). Moreover, he fails to consider the suggestion that there might be some other way in which they could be said to 'move themselves'. Instead his solution is that they are moved by whatever

---

[41] The arguments at 253^b6–26, against the claim that everything is always changing, are of some interest. They appear to conflict with the claim of Book VI that all change is continuous.

it is that brings them into existence in the first place (e.g. whatever it is that turns water into air), or by whatever it is that subsequently removes an obstacle to their motion (255^a18–256^a3). It emerges from this that Aristotle's real claim is that everything that changes is changed by something *other than itself*, for although a living thing is said to 'change itself', Aristotle thinks that what happens in this case is that one part of it changes another. (He gives reasons for this in the next chapter.)

It is important to notice at this point that one very relevant class of things has been omitted from consideration, namely the heavenly bodies (or rather, the spherical shells that carry them). According to the theory of the *De Caelo*, whereas earth moves naturally to the centre of the universe, and fire moves naturally away from it, these heavenly spheres are made of a stuff which moves naturally round the centre. Now the reasons which Aristotle gives for saying that earth and fire do not 'move themselves' as living things do will equally show that the heavenly spheres do not 'move themselves' in this way either. At any rate we can certainly say (i) that only the one motion is natural to them, and that (in Aristotle's view) they cannot stop themselves moving in this way. Whether we can add (ii) that they do not consist of two parts of a different kind, one moving the other, is less clear, but at any rate the *De Caelo* did not think of them in this way. So it seems fair to say that they do not 'move themselves' in the one way which Aristotle accepts. But it is also true that they are not 'moved by others' in the way that air and water are, for they exist eternally, and so never were created, and equally there never have been any obstacles to their motion. Consequently we have here a counter-example to the thesis of this chapter, that is, some things that undergo change (namely change of place) but are not *changed by* anything at all. Moreover, the case is crucial to Aristotle's argument, as we shall see.

Let us come back to that argument, which continues in Chapter 5. There Aristotle first claims that if we start with any change and look for its cause, and then for the cause of that in turn, and so on, we shall eventually come to a first cause which is either an unchanging changer or a self-changer. He offers three main arguments for this (256^a4–21, 256^a21–$^b3$, 256^b3–257^a14), together with a coda which is mainly a repetition of the previous

argument (257ᵃ14–27).⁴² An aside in the first argument rules out
what would seem to us to be the obvious alternative, namely
that every change is caused by another change, so that we have
an infinite regress of things which cause change by being changed
themselves (256ᵃ17–18). But this seems reasonable to us because
we think of one change being caused by a *previous* change, so
that the regress takes us to ever earlier times. To appreciate
Aristotle's position, however, one must remember that he naturally thinks of the cause (i.e. agent) of a change as operating all
the time that the change is going on, for the change would not
continue if the agent ceased to operate. To apply this to motion
in particular, one must remember that he does not share our
concept of inertial motion. I digress here to put in a word on this
topic.

An obvious objection to Aristotle's position is the case of
projectiles. He himself recognizes that this is an objection to his
argument in this book, and he attempts to deal with it in Chapter 10 at 266ᵇ27–267ᵃ20. From his point of view, the difficulty
is to explain why a stone thrown upwards continues to move
even after it has left the hand. His answer is that the stone is
pushed on by the air, one piece of air after another taking on the
motion and continuing to push the stone. But in a clear-sighted
way he has remarked at the beginning that this suggestion is
already problematic; for, to conform to what seem to him to be
obvious principles governing motion, we should say that once
the hand has stopped moving, *nothing at all* will continue to
move, neither the stone *nor the air* (266ᵇ30–267ᵃ1). Hoping to
avoid this suggestion he therefore says that a mover may continue to move something even after it has stopped moving itself,
and he is thinking here of one piece of air continuing to move
the next for a short time after it has itself stopped moving (267ᵃ5–
8). But of course this is a breach of the general principle that
motion occurs only while a force is being applied, though Aristotle intends it as a very *small* breach of that principle. But if the
principle can be breached at all, then why not a larger breach?
Why not just say that the hand continues to move the stone long

⁴² The second argument apparently concludes by saying that the first cause must
always be a self-changer (256ᵇ1–3). This appears to be a slip on Aristotle's part.

after it has itself ceased to move? Aristotle started by finding this an impossibility. In that case, he should find that his own solution is equally an impossibility, for the principle is the same in either case.

In our eyes, then, Aristotle's general position on how change is caused is quite unacceptable, but if we look through his eyes it is perfectly reasonable that he should pay no serious attention to the possibility of an infinite regress of things that cause change by being changed themselves. For the members of this regress would all have to be simultaneous with one another, and that cannot happen in a finite universe.[43] For the sake of argument, then, we may grant that, when tracing the causes of any change, one must eventually come to some first agent that causes change but not because it is changed by something else. The first agent then will cause change either by being changed by itself, or without being changed at all.

In the second half of Chapter 5 (from 257^{a}31) Aristotle then argues that a thing that is changed by itself must have two parts, one of which is changed by the other *without* being changed itself (except coincidentally, as is admitted later at 259^{b}16–20. The idea is that if some part of an animal causes it to move from A to B, then that part is itself moved from A to B, but only coincidentally, i.e. only because it remains a part of the animal throughout the movement.) Now Aristotle does give a reason for saying that it cannot be literally true that x changes x, but the reason applies only to some cases, namely those where a changer operates by imparting some property (such as heat) that the changer already has and the changed object does not yet have (257^{b}2–13). This would not prevent us supposing that in some *other* cases (in particular, in the case of the motion of the heavenly bodies) it is x itself that causes x to change. This point is important for the argument overall. But a different objection, applying to the present stage of the argument in particular, is that Aristotle gives us no good reason for his claim that where one part x of a self-changer causes another part y to change, the part x must itself be something unchanging. It *appears* that his thought must be that the part x which causes change will be the mind (or

---

[43] Or so it would seem. I discuss the point more closely in §17, on Book VII.

soul, *psyche*), and the part *y* which is caused to change will be the body (though it is to be observed that our text never quite says this). Moreover, one notes that elsewhere Aristotle is prepared to argue that in a sense the mind (soul) is never changed,[44] though here he is prepared to grant (at least for the sake of argument) that it does both begin to exist and cease to exist (258^b16-22). But the proposition that the mind never changes is one that holds little appeal for us.

After these preliminaries, Aristotle comes in Chapter 6 to his main claim that there is an eternal cause of change, offering three arguments. The third of these, however ($259^a20-{}^b21$), adds little that is not already in the first two, so I here ignore it. The first ($258^b16-259^a6$) points out that the familiar self-changers (i.e. animals) come to be and cease to be, and it claims that this succession of self-changers is an eternal succession.[45] But, says Aristotle, an eternal succession requires an eternal cause to account for it. However, he fails to make clear to us why an alternative that he himself considers, namely an eternal succcession of different causes, would not do equally well. One can only say that at this stage his case is not proven. The second argument (259^a13-20) seeks to evade this objection by strengthening the premiss. We have proved that always there is some change, and Aristotle now construes this as stating that there is some one change which exists always. (He says: 'since change is eternal it is continuous, and since it is continuous it is single'.) But, he claims, a single change requires a single cause, and if the change is eternal then the cause must be eternal.

There are two serious objections to this argument. The first is the obvious point that Aristotle is not entitled to assume that there is some one eternal change, for it certainly does not follow from what we have had so far. But it is a belief that must have played a crucial role in his own thinking. In Chapters 7–9 he will argue that the primary kind of change is movement, that the primary kind of movement is movement in a circle, and that there can be a single and eternal circular movement, whereas no

---

[44] Cf. *De Anima* i. 4, $408^a30-{}^b31$.

[45] One might here cavil that this point has not been proved. From the fact that there is always some change it does not follow that there is always some change that is caused by a self-changer.

other change can be eternal. He is referring, of course, to the movement of the heavenly bodies, which he believes to be single and eternal. But he has no way of proving that this movement is eternal, and the best that he could do would be to offer a probable argument, appealing to experience. Thus he might say that records of the heavenly bodies have been kept for many centuries, and those records reveal regular patterns that repeat themselves over and over again without alteration; the most likely explanation of this is that the movement of these bodies is eternal. But even if we grant to Aristotle that this is an example of a single and eternal movement, still his conclusion will not follow. For I have pointed out that his argument in Chapter 4 fails to show that the movement of these bodies is caused by anything other than their own nature. Consequently the argument of Chapter 5, to show that every movement is ultimately caused by something that cannot change, equally fails to apply in their case.

In the remainder of this book we do not learn very much more about the nature of this first and unchanging cause of all changes. At the end of Chapter 6 it is confirmed that the change that it directly causes is the rotation of the heavenly bodies, which is always the same. (But this in turn causes other changes which are not always the same, because they are changes in things differently related to the heavenly bodies ($259^b32-260^a19$).) In Chapter 10 Aristotle argues that the first cause must have 'an infinite power', and hence that it cannot have any magnitude; in other words, that it is not a material thing. He adds that it is located at the periphery of the world, at the same place as the sphere of the fixed stars, since that is the sphere that it moves most quickly. But he tells us nothing here about *how* it causes this movement, or what kind of 'power' we are to think of it as having. A natural guess might be that it acts in whatever way a human mind acts, when it causes movement in a human body.

We learn some more from elsewhere, in particular from Book *Λ* of the *Metaphysics*, which must be later than Book VIII of the *Physics*. Briefly, in chapter 7 of that book Aristotle tells us that this first cause of all movement acts as a final cause, by being an object of love. (To fulfil this role, it hardly seems necessary that it should have 'power' in any more than a metaphorical sense.)

He adds further that it is divine, and that it engages continually in its own proper activity, which is contemplative thought. But whereas in our discussion Aristotle had supposed that one 'first mover' would be enough to explain the phenomena (259^a8–13), in chapter 8 of Book Λ he tells us that there must in fact be *many* eternal 'first movers', to account for the variety of movements exhibited by the various heavenly bodies.[46] But he still counts all of them as 'Gods'.

17. Appendix: Book VII

I have relegated a comment on Book VII to the end, because I believe that, if Aristotle had himself prepared his *Physics* for publication, he would have omitted it altogether. First, it is a short book, with little connection between its five chapters. (One can see a link between Chapters 1 and 2, and again between Chapters 2 and 3, but nevertheless they do not form a single continuous discussion. Chapters 4 and 5 have no links either with the earlier chapters or with one another.) Second, most of its arguments are unpersuasive, and they often strike one as immature. Third, although Book VIII freely refers back to other books of the *Physics*—in fact to each of Books II, III (twice), V, and VI (twice)[47]—it never refers back to Book VII, even though there are two places where you would certainly have expected such a reference. (Chapter 1 of Book VII has argued at length that there cannot be an infinite regress of causes of a single change (242^a49–243^a31), but when Book VIII introduces this thesis without argument it gives no reference back (256^a17–19). Similarly, Chapter 5 of Book VII has devoted some attention to the proportion:

$$(\text{Power acting}) = \frac{(\text{Weight moved}) \cdot (\text{Distance covered})}{(\text{Time taken})}$$

but when this thesis is put to use in the arguments of Chapter 10 of Book VIII there is again no reference back.) The most obvious explanation for this is that, just as the opening argument

---

[46] For this variety, see note on 259^b30. (*Physics* 259^b28–31 apparently reflects the later view that there must be many eternal 'first movers'.)

[47] 251^a9 (cf. 257^b6–9), 253^b7–9, 257^a34, 262^a1–2, 263^a11–12, 267^b21.

of Book VII (i.e. 241^{b}34–242^{a}49) is very clearly superseded by the much fuller argument of Chapters 4 and 5 of Book VIII, so Aristotle intended the *whole* of Book VII to be superseded by the whole of Book VIII. Finally, I should observe that there is the odd situation that we have two versions of Chapters 1–3 of Book VII, differing from one another in a large number of small ways,[48] and both *seeming* to be by Aristotle (though this is disputed). There is nothing else quite like this with any of Aristotle's other writings, but it is not clear what hypothesis would best explain the anomaly.

Chapter 1 opens with an argument to show that everything that changes is changed by something other than itself, but the argument is evidently fallacious, and is clearly superseded by Chapters 4 and 5 of Book VIII. The bulk of the chapter (from 242^{a}49) then attempts to deduce a contradiction from the assumption of an infinite chain of causes, confining its attention to the case where the causes and their effects are all movements. The argument is this. If A moves B by being in motion itself, then the movements of A and of B are simultaneous. Hence, if there is an infinite chain of moved movers, each moving the next, but with no first mover, then all these infinitely many movements are simultaneous. Further, each member of the chain is a finite movement, with definite termini, occurring in a finite time. So the sum of all the infinitely many movements also takes place in a finite time. But if one movement causes another, the first must be at least as great as the second, so the sum of all the infinitely many movements is an infinite movement. Moreover, it is a single movement, for each moved body must be in contact with the one that moves it, and so they can all be regarded as together forming a single body, unified by contact, which has that movement. But it is impossible that there should be a single infinite movement in a finite time.

The most obvious comment to make on this argument concerns the principle that if one thing causes another to move, by being in contact with it, then the movement of the first must be 'at least as great' as the movement of the second. The question

---

[48] The version translated here is called version α. In the notes I have mentioned the alternative provided by version β only in the one case where it makes a significant difference to the argument. See note on 241^{b}34–242^{a}49.

is: how is one to compare the 'greatness' of two movements? I give two ways of bringing out the force of this question.

First, suppose that, by pushing on one end of a pencil, I cause the whole pencil to move in the direction of its length. Then it seems correct to say that the leading half of the pencil moves because it is being pushed by the rear half. (It would be even more clear that this was a correct thing to say if the pencil had actually been cut in half, and the one half was merely in contact with the other. But the point still seems to hold even if the halves are unseparated. Cf. note on 242^{b}60.) But now, considering just the rear half of the pencil, can we not say in the same way that *its* leading half moves because it is being pushed by *its* rear half? Then (*a*) is this consistent with the principle that the causing movement is 'at least as great' as the movement caused? It *seems* not, for we seem to be implying that the movement of the rear quarter is at least as great as that of the next quarter, which in turn is at least as great as that of the leading *half*. But (*b*), supposing that the principle is still satisfied (e.g. because the 'greatness' of a movement is just its velocity), can we not use this thought to generate a counter-example to Aristotle's argument? For by continuing (in thought) the division of the pencil into ever smaller parts, it seems that we may regard it as consisting of infinitely many parts, each of which moves only because it is pushed by its smaller predecessor. If so, then there are infinitely regressive causal chains, even within a finite universe.

Second, consider the simple example of a horse towing a barge up a river, by walking along the tow-path. Since horse, tow-rope, and barge are in contact with one another, it seems that Aristotle's principle should cover this case. But in what sense is the movement of the horse 'at least as great' as that of the barge? There need not be any simple relation between their speeds. For example, if the river winds then the horse, when it is on the inside of a bend, travels more slowly than the barge. And even if we do confine attention to the simple case when the river is straight, and the speeds are equal, still we are likely to be impressed by the fact that the barge has a much greater *momentum* than the horse, so we would need much more force to stop the barge than to stop the horse. Is this not a reason for saying that the movement of the barge is 'the greater'?

I raise these questions just to make it clear that they do need consideration, and I think one can be sure that if Aristotle had stopped to consider them, then he would have to have realized that his dynamics were oversimple. (And I mean that he should have realized this, even without taking into account the case of projectiles.) But as it is he assumes, without stopping to think about it: (*a*) that there is a general principle that causes must always be 'at least as great' as their effects, and (*b*) that the 'greatness' of a movement is an entirely straightforward notion.

The remaining chapters of Book VII are of less interest, and I deal with them more briefly. Chapter 2 has a connection with Chapter 1, for Chapter 1 had simply assumed that the mover and the object moved must be in contact, and Chapter 2 attempts to reinforce this claim and to generalize it to all kinds of changes, and not just movement. But (*a*) in another way Chapter 2 contradicts Chapter 1, since it casually assumes that there are things which move themselves (243^a11-15), whereas Chapter 1 had argued that this is impossible ($241^b34-242^a49$). Further (*b*) Chapter 2 appears to describe the movement of projectiles (243^a20-^b2), and another somewhat obscure change involving fire (244^a11-14), as movements in which there is contact, but *not* throughout the movement. This, however, is not what the argument of Chapter 1 requires, for it demands that the mover and the object moved should be all the time in contact with one another.

We have a similar situation between Chapters 2 and 3. There is a link, because Chapter 2 has claimed that alteration applies only to change in perceptible qualities and is always caused by perceptible qualities (244^b5-6). Chapter 3 then aims to support this claim by arguing that various other changes, which might *seem* to be alterations, are not really so. But at the same time Aristotle fails to consider the question of whether, in these other changes which he denies to be alterations, the contact condition is still satisfied. So he has apparently lost sight of the overall purpose of Chapter 2, and of its connection with Chapter 1. I remark also that the doctrine of Chapter 3 is implausible in itself, and is often contradicted by Aristotle elsewhere. In particular, Chapter 3 claims that becoming healthy is not an alteration

(246^b4), but elsewhere it is very often cited as a standard example of an alteration. In fact, it figures in this role in the very next chapter of Book VII (249^a29-^b11).

Chapter 4 asks when two changes are comparable in respect of speed. In the course of discussing this Aristotle makes an astonishing geometrical mistake, when he assumes that straight and curved lines cannot be compared in length (248^a12-13, $^a24-5$, $^b4-7$). Setting this aside, the main interest of the chapter is that it shows Aristotle coming to recognize that one cannot just assume that 'faster than' makes sense between any two changes, or even (on his account) between any two movements. One wishes that he had shown the same caution over 'greater than'. Finally, Chapter 5 is concerned to put forward the proportion already mentioned on p. lxix, but is mainly interested in a certain range of *exceptions* to it. That is, when the power is small in comparison to the weight to be moved, it may not move the weight at all, no matter how long it is applied. (This thought recurs in a different form at 253^b14-23 of Book VIII.)

SELECT BIBLIOGRAPHY

The interpretation and evaluation of an Aristotelian work is always controversial. In the Introduction and the Explanatory Notes to this volume there has not been space to explore the many controversies that exist over various aspects of the *Physics*, and I have had to state my own views somewhat dogmatically. This list of suggestions for further reading aims mainly to introduce the chief points of dispute. The list is confined to works written in English, and is *very* selective.

Introduction to Aristotle

Many general introductions exist. I mention here two that are fairly recent and one that is now a classic.

[1] Ackrill, J. L., *Aristotle the Philosopher* (Oxford: Oxford University Press, 1981).

[2] Lear, J., *Aristotle: The Desire to Understand* (Cambridge: Cambridge University Press, 1988).

[3] Ross, W. D., *Aristotle* (London: Methuen, 5th edn., revised 1949; first edn., 1923).

Ackrill's introduction is the most straightforward, Lear's is interestingly controversial, Ross's contains the most information on Aristotle's writings. All of them give a substantial amount of space to the *Physics*. There is a complete translation of all Aristotle's works in:

[4] Barnes, J. (ed.), *The Complete Works of Aristotle: The Revised Oxford Translation*, 2 vols. (Princeton and Oxford: Princeton University Press, 1984).

General Works on the Physics

A translation which follows the phrasing of the Greek text more closely than the present one is:

[5] Hardie, R. P., and Gaye, R. K., *Aristotle's Physics* (Oxford: Clarendon Press, 1930); repr. in [4].

By far the most important edition of the *Physics* is:

[6] Ross, W. D., *Aristotle's Physics*, text, introduction, analysis, and commentary (Oxford: Clarendon Press, 1936).

The present translation is based upon the above text. Even for those who know no Greek, the book contains much that is extremely useful. There are two volumes in the Clarendon Aristotle Series devoted to parts of the *Physics*, namely:

[7] Charlton, W., *Aristotle's Physics, Books I and II*, tr. with notes (Oxford: Clarendon Press, 1970).

[8] Hussey, E., *Aristotle's Physics, Books III and IV*, tr. with notes (Oxford: Clarendon Press, 1983).

These translations stay close to the Greek phrasing, and the notes give full discussions of questions of philosophical interest. There is also a special study of Book VII in:

[9] Wardy, R., *The Chain of Change* (Cambridge: Cambridge University Press, 1990).

This contains a text and translation, as well as a very full discussion. (Wardy's general view of Book VII is very different from mine.) I also mention here four books which concentrate upon particular aspects of Aristotle's *Physics*, namely:

[10] Solmsen, F., *Aristotle's System of the Physical World* (Ithaca, NY: Cornell University Press, 1960).

[11] Waterlow, S., *Nature, Change and Agency in Aristotle's Physics* (Oxford: Clarendon Press, 1982).

[12] Sorabji, R., *Time, Creation and the Continuum* (London: Duckworth, 1983).

[13] White, M. J., *The Continuous and the Discrete* (Oxford: Clarendon Press, 1992).

Solmsen covers not only the *Physics* but also the *De Caelo*, *De Generatione et Corruptione*, and *Meteorologica*; he also pays particular attention to Aristotle's relation to his predecessors. Waterlow is closely focused on the *Physics* itself, and in particular on the topics in it that her title picks out. White is concerned instead with Aristotle's theories of time, space, and motion; and he discusses also the rival theories of Aristotle's immediate successors, the Stoics and the Epicureans. Sorabji's sphere of interest is much the same as White's—time, space, and motion— but he has a much broader canvas; his object is to trace the story from its beginnings in the pre-Socratic philosophers through classical, hellenistic, and Roman times, and on through the Arabs, until the philosophers of the Latin West in the thirteenth and fourteenth centuries AD.

It is also convenient to mention here two recent collections of articles, namely:

[14] Barnes, J., Schofield, M., and Sorabji, R. (eds.), *Articles on Aristotle* (London: Duckworth; vol. i, 1975 and vol. iii, 1979).

[15] Judson, L. (ed.), *Aristotle's Physics: A Collection of Essays* (Oxford: Clarendon Press, 1991).

Both of these also contain useful bibliographies. Finally, I add at this point (because it does not fit under any of my later headings):

[16] Owen, G. E. L., 'Tithenai ta phainomena', in S. Mansion (ed.), *Aristote et les problèmes de méthode* (Louvain: Publications Universitaires de Louvain, 1961), 83–103; repr. in J. Moravcsik (ed.), *Aristotle: A Collection of Critical Essays* (New York: Anchor Books, 1961); and in G. E. L. Owen, *Logic, Science, and Dialectic*, ed. M. Nussbaum (London: Duckworth, 1986).

This is a classic account of Aristotle's method in the *Physics* (but it is contradicted by [19]).

The Principles of Change (Book I)

The main issues in Book I are treated in:

[17] Weiland, W., 'Aristotle's *Physics* and the Problem of Enquiry into Principles', in [14], i. 127–40. (Translated from *Kant-Studien*, 52 (1960/1), 206–19.)

[18] Bostock, D., 'Aristotle on the Principles of Change in *Physics* I', in M. Schofield and M. Nussbaum (eds.), *Language and Logos* (Cambridge: Cambridge University Press, 1982), 179–96.

[19] R. Bolton, 'Aristotle's Method in Natural Science: *Physics* I', in [15], 1–29.

A question which at once arises from Book I is Aristotle's commitment to the traditional notion of *prime matter*. Much has been written on this. I select in particular:

[20] King, H. R., 'Aristotle without *Prima Materia*', *Journal of the History of Ideas*, 17 (1956), 370–87.

[21] Solmsen, F., 'Aristotle and Prime Matter', *Journal of the History of Ideas*, 19 (1958), 243–52.

[22] Charlton, W., appendix to [7].

[23] Robinson, H. M., 'Prime Matter in Aristotle', *Phronesis*, 19 (1974), 168–88.

[24] Williams, C. J. F., *Aristotle: De Generatione et Corruptione*, tr. with notes (Oxford: Clarendon Press, 1982), appendix.

The Concept of Nature (Book II)

There is an interesting discussion in Waterlow [11], chs. 1–2. Apart from this, it is also useful to consider:

[25] Thayer, H. S., 'Aristotle on Nature', *Review of Metaphysics*, 28 (1975), 725–44.

Explanation (Book II)

On the topic of chance one should compare Aristotle's *Metaphysics E3*. A convenient source for this is:

[26] Kirwan, C. A., *Aristotle's Metaphysics, Books ΓΔE*, tr. with notes (Oxford: Clarendon Press, 1971).

There are also helpful discussions in:

[27] Sorabji, R., *Necessity, Cause and Blame* (London: Duckworth, 1980), ch. 1.

[28] Lennox, J. G., 'Aristotle on Chance', *Archiv für Geschichte der Philosophie*, 66 (1984), 52–60.

[29] Judson, L., 'Chance and "Always or for the Most Part" in Aristotle', in [15], 73–99.

On the four causes more generally, it is useful to read:

[30] Hocutt, M., 'Aristotle's Four Becauses', *Philosophy*, 49 (1974), 385–99.

[31] Moravcsik, J. M. E., 'Aristotle on Adequate Explanations', *Synthèse*, 28 (1974), 3–17.

[32] Schofield, M., 'Explanatory Projects in *Physics* II, 3 and 7', *Oxford Studies in Ancient Philosophy, Suppl. Vol. 1991*, 29–40.

On the particular question of final causes, i.e. of Aristotle's views on teleological explanation, a great deal has been written. I select in particular:

[33] Balme, D. M., *Aristotle's De Partibus Animalium I and De Generatione Animalium I* (Oxford: Clarendon Press, 1972), esp. 76–85, 93–100.

[34] Gotthelf, A., 'Aristotle's Conception of Final Causality', *Review of Metaphysics*, 30 (1976/7), 226–54; repr. in Gotthelf and Lennox (eds.), *Philosophical Issues in Aristotle's Biology* (Cambridge: Cambridge University Press, 1987), 204–42.

[35] Sorabji, R., *Necessity, Cause and Blame* (London: Duckworth, 1980), chs. 9–11.

[36] Cooper, J. M., 'Hypothetical Necessity and Natural Teleology', in Gotthelf and Lennox (see [34]), 243–74.

[37] Lewis, F. A., 'Teleology and Material/Efficient Causality in Aristotle', *Pacific Philosophical Quarterly*, 69 (1988), 54–98.

[38] Charles, D., 'Teleological Causation in the *Physics*', in [15], 101–28.

The Definition of Change (Book III)

There are discussions in Waterlow [11], ch. 3, and in White [13], 96–132. In addition it is worth consulting:

[39] Kosman, L. A., 'Aristotle's Definition of Motion', *Phronesis*, 14 (1969), 40–62.

[40] Graham, D. W., 'Aristotle's Definition of Motion', *Ancient Philosophy*, 8 (1988), 209–15.

Select Bibliography

Infinity (Book III)

There are good treatments in Sorabji [12], ch. 14, and in White [13], esp. ch. 4. In addition, see:

[41] Hintikka, J., 'Aristotelian Infinity', *Philosophical Review*, 75 (1966), 197–212; repr. in J. Hintikka, *Time and Necessity* (Oxford: Clarendon Press, 1973), ch. 6.

[42] Lear, J., 'Aristotelian Infinity', *Proceedings of the Aristotelian Society*, 80 (1979/80), 187–210.

The readings listed below under Zeno's Paradoxes, [55] and [56], are also relevant here.

Place and Void (Book IV)

A good starting-point is Solmsen [10], ch. 6. Other useful readings are:

[43] Furley, D. J., 'Aristotle and the Atomists on Motion in a Void', in P. K. Machamer and R. J. Turnbull (eds.), *Motion and Time, Space and Matter* (Columbus, Oh.: Ohio State University Press, 1976), 83–100; repr. in D. J. Furley, *Cosmic Problems* (Cambridge: Cambridge University Press, 1989).

[44] Machamer, P. K., 'Aristotle on Natural Place and Natural Motion', *Isis*, 69 (1978), 377–87.

[45] Mendell, H., 'Topoi on Topos: The Development of Aristotle's Concept of Place', *Phronesis*, 32 (1987), 206–31.

Time (Book IV)

There are treatments in Sorabji [11], chs. 1 and 6–7, and in White [13], ch. 2. From many other contributions I select:

[46] Annas, J., 'Aristotle, Number, and Time', *Philosophical Quarterly*, 25 (1975), 97–113.

[47] Owen, G. E. L., 'Aristotle on Time', in Machamer and Turnbull (see [43]), 3–25; repr. in [14], iii. 140–58; and in his *Logic, Science and Dialectic* (see [16]).

[48] Bostock, D., 'Aristotle's Account of Time', *Phronesis*, 25 (1980), 148–69.

[49] Waterlow, S., 'Aristotle's Now', *Philosophical Quarterly*, 34 (1984), 104–28.

[50] Inwood, M., 'Aristotle on the Reality of Time', in [15], 151–78.

Continuity and Atomism (Book VI)

There are treatments in Sorabji [12], chs. 22–4, and in White [13], ch. 1. Other useful reading is:

[51] Furley, D. J., *Two Studies in the Greek Atomists* (Princeton: Princeton University Press, 1967), Study I, esp. ch. 8.

[52] Waterlow, S., 'Instants of Motion in Aristotle's *Physics VI*', *Archiv für Geschichte der Philosophie*, 65 (1983), 128–46.

[53] Bostock, D., 'Aristotle on Continuity in *Physics VI*', in [15], 179–212.

Zeno's Paradoxes (Books VI and VIII)

For some account of Zeno himself, see:

[54] Vlastos, G., 'Zeno of Elea', in P. Edwards (ed.), *The Encyclopaedia of Philosophy* (London and New York: Macmillan, 1967), viii. 369–79.

For discussions which concern Aristotle's treatments of Zeno, see Sorabji [11], ch. 21, and:

[55] Owen, G. E. L., 'Zeno and the Mathematicians', *Proceedings of the Aristotelian Society*, 58 (1957/8), 199–222; repr. in his *Logic, Science and Dialectic* (see [16]).

[56] Bostock, D., 'Aristotle, Zeno, and the Potential Infinite', *Proceedings of the Aristotelian Society*, 73 (1972/3), 37–51.

[57] Lear, J., 'A Note on Zeno's Arrow', *Phronesis*, 26 (1981), 91–104.

There are many modern discussions which pay little or no attention to the ancient context. I mention in particular:

[58] Thomson, J. F., 'Tasks and Super-Tasks', *Analysis*, 15 (1954/5), 1–13.

[59] Benacerraf, P., 'Tasks, Super-Tasks, and the Modern Eleatics', *Journal of Philosophy*, 59 (1962), 765–84.

Both of these are reprinted in a useful anthology:

[60] Salmon, W. C. (ed.), *Zeno's Paradoxes* (New York: Bobbs-Merrill, 1967).

Aristotle's Laws of Physics

By this heading I mean Aristotle's *mathematical* laws, i.e. his statements of exact proportionality. The debate is on how seriously they are intended.

[61] Carteron, H., 'Does Aristotle Have a Mechanics?', in [14], i. 161–74. This is translated from his *La Notion de force dans le système d'Aristote* (Paris, 1923).

[62] Drabkin, I. E., 'Notes on the Laws of Motion in Aristotle', *American Journal of Philology*, 59 (1938), 60–84.

[63] Owen, G. E. L., 'Aristotelian Mechanics', in A. Gotthelf (ed.), *Aristotle on Nature and Living Things* (Bristol: Bristol Classical Press, 1985), 227–45; repr. in his *Logic, Science and Dialectic* (see [16]).

[64] Hussey, E., 'Aristotle's Mathematical Physics: A Reconstruction', in [15], 213–42.

The Unchanging Cause of All Change (Book VIII)

(In the discussions this is generally referred to as 'the unmoved mover' or 'the prime mover'.) Aristotle's treatment in the *Physics* leaves many questions unanswered. There are also treatments in the *De Caelo* (which appears to be inconsistent on this topic), in Book *Λ* of the *Metaphysics*, and in the *De Motu Animalium*. A classic treatment of Aristotle's apparently changing views is:

[65] Guthrie, W. K. C., 'The Development of Aristotle's Theology', *Classical Quarterly*, 27 and 28 (1933 and 1934), 162–71 and 90–8.

For *Metaphysics Λ*, and for the *De Motu Animalium*, see:

[66] Judson, L., *Aristotle's Metaphysics, Book Λ*, tr. and commentary (Oxford: Clarendon Press, forthcoming).

[67] Nussbaum, M., *Aristotle's De Motu Animalium*, text, tr., commentary, and essays (Princeton: Princeton University Press, 1978; repr. with corrections, 1985). See esp. Essay 2.

For a debate on the way in which animals are self-movers, see:

[68] Furley, D. J., 'Self-Movers', in G. E. R. Lloyd and G. E. L. Owen (eds.), *Aristotle on Mind and the Senses* (Cambridge: Cambridge University Press, 1978, 165–79); repr. in Furley, *Cosmic Problems* (see [43]).

[69] Sorabji, R., *Matter, Space and Motion* (London: Duckworth, 1988), ch. 13.

There is a provocative account of the significance of Aristotle's theory in:

[70] Kahn, C. H., 'The Place of the Prime Mover in Aristotle's Teleology', in Gotthelf (see [63]), 183–205.

ARISTOTLE
PHYSICS

CONTENTS

BOOK I

BOOK II

Contents

4

Contents

Contents

BOOK VI

BOOK VII

Note

Note that the headings to the various books and chapters or sections are not in the Greek, but have been supplied to delineate the structure of the main arguments of the book.

The letters and numbers in the margin of the translation give the page number, the column (ᵃ or ᵇ) and the line number of the edition of the Greek text by I. Bekker (Berlin, 1831). This is the standard means of precise reference to Aristotle's works. Because the lengths of line of the English translation and the Greek text inevitably differ, and because the word-order and even occasionally the clause-order of the English does not necessarily correspond to those of the Greek, it seems best simply to mark the beginnings of paragraphs and of other significant sentences.

The Greek text used for this translation has been that of W. D. Ross in his 1936 edition (*Aristotle's Physics* (London: Oxford University Press)), rather than that of his Oxford Classical Text. The places where a reading has been adopted other than that to be found in Ross's edition have been marked with an obelisk (†) in the translation and explained in the Textual Notes.

Asterisks (*) in the text refer to the Explanatory Notes.

I

THE PRINCIPLES OF NATURE

1. *The importance of distinguishing the principles of nature*

In any subject which has principles, causes, and elements, 184ª10
scientific knowledge and understanding stems from a grasp of
these, for we think we know a thing only when we have
grasped its first causes and principles and have traced it back
to its elements. It obviously follows that if we are to gain
scientific knowledge of nature as well, we should begin by
trying to decide about its principles.

The natural way to go about this is to start with what is ª16
more intelligible and clear *to us* and move from there to what
is clearer and more intelligible *in itself*. For the fact that
something is intelligible to us does not mean that it is intelligible *tout court*. So we have to proceed as I have said: we
have to start with things which are less clear in themselves,
but are clearer to us, and move from there to things which
are clearer and more intelligible in themselves. The things
which are immediately obvious and clear to us are usually
mixed together; their elements and principles only become
intelligible later, when one separates them. That is why we
have to progress from the general to the particular; it is because it is whole entities that are more intelligible to the
senses, and anything general is a kind of whole,* in the sense
that it includes a number of things which we could call its
parts. In a way, the same relationship also obtains between ª26
names and definitions: a word means an undifferentiated whole
(a circle, for instance), whereas the definition separates it into
particulars. And little children initially call all men 'father'
and all women 'mother', and only later distinguish who their
fathers and mothers are.

2. How many principles are there?

184ᵇ 15 Inevitably, there is either just one principle or there are more than one. If there is a single principle, it either does or does not change. The latter is what Parmenides and Melissus say; the former is the view of the natural scientists,* some of whom say that air is the principle and some of whom say it is water. If there is a plurality of principles, they are either finitely or infinitely many. If the number is finite but larger than one, there are two, three, four, or some other determinate number of principles. If the number is infinite, they are either like Democritus' principles,* which are of a single kind but differ in shape or form, or they are different in kind,† or they are even opposites.

ᵇ22   This is pretty much the same as asking how many things there are in the world, since to ask this question is actually to look into the original constituents of things. Those who ask it are trying to find out whether there is just one original constituent or whether there are more than one, and if there are more than one, whether they are finitely or infinitely many. In other words, they are trying to find out whether there is a single elementary principle or whether there are a number of elementary principles.

Criticism of the view that there is just one principle, and that it never changes

ᵇ25 Now, to enquire whether being is single and unchanging* is no part of an enquiry into nature. A geometer can no longer carry on a discussion with someone who denies the principles of geometry; such a discussion belongs to some other branch of knowledge, or to what is common to all branches. The same goes for an enquiry into principles: if there is only one thing, and a thing of this sort, the notion of a 'principle' is redundant, since a principle is a principle *of* some thing or things. So to enquire whether being is this sort of unity is no different from addressing any other thesis of the kind which is advanced just for the sake of argument—the Heraclitean thesis,* for instance, or the idea that being is a single person.

185ᵃ7 Alternatively, it is no different from resolving a sophistic argument, because both Melissus and Parmenides argue sophistically; indeed, their premisses are false and their conclusions do not follow. Or rather, Melissus' argument is crude and presents no problems: there is nothing difficult about deducing a whole string of absurdities once a single absurdity has been conceded.

We can assume that some or all natural things are chang- ᵃ12
ing; a survey of instances makes it clear that this is the case. At the same time, it is not our business to correct all mistakes, but to do so only where someone has drawn false inferences from principles, and not otherwise. Similarly, it is a geometer's job to refute the attempt to square the circle by means of segments, but it is not up to a geometer to refute Antiphon's method of squaring the circle.* All the same, although these people are not concerned with nature, they do incidentally address some problems which are relevant to the study of nature, so it might perhaps be a good idea to discuss them a bit, since the enquiry does involve some philosophy.

The most suitable place to start is to note that *being* means ᵃ20
different things and to ask what they mean when they say that all things are one.* Do they mean that all things are substance, or quantities, or qualities? And again, do they mean that everything is a single substance—a single person, as it were, or a single horse or a single mind? Or is everything a single quality—pale, for instance, or hot or something like that? These ideas may be equally indefensible, but there is a great deal of difference between them. If substance, quality, ᵃ27
and quantity all exist, there is a plurality of existing things, whether or not they are separate from one another. On the other hand, if everything is quality or quantity, then whether there is or is not such a thing as substance, the situation is absurd—if an impossibility can be called absurd. Why is it absurd? Because nothing except substance can exist by itself: everything else is an attribute of an underlying substance.

Melissus says that being is infinite. It follows that being is ᵃ32
a quantity, because infinity is in the category of quantity. It is impossible for a substance or quality or affection to be infinite, except coincidentally—that is, if they also possess some quantity. The point is that the concept of quantity, but

not of substance or of quality, is needed to explain infinity. So either there is substance as well as quantity, in which case being is twofold and not merely single, or there is only substance, in which case it is not infinite and in fact does not have any magnitude at all, because that requires quantity.

185^b5 That is not all. Oneness means just as many different things as being does, so we had better consider what they mean when they say that all is *one*. Now, a thing is said to be single when it is continuous or when it is indivisible, and things are also said to be single when the definition of what they are is one and the same ('ale' and 'beer', for instance).

^b9 If, first, they mean their one to be continuous, then their one is many, because anything continuous is infinitely divisible.

^b11 (There is a difficulty relating to parts and wholes, but it probably needs taking on its own and would be out of place here. It is the question whether the part and the whole together constitute a single unit or a plurality, and in what sense they constitute a single unit rather than a plurality, or a plurality (if that is what they are). These questions apply to non-continuous parts as well. And there is also the difficulty that if each of two parts is the same as the whole in the sense of being indivisible from it, they must also be indivisible from each other.)

^b16 Next, if they mean their one to be indivisible, there will be no such things as quantity or quality,* and being will not in fact be unlimited, as Melissus claims. It will not be limited either, as Parmenides claims, because indivisibility is a property of limits, not limited things.

^b19 Furthermore, if they mean that everything in the world is one in definition (like 'mantle' and 'cloak'), they will find that they are committed to the Heraclitean thesis: there will be no difference between what it is to be good and what it is to be bad, what it is to be good and what it is to be not good. Good and not good will end up identical, and so will man and horse, and their doctrine will not be about things being one, but about things not being anything at all. And there will be no difference between what it is to be such-and-such a quality and what it is to be such-and-such a quantity.

^b25 Even our more recent predecessors were anxious to avoid

making the same thing one and many at the same time. This is why some of them, like Lycophron,* eliminated the word 'is', and others tried to alter the way we speak, by saying 'He pales' instead of 'He is pale', 'He walks' instead of 'He is walking'. They wanted never to add 'is' in case they turned what is one thing into a plurality. This is to assume, however, $^{b}31$ that there is only one way in which 'one' or 'is' may be used, whereas in fact things can be many either in definition (so, for instance, being pale is different from being educated, but the same person can still be both, so what is one thing can be many) or by division, as a whole is divisible into many parts. This last point did in fact create difficulties for these more recent thinkers, and they conceded that the one was many. And why should the same thing not be one and many, as long as it is not so in conflicting ways? After all, something may be one either potentially or actually.*

3. *The same subject continued*

If we look at the matter in this way, then, it seems impossible $186^{a}4$ for all things to be one. Nor is it difficult to deal with the arguments brought up in support of the thesis, because both of them—Melissus and Parmenides—argue sophistically.

It is obvious that Melissus' argument* is invalid. He thinks $^{a}10$ it follows from the idea that every created thing has a beginning that every uncreated thing does not have a beginning. And a second absurdity is that every created thing should have a beginning—that there should be a beginning of the thing, not of the time—and that this applies not only to simple coming to be but also to alteration, as if change does not happen all over. Next, why does it follow from there only being one thing that it is unmoving? Why can the universe not move within itself, as its components can? For example, this water moves internally while still remaining one. Next, $^{a}18$ why can alteration not occur? (As a matter of fact, it is not even possible for everything to be one in species, because a man is specifically different from a horse, and opposites are specifically different from one another. But it may be possible for everything to be made of the same stuff; this, rather than the former alternative, is the sense in which some natural scientists say that everything is one.)

^a22 Parmenides is also liable to the same kinds of objection, besides others which apply particularly to him. The solution in his case is to point out, first, that he is mistaken and, second, that his conclusions do not follow. He is mistaken in his assumption that 'being' has just one meaning, when in fact it is equivocal; his conclusion does not follow because if we take just pale things, and agree that 'pale' has a single meaning, that still would not alter the fact that there is a plurality of pale things, not just one. For pallor will not be one either *qua* continuous or *qua* having a single definition, because what it is to be pallor will be different from what it is to be something with pallor. There still will not be anything separate from pallor; the reason that pallor and something with pallor are different is not because they are separate

^a31 things, but because they are different in definition. However, Parmenides was not in a position to appreciate this, and so he was bound to assume not only that being has only one meaning, whatever it is predicated of, but also that it means *just* being* (and *just* oneness).

^a34 The point is that coincidental attributes are predicated of something underlying, and consequently anything of which being is merely a coincidental attribute will not be, since it is different from being; it follows that there will be something without being. What just is being will not, then, be an attribute possessed by anything else, since it will be impossible for anything else to be a being, unless being means many things in such a way that each of them can be something. But *ex hypothesi* being means only one thing.

186^b4 Therefore, if what just is being is not an attribute of anything, but other things are attributed to *it*,[†] why does just being mean being rather than non-being? Suppose just being is also pale. Since nothing has being except just being, being cannot be attributed to being pale, and therefore what it is to be pale is not just being. Therefore, what is pale has no being*—not in the sense of not being this or not being that, but in the sense that it lacks being altogether. It follows that what is just being has no being, since it is true to say that it is pale, but we found that to say that something is pale is to say that it has no being. Consequently, 'just being' means 'pale' as well as 'just being',[†] and therefore 'being' means more than one thing.

Besides, if being is what just is being, then it will have no ᵇ12 magnitude either, because if it did the being which each of its parts have would be different.

Definition also shows that what just is a being[*] is divisible ᵇ14 into something else which just is a being. Suppose, for instance, that to be a man is the same as what just is some being; in that case to be an animal and to be two-footed are also necessarily just some beings. If each of them is not just some being, then they are coincidental attributes, and they would have to be attributes either of man or of some other subject, but that is out of the question, for the following reasons.

A thing is said to be coincidental if it may, but equally may ᵇ18 not, be an attribute of something, or if,[*] in order to define what it is, you have to mention what it is an attribute of. Being seated illustrates the first alternative, because it is separable from the subject; snubness illustrates the second alternative, because it includes the definition of the nose which the snubness is being said to be a coincidental attribute of. Moreover, the definition of the whole is not included in the ᵇ23 definition of the terms which occur in or form part of its definition. For example, the definition of a man does not occur in the definition of what it is to be two-footed, and the definition of a pale person does not occur in the definition of what it is to be pale. Assuming that all this is so, if two-footedness is a coincidental attribute of man, either it must be separable (in which case it would be possible for there to be a man who was not two-footed) or the definition of what it is to be two-footed will include the definition of what it is to be a man—but this is out of the question, because in actual fact it is the other way round.

If, on the other hand, two-footedness and animalness are ᵇ31 coincidental attributes of something other than man, and neither of them is what just is some being, then man must also be a coincidental attribute of something else. But let it be granted that what is just a being cannot be an attribute of anything else, and that when something has a pair of attributes, it must also have the attribute which the pair forms as well. Does it then follow that the universe is made of indivisible entities?*

187ª1 Some people* have capitulated to both arguments. Faced with the argument that if being is univocal, then all is one, they claimed that non-being has being; faced with the argument from dichotomy, they came up with the idea of indivisible magnitudes. But to say that if being is univocal, and the contradictory of being is ruled out, then there will be nothing which is not, is another patent falsehood. For there is nothing to stop what is not from being what is not something or

ª6 other, rather than what is not *tout court*. And to say that all will be one if there is nothing other than being itself is absurd, because the obvious sense of 'being itself' is what just is some being. Therefore, there is still no reason why there cannot be a plurality of things in the world, as I have already said. Anyway, it is clear that it is impossible for being to be single in the way they want it to be.

4. *Other views of earlier thinkers*

ª12 What about the natural scientists, though? They fall into two schools of thought. Some make the underlying stuff single, and identify it either with one of the three* or with some other stuff which is more condensed than fire and more refined than air. Then they have condensation and rarefaction generate everything else, and so they arrive at a plurality of objects. Now, density and rarity are opposites, and fall under the general class of excess and defect, which is how Plato describes his 'great and small',* except that according to him the great and the small constitute matter and it is the one that is form, whereas according to the natural scientists the one is the underlying matter, and the opposites constitute the differences between things, which is to say their forms.

ª20 Others, however, claim that the one contains oppositions, which are then separated out. This is the view of Anaximander and of those like Empedocles and Anaxagoras whose underlying stuff is simultaneously one and many. They belong to this school of thought because they too separate everything else out from the mixture. The differences between Empedocles and Anaxagoras are that according to Empedocles mixture and separation occur in cycles, while according to Anaxagoras the separation was a unique event, and that Anaxagoras separates out an infinite number of things—the homoeomerous substances* and the opposites—while Empedocles separates out only the familiar elements.

Criticism of the view that there are infinitely many principles

It seems likely that Anaxagoras posited an infinite number of ᵃ26 things in this way because he assumed the truth of the view held by all the natural scientists that nothing comes into being from non-being. That is why they make statements* like 'Everything was originally mixed together', and 'This is the kind of thing that coming into being is—alteration', though others talk in this context of combination and separation. They also thought that since the opposites come from each ᵃ31 other, they must have been present in each other. They reasoned as follows: necessarily, everything which comes into being comes either from things with being or from things without being; but it is impossible for anything to come into being from non-being (all the natural scientists are unanimous on this point); therefore, the only remaining possible conclusion, they thought, was that anything which comes into being* comes from things with being, which are already present in the source, but which are too small for us to detect with our senses. So the reason they say that everything is mixed in 187ᵇ1 everything is because, in their view, everything comes from everything; and they explain the fact that although everything is a mixture consisting of an infinite number of ingredients, things still look different from one another and are called one thing rather than another, by saying that this depends on which ingredient is numerically predominant within the mixture. There is nothing, they say, which is wholly and purely pale or dark or sweet or flesh or bone; people assess the nature of an object according to whichever ingredient there is most of within that object.

b7   Now, the infinite, as such, is unknowable. The measure of something which is infinite in number or in magnitude is unknowable, and the nature of something which has an infinite variety of forms is unknowable. And where the principles are infinite in quantity and in form, things composed of these principles are unknowable. The point is that knowledge of a compound is taken to depend on knowing what things, and how many things, it is composed of.

b13   Secondly, if a part of a thing can be arbitrarily big or small, the thing itself must also be able to be arbitrarily big or small. (By 'part' here I mean something which is a component of the whole such that the whole can be divided into it.) But since it is impossible for an animal or a plant to be arbitrarily big or small, clearly none of its parts can either (because then by the same token the whole could too). Now, flesh and bone and so on are parts of animals, and fruits are parts of plants. Obviously, then, it is impossible for flesh or bone or anything else to be of an arbitrary size—either arbitrarily large or arbitrarily small.

b22   Thirdly, if flesh, bone, and so on are all present in one another (so that they do not come into being but are separated out instead, as already present ingredients, and things get their names from their predominant ingredient), and if anything can come from just anything (for example, water can be separated out from flesh, and flesh from water), and if every finite body is exhausted by the repeated abstraction of a finite body, then we can easily see that it is impossible
b27 for everything to be present in everything. Suppose flesh is extracted from water, and then more flesh is generated out of the remaining water by separation. Even if the quantity of flesh being separated out is constantly diminishing, still it will not become smaller than some definite size.* Consequently, either the process of separating out will stop, in which case the remaining water will contain no flesh and there will not be 'a portion of everything in everything', or alternatively it will not stop, and there will always be the possibility of further extraction, in which case contained within an object

of finite size there would have to be an infinite number of
equal parts of finite size, which is impossible.

Fourthly, every material body is bound to become smaller ᵇ35
when some part of it is removed, and flesh cannot be arbitrarily large or small. It clearly follows that once you have the
smallest possible piece of flesh, nothing material can be extracted from it, because then there would have to be a piece
smaller than the smallest possible piece.

Fifthly, the infinite number of bodies would have to contain 188ᵃ2
an infinite amount of flesh, blood, and brain, each of which
would have to exist, and to exist in infinite quantity, without of
course being separated from one another. This is unthinkable.

Anaxagoras was right to say that the separating out will ᵃ5
never be over, but he failed to understand why. The point is
that attributes are inseparable from what they are attributes
of. Suppose, then, that colours and states are included in the
mixture. If they are separated out, there will be a 'pale' and
a 'healthy' which is nothing but what it is, in the sense that
it is not predicated of some subject. This puts Anaxagoras'
intelligence into the absurd position of trying to achieve the
impossible: it wants things to be separated out when this is
impossible, and is impossible not only in the domain of quantities but also in the domain of qualities. It is impossible in
the domain of quantities because there is no smallest magnitude, and it is impossible in the domain of qualities because
attributes are incapable of separate existence.

Anaxagoras is not even right about the origin of homogene- ᵃ13
ous substances, since although there is a sense in which mud
breaks down into lumps of mud, there is also a sense in which
it does not. And the ways in which bricks come 'from' a house
and a house comes 'from' bricks are not equivalent to the way
in which water and air are made of each other* and come from
each other. Empedocles takes there to be fewer principles—
that is, a finite number—and this is a better approach.

5. *Some opposites must be principles*

That the opposites are principles is agreed by everyone, includ- ᵃ19
ing those who say that the universe is single and motionless

(even Parmenides* regards hot and cold—or fire and earth, as he calls them—as principles), as well as those who talk of rarity and density, and Democritus too with his talk of the full and the void, which he says exist as being and non-being. Moreover, he relies on position, shape, and arrangement,* and these are genera containing opposites; for instance, under 'position' come above and below, in front and behind, and under 'shape' come angular and smooth, straight and rounded.

ᵃ26   It is clear, then, that in one way or another everyone regards the opposites as principles. This is a reasonable position to hold, because for things to qualify as *principles* they must not consist of one another or of other things, and everything must consist of them. Primary opposites fulfil these conditions. Because they are primary, they do not depend on other things, and because they are opposites, they do not depend on one another.

ᵃ30   However, we should also try to argue the matter through and see what conclusions follow. We had better start by assuming that nothing in the universe is such that it affects or is affected by anything else at random, nor does anything come from just anything, except coincidentally. How could anything become pale from being educated unless 'being educated' was coincidentally an attribute of what was not pale (i.e. what was dark)? No, something becomes pale from being not pale—not from being not pale in the sense in which just anything is not pale, but from being dark or something between pale and dark—and something becomes educated from being not educated, not in the sense in which just anything is not educated, but from something uneducated or from something between educated and uneducated (if such a 188ᵇ3 state exists). And when a thing ceases to be it does not turn into the first thing that just happens along. What is pale, for instance, does not turn into something educated (except on occasions coincidentally); it turns into what is not pale, and not into something which just happens to be not pale, but into something dark or something between pale and dark. The same goes for what is educated: it turns into something not educated, and not into something which just happens to be not educated, but into what is uneducated or something in an intermediate state (if there is one).

The same goes for everything else as well: the same account ᵇ8
holds for things which are complex rather than simple, but it
is just that we do not realize that this is what is happening,
because the opposite states have not got names. But the point
is that something structured must come from something unstructured and something unstructured from something structured, and when structure ends it is bound to become lack of
structure, and not just any lack of structure, but the corresponding opposite. It makes no difference whether we talk of
'structure' or 'order' or 'composition'; the same account clearly
holds good. Houses too, and statues—in fact, anything and ᵇ16
everything—come to be in the same way. A house, for instance, comes from this particular set of things in this particular state of separation rather than combination; a statue
or any other shaped object comes from a state of lacking
shape. Each of these things is either a certain arrangement or
a certain combination.

If all this is true, then everything which comes into being ᵇ21
comes from its opposite or from some intermediate between
the two extremes, and everything which ceases to be turns
into its opposite or into some intermediate between the two
extremes. The intermediates too are formed from the opposites: the various hues, for instance, are formed from pale and
dark. Consequently, everything that naturally comes to be is
either an opposite or consists of opposites.

Now, as I have already said, most other thinkers would ᵇ26
pretty much agree with what we have been saying so far,
because they all regard their elements and what they identify
as their principles as opposites; they may offer little argument
for their position, but they still make them opposites—it is as
if the truth itself left them no choice. There are differences
between them, however. The opposites they select may be
more or less primary, and more or less intelligible to reason
rather than to the senses. Some, for instance, consider hot
and cold to be responsible for coming into being, or wet and
dry, while others choose odd and even or hatred and love; the
differences in these cases are as stated.

The upshot is that in a sense they are saying the same ᵇ36
things, and in a sense they are disagreeing with one another;

their disagreements are in fact what strike most people, but by analogy they are in agreement with one another. The point is that they draw on the same list,* in which some opposites are more inclusive, and some less. This is what makes their ideas identical as well as different, and also what makes some

189ª4 of them better and some worse. It also explains what I said earlier—that some speak of opposites which are more intelligible to reason, while others take opposites which are more intelligible to the senses. The point is that reason grasps generalities, while the senses grasp particulars, because reasoning concerns generality, while what one perceives is particular. So, for instance, the great and the small are accessible to reason, while the rare and the dense are accessible to the senses.

ª9 Anyway, it is plain to see that the principles must be opposites.

6. *There are either two principles or three, but not more*

ª11 We should next state whether there are two of them, or three, or more than three. There cannot be only one, because the opposites are not a single thing. At the same time, there cannot be infinitely many. Why? Because, first, being would then be unknowable; second, because every genus contains just one opposition* and substance is a genus; third, because we can make do with a finite number, and it is better to rely on a finite number, as Empedocles does, than on an infinite number. (Empedocles thinks he can explain everything which Anaxagoras uses an infinite number of principles to explain.) Also, some pairs of opposites are prior to others, while some come from others† (as do sweet and bitter, and pale and dark), but principles should be constant.

ª20 This makes it clear that there cannot be just one principle, and there cannot be infinitely many either. There are a finite number, then. But if so, there is some reason* not to restrict them to being only two, because it is difficult to see how density could make rarity into something, or vice versa. And the same goes for any other opposition. Love does not combine hatred and make something from it, and hatred does not act on love in an equivalent fashion either: both of them act on some third thing. Some thinkers, in fact, think they need more than three principles to construct the way things are.

Another difficulty we might encounter if we do not posit ^{a}27 the existence of a third principle which underlies the opposites is that the opposites never seem to constitute the substance of anything. This is a problem, because a principle should not be predicated of something else which underlies it; if that were the case, the principle would have a principle, because if *A* underlies *B*, then *A* is a principle of *B* and is taken to be prior to *B*.

Also, substances are not, in our opinion, opposed to one ^{a}32 another. So how could a substance depend for its existence on things which are not substances? And how could non-substance be prior to substance?

It follows that if we find both arguments* valid—the ear- ^{a}34 lier one as well as the present one—and want to preserve them, we have to claim that there is a third, underlying thing. This is the claim made by those who say that the whole universe consists only of a single stuff, like water or fire or something intermediate between them. The most plausible of these ideas is that it is some intermediate stuff, because fire, earth, air, and water are already intrinsically connected with various oppositions. So it was quite sensible for some thinkers to make the underlying stuff different from these four, or at least to choose air, which has fewer perceptible differentiating qualities than the others. The next-best choice is water. They all, however, make opposites (such as rarity and den- 189^{b}8 sity, and more and less) the means by which they give shape to their single stuff. Now, all these opposites of theirs can obviously be covered by the concepts of excess and defect, as I have already said.* So the idea that oneness and excess and defect are the principles of being also seems to have a long history, though it does not always appear in the same way: early thinkers made the two active and the one acted on, whereas some more recent thinkers* say instead that the one is active and the two are acted on.

Anyway, these considerations and others like them might make ^{b}16 the idea that there are three elements seem rather plausible, as I have said, but the idea that there are more than three no longer seems plausible. In the first place, we do not need more than just the one to be acted on. In the second place, if there are four, forming two oppositions,* then each pair will need a separate extra thing as an intermediate; alternatively, if there are two pairs and they are each capable of generating out of the other, then one or the other of the oppositions

b22 is redundant. Moreover, there cannot be more than one *primary* opposition, because substance is a single genus of being, and it follows from this that the principles do not differ from one another in genus, but only in that one is prior to another. For within a single genus there is only ever a single opposition, to which all oppositions are apparently reducible.

b27 It is clear, then, that the number of elements is neither one, nor more than two or three. But it is very problematic to decide whether it is two or three, as I have indicated.

7. *The true view of the principles*

b30 Here is my own account. The natural procedure is to start by discussing what is common to the whole area, before considering the distinctive features of individual cases, so I will first discuss coming to be in general.

b32 The formula 'A comes from B'—A and B being different from each other—covers both simple and complex cases. In other words, a person can become educated, and what is not educated can become educated, and a non-educated person can become an educated person. What I am saying, then, is that in the first two cases, what comes to be something— namely, the person and what is not educated—are simple, and what it comes to be, namely something educated, is also simple. However, when we say that a non-educated person becomes an educated person, both the thing which comes to be and the thing which it becomes are complex entities.

190a5 Now, in some cases we can say not only that what is A comes to be B, but also that it comes to be B from being A; for instance, we say that what is educated comes from what is not educated. In other cases, however, we cannot speak in

24

this way: what is educated does not come from being a person, even though it is a person who has become educated.

Another point: of the things which come to be in the way ^{a}9 we are saying that simple things do, some persist and others do not persist. For instance, a person persists—he is still a person when he has become educated—but what is not educated (or what is uneducated) does not persist* either on its own or in conjunction with the subject.

Bearing these distinctions in mind, what emerges from ^{a}13 considering all cases of coming to be in the way I have been suggesting is that there always has to be some underlying thing which is what comes to be, and that even if this is numerically one, it is not one in form (which is just another way of saying 'in definition'), because what it is to be a person is not the same as what it is to be uneducated. More- ^{a}17 over, only one of these two things persists: the one which is not an opposite persists (i.e. the person persists), but what is not educated, or what is uneducated, does not persist, and neither does the complex which the two things together constitute (i.e. the uneducated person).

We typically say that *B* comes from *A* (as distinct from ^{a}21 saying that *A becomes B*), in cases where *A* does not persist. For example, we talk of something educated coming from something uneducated, but we do not say that what is educated comes from a person. (Nevertheless, this way of speaking is occasionally used even when things which persist are involved. For example, we talk of a statue coming from bronze and not of bronze becoming a statue.*) However, both expressions are used when a thing comes from something which is its opposite and which does not persist: we talk in these cases not only of *B* coming from *A*, but equally of *A* becoming *B*. For instance, we talk of something educated coming from something uneducated and also of something uneducated becoming something educated. And so the same goes for complexes: we say both that an uneducated person becomes educated, and that he does so from being an uneducated person.

'Coming to be' is ambiguous. In some cases a thing is not ^{a}31 said to come to be; rather, something is said to come to be

it. But substances alone are said to come to be *tout court*.
Now, where things other than substances are involved, it is
obvious that there has to be some underlying thing which
comes to be. A quantity, for instance, comes to be the quantity *of* an underlying something, and the same goes for quality, relation, time,[†] and place, because the only thing which
is not predicated of some underlying thing other than itself is

190^b1 substance, while everything else is predicated of it. However,
reflection on the matter shows that the same goes for substances too, and for anything else[†] which simply *is* with no
further qualification: there is something underlying them too,
which they come from. Plants and animals, for instance, come
from seeds. There is always *something* underlying substances,
something for them to come from. Things which come to be
without further qualification do so either by change of shape
(as a statue does) or by addition (as things that grow do) or
by subtraction (which is how a herm emerges from stone) or
by composition (as a house does) or by alteration (as things
whose matter alters do). Clearly, in each of these cases, there
is some underlying thing which they are coming from.

^b10   All this makes it clear, then, that everything which comes
to be is composite: first, there is that which comes into being,
and second, there is that which comes to be this first thing.
It is also clear that there are two possibilities as to what this
second thing might be: it is either an underlying object or an
opposite. By 'opposite' I mean what is uneducated, by 'underlying object' I mean the person; or again, shapelessness, form-
lessness, and disorder are opposites, and bronze or stone or
gold are underlying things.

^b17   It is clear, therefore, that if naturally existing things have
causes and principles, which are the sources of their being
and from which each thing has come to be what we say it
is when we are describing its substance rather than its coincidental attributes, then everything comes from an underlying
thing and a form. An educated person, for instance, is in a
way composed of 'a person' and 'educated'—in the sense that
its analysis depends on explaining these two terms. So it is
clear that things which come to be must come from these
two.

Now, the underlying thing is one in number, but two in $^{b}23$
form. First, there is the person and the gold, and in general
the matter (which is subject to number because it is closer to
being an identifiable thing and because it is not coincidental
that what comes to be comes from it); second, there is the
privation or opposite of form, which is a coincidental attribute. The form, however, is single. It is the order, for
example, or the education, or anything else which is predicated in this way.

From one point of view, then, we have to say that there are $^{b}29$
two principles, but from another point of view that there are
three; and from one point of view we have to say that they
are the opposites, but from another point of view that they
are not. They are the opposites in the sense that 'educated'
and 'uneducated', 'hot' and 'cold', 'structured' and 'unstructured', are principles; they are not opposites in the sense that
opposites are incapable of acting on one another. The solution to this problem too lies in differentiating the underlying
thing from the opposites: it is not an opposite.

The upshot of all this is that there is a sense in which there $^{b}35$
are no more principles than there are opposites—in other
words, one could say that there are two principles—but at
the same time they are not in all respects two, but rather
three, because they are different in definition: what it is to be
a person is different from what it is to be uneducated, and
what it is to be shapeless is different from what it is to be
bronze.

We have now said, then, how many principles are required $191^{a}3$
for natural things to come to be, and why they are that many.
It is clear that there has to be something to underlie the
opposites, and that the opposites are two in number. (From
another point of view, however, there need not be three: the
absence or presence of one or the other of the opposites will
be enough to effect the change.) The underlying nature can be
understood by analogy. Its relation to substance—that is, to
a particular existing object—is analogous to the relation of
bronze to a statue, of wood to a bed, or in general of matter
(which is to say,† something shapeless), before it gains shape,
to something with shape.

ᵃ12   This, then, is one principle, although it is not single in the way in which a particular identifiable thing is, nor does it exist in the way in which a particular identifiable thing exists; another principle is that which enables us to define what a thing is; and then there is this one's opposite or privation. I have already explained how it makes sense to describe them as two and as more than two from different points of view.

ᵃ15   We started by arguing that only the opposites are principles; then we went on to argue that there has to be something else to underlie the opposites, and that there were accordingly three principles; and now we can see what the differentiating feature of the opposites is, how the principles are related to one another, and what the underlying thing is. Whether the form or the underlying thing is substance is not yet clear.* However, what is clear is *that* there are three principles, *in what way* they are three, and *what kind* of principle each of them is. This, then, should be enough to clarify my position as regards how many principles there are and what they are.

8. *This view removes the difficulties felt by earlier thinkers*

ᵃ23   Next, let us explain why this is also the only way of resolving the problem which faced our predecessors. Those who first tried to discover the truth and to understand the nature of things in a philosophical manner were deflected by their inexperience down a side-alley, so to speak. They claimed that nothing comes to be or ceases to be, on the grounds that for anything to come to be it would have to come either from what is or from what is not, but that neither of these is possible. What is cannot come to be because it already is, and nothing can come from what is not because there must be some underlying thing. And then they extrapolated from this to conclude that there cannot be a plurality of things, but only being itself.

ᵃ33   Anyway, this is why *they* held the view that neither option was possible. But what *we* are saying is that in a way the idea that something might come from what is or what is not, or that what is or what is not might have an effect or be affected

or come to be some particular identifiable object, is no different from the idea that a doctor might have an effect or be
affected, or might be or come to be anything from being a
doctor. There are two ways in which we could take this   191^{b}2
proposition about the doctor, and so there are obviously two
ways in which we could take the idea of things coming from
what is, or of what is having an effect or being affected. If a
doctor builds a house, he does so not as a doctor but as a
builder; if he becomes pale, he does so not as a doctor but as
a person with a dark complexion. If he heals, however, or
becomes ignorant of medicine, he does so as a doctor. Now,
it is particularly appropriate to say that a doctor acts or is
affected in a certain way, or that he comes to be something
from being a doctor, if it is *as a doctor* that he acts or is
affected, or becomes something. It obviously follows that the
same goes for talk of coming to be from what is not: this
means 'coming to be from what is not *as what is not*'.

The reason why earlier thinkers went astray is that they ^{b}10
failed to make this distinction. And having failed to understand this point, they sank even further into incomprehension,
until they imagined that nothing comes to be and that nothing else is,* and ruled out coming to be altogether. Now, we
agree that nothing comes in an unqualified sense from what
is not, but we maintain that there still is a sense in which
things do come from what is not—that is, coincidentally: they
come to be something from the privation, which is in its own
right something that is not, and which does not remain. This
is what puzzles people, so that it has been taken to be impossible for something to come from what is not.

By the same token, it is impossible for anything to come ^{b}17
from what is, or to come to be what is, except coincidentally.
That is how this happens too. An analogy for how it happens
is provided by the way in which an animal comes from an
animal, and an animal of a specific kind comes from an animal
of a specific kind—a dog from a dog, for instance, or a horse
from a horse. In the first place, a dog comes not only from
an animal of a specific kind but also from an animal; but it
already has the property of being an animal,* so it is not *as
an animal* that a dog comes to be. If something is going to

become an animal (where 'animal' is not just a coincidental attribute), it will not do so from being an animal, and if something is going to become something that is, it will not do so from being something that is. Of course, it cannot do so from being something that is not either, because (as we have already said) the meaning of 'coming to be from something that is not' is 'coming to be from something that is not *as something that is not*'. We have also left intact the principle that everything either is or is not.

b27   This is one way of resolving the problem. Another way takes into consideration that it is possible to describe the same things in terms of potentiality and in terms of actuality. But I have developed this distinction in greater detail elsewhere.*

b30   The upshot (to continue) is that we have resolved the difficulties which forced people to rule out some of the things we have been talking about. For we have seen why earlier thinkers too were so thoroughly deflected from understanding coming to be and ceasing to be and change in general. If they had understood the underlying nature, all their misunderstandings would have fallen away.

9. *Criticism of the Platonist theory*

b35 Now, others* have touched on the underlying nature, but not clearly enough. In the first place, they concede the truth of Parmenides' argument that anything which simply comes into being comes from what is not; secondly, they imagine that if the underlying nature is numerically one, it must be one in potential too. But there is a great deal of difference between these two propositions. What we say is that matter and privation are two different things, one of which—matter—is a thing that coincidentally is not (and it is in a sense very close to being substance), while privation is a thing that in its own right is not (and is nothing like substance). In their view, however, the great and the small, together or separately, equally are what is not.

192ª8   In other words, their set of three factors and ours are completely different. They got as far as seeing that there has to be some underlying nature, but they made it one. Even describing it as a pair—great and small—does not stop it being single, because it remains the case that no account is taken of the privation. According to them, what persists is responsible, along with form, for things coming to be: it is the mother, as it were, of creation. As for the other aspect of ^{a}14 the opposition,* however, focusing on its pernicious features might almost make it seem not to exist at all. The point is that while *our* view, in the context of there being something divine and good and desirable, is that the opposite to this also exists, as does that which by its own nature desires and longs for it, *they* are committed to the view that the opposite longs for its own destruction. In fact, however, the form ^{a}20 cannot desire itself, because it is not in need of anything, and the opposite cannot desire the form, because opposites are mutually destructive. It is the matter which does the desiring. You might liken it to a woman longing for a man,* or what is ugly longing for what is beautiful, if it were not for the fact that matter is not in its own right something that is either ugly or female, except coincidentally.

Matter is not subject to generation and destruction

There is also a sense in which it comes to be and ceases to ^{a}25 be, and a sense in which it does not. Considered as the subject of attributes, it does in its own right cease to be, since it is the subject of the privation, which is what ceases to be. Considered as potential, however, and not in its own right, it does not cease to be, but is bound *not* to be liable to either coming to be or ceasing to be, because if it came into being, there would have to be a prior underlying thing from which it would come and which would be a component of it. But since that is precisely what it itself is, this would entail that it *was* before it *came into being*. For matter *ex hypothesi* is what ultimately underlies a thing; it is that from which something comes to be and which remains as a non-coincidental component in the thing's make-up. And if it ceased to be, the final destination of this process would be precisely this underlying thing, with the consequence that it would have ceased to be before it ceased to be.

a34   Deciding in detail how many principles of form there are—
one or more than one—and of what kind it is or they are, is
the job of first philosophy,* so we had better defer that issue
until then. In the following expositions* we will discuss natural forms which *are* subject to destruction.

192b2   Anyway, so much for our arguments for the existence of
principles, what they are, and how many there are. Now let
us make a fresh start in our discussion.

II

THE STUDY OF NATURE

1. *A natural object has a nature; explanation of this*

Some things exist by nature, others are due to other causes. _{192^b8} Natural objects include animals and their parts, plants and simple bodies like earth, fire, air, and water; at any rate, we do say that these kinds of things exist naturally. The obvious difference between all these things and things which are not natural is that each of the natural ones contains within itself a source of change and of stability, in respect of either movement or increase and decrease or alteration. On the other _{b16} hand, something like a bed or a cloak has no intrinsic impulse for change—at least, they do not under that particular description and to the extent that they are a result of human skill, but they do in so far as and to the extent that they are coincidentally made out of stone or earth or some combination of the two.

The nature of a thing, then, is a certain principle and cause _{b20} of change and stability in the thing, and it is *directly* present in it—which is to say that it is present in its own right and not coincidentally. By 'not coincidentally', I mean that if a doctor, say, is responsible for curing himself, this does not alter the fact that it is not *qua* being cured that he possesses medical skill: it is just a coincidence that the same person is both a doctor and is being cured, and that is why the two things are separable from each other. The same considera- _{b27} tions apply to all other products. None of them intrinsically contains the source of its own production. Some rely on external objects other than themselves (I am thinking here of houses and other products of manual labour), and even if others do contain the source of their own production, they do not do so in their own right, but are only coincidentally responsible for themselves.

ᵇ32   Nature, then, is as stated. The things which have a nature are those which have the kind of source I have been talking about. Each and every one of them is a substance, since substance is an underlying thing, and only underlying things can have a nature. They are all natural, and so is any property they have in their own right, such as the property fire has of moving upwards. This property is not a nature, and does not have a nature either, but it is due to nature and is natural.

193ᵃ1   Now, we have said what nature is and what 'natural' and 'due to nature' mean. It would be absurd, however, to try to *prove* that nature exists, since it is evident that there do exist many things of this sort. To rely on the non-obvious to establish the obvious is a sign of being incapable of distinguishing between what is and what is not intelligible in itself. This is a situation it is quite possible to be in: someone born blind, for instance, might reach conclusions about colours by a process of reasoning. Inevitably, then, people in this kind of situation argue only at the verbal level, but do not understand anything.

Is the nature of a thing its matter or its form?

ᵃ9   Some people take the nature and substance* of any natural thing to be its primary component, something which is unformed in itself. They say, for instance, that wood is the 'nature' of a bed, bronze the 'nature' of a statue. Antiphon* cites as evidence the fact that if you bury a bed and, as it rots, it manages to send up a shoot, the result is wood, not a bed. He concludes from this that the arrangement and design of the bed, which are due merely to human convention, are coincidental attributes, and that the substance is that which
ᵃ17   persists throughout, however it is affected. If, on the other hand,* any of these materials stands in the same relation to something *else*—if, for instance, bronze and gold stand in this relation to water, and bones and wood to earth—then *this*, they say, is their nature and substance. And so fire or earth or air or water, or some of them, or all of them, have been named by various thinkers as the nature of things. Whichever candidate or candidates they select, it or they are said to comprise all the substance there is, while everything else is an

affection, state, or disposition of this. And each of them is everlasting (since it is impossible for it to change from what it is), while everything else comes to be and ceases to be countless times.

This is one way in which people think of a thing's nature, ᵃ28 as the first matter* underlying anything which has its own source of motion and change. An alternative is to think of it ᵃ30 as the shape and form which enables us to define what an object is. The point is that we speak of nature where things happen by nature and are natural, just as we speak of skill where things happen by skill and are designed. We would not say that skill has played the slightest part, or talk of skill, when a thing is only potentially a bed and does not yet have the form of a bed, and the same goes for things which are constituted by nature. That which is potentially flesh or bone has not yet gained its own nature, and is not a natural object, until it has acquired the form which enables us to define what the thing is and to define it as flesh or bone. Consequently, 193ᵇ3 the alternative is to think of the nature of a thing as the shape and form of that which has in itself its own source of motion and change, where this shape or form is not separable from the thing itself except in definition. And anything which is a compound of the two, such as a person, is a natural object, rather than a nature.

Also, form is a more plausible candidate for being nature ᵇ6 than matter is because we speak of a thing as what it actually is at the time, rather than what it then is potentially.

Moreover, men come from men, but beds do not come ᵇ8 from beds. That is why people say that the wood, not the shape, is the bed's nature, because any offshoot that occurred would be wood, not a bed; but if the wood is its nature, the fact that men come from men shows that form too is nature.*

Moreover, 'nature' in the sense of a process* is a passage ᵇ12 towards nature. It is not like doctoring, which we say is a process which ends not in the skill of doctoring but in health, since in the case of doctoring the skill is necessarily the starting-point rather than the end-point. The relation between nature as a process—that is, growth—and nature is not like this: that which is growing is proceeding from something to something—that is what it means to be growing. What, then, is the end-point of growth? It is not that which the growing is from, but that which the growing is into.† From which it follows that form is nature.

^b18 'Form' and 'nature' are ambiguous, since from one point of view the privation is form as well. But we had better consider later* whether or not cases of simple coming into being involve a privation and an opposite to form.

2. *The scope of natural science*

^b22 Now that we have sorted out the different views about nature, we had better go on to consider what the difference is between a mathematician and a natural scientist. The point is that natural bodies have surfaces and solidity, lengths and points, and these are the subjects of mathematical investiga-^b25 tion. We also have to try to find out whether astronomy is different from or an aspect of natural science. It would be strange for a natural scientist to know what the sun and the moon are, but to be completely ignorant about their necessary attributes, especially since writers on nature obviously do discuss the shape of the moon and the sun, as well as whether or not the earth and the universe are spherical.

^b31 Now, it is true that the matters I mentioned are the concern of mathematicians as well, but they are not interested in the fact that surfaces and so on form the limits of natural bodies, and they also do not consider their properties as the properties of natural bodies. The reason why they abstract these properties, then, is that the properties are conceptually separable from the world of change. It makes no difference if you treat them as separate, in the sense that it does not ^b35 result in error. Albeit unconsciously, those who say that there are forms* do the same: they abstract natural properties, even though these are less separable than mathematical properties. This would become clear if you were to try in both cases to define the objects themselves and their attributes. One can conceive of odd and even, and straight and curved, in isolation from change, and similarly number and line and shape; but this is impossible in the case of flesh and bone and man,

which are defined like a snub nose* rather than a curved thing. Further clarification comes from the branches of math- 194ª7 ematics which are closest to natural science (such as optics, harmonics, and astronomy), since they are in a sense the converse of geometry: where geometry studies naturally occurring lines, but not as they occur in nature, optics studies mathematical lines, but as they occur in nature rather than as purely mathematical entities.

Since 'nature' refers to two things—that is, both to form ª12 and to matter—our investigation had better imitate an enquiry into what it is to be snubness, or something else which should not be considered in isolation from matter, but should not be restricted to matter either. In fact, an awkward question arises in this context: since there are two kinds of nature, which one does the natural scientist investigate? Or does he investigate something compounded from the two? But if his subject is a combination of the two, he is concerned with both natures, and we therefore have to ask whether it takes the same or a different branch of knowledge to understand them both.

One gets the impression, from considering the early natural ª18 scientists, that the study of nature is the study of matter. (Empedocles and Democritus did deal with form and with what it is to be a thing, but only to a minimal extent.) But design imitates nature, and it is the job of the same branch of knowledge to know about both form and, to a certain extent, matter: it is a doctor's job, for instance, to know not just about health, but also about bile and phlegm, in which health is found; likewise, it is the builder's job to know not just about the form of the house, but also about the bricks and planks which constitute its matter; and the same goes for all other branches of knowledge. It follows that it is also a natural scientist's job to understand both kinds of nature.

Moreover, it takes a single branch of knowledge to know ª27 the purpose or end of something and the way in which the purpose is achieved. Now, the nature of a thing is its end and its purpose, since in any case of continuous change which comes to an end, this concluding point is also the purpose of the change. This is why the poet said, absurdly, 'Now he has

37

reached the end for which he was born.'* In fact, of course, not every final stage has a claim to be called an end; only the best is an end.

ᵃ33   It is also† relevant that artisans make their matter—some simply make it, while others make it workable—and that we make use of things as if they existed for our sake. (From one point of view we too are ends. What a thing is for is ambiguous, as I explained in my dialogue *On Philosophy*.*) There are two kinds of skill, then, to which matter is subordinate and which have knowledge of it: one makes use of matter 194ᵇ2 and the other directs its making. The one which makes use of matter is in a way directive as well, but the difference is that it involves knowing about the form, while the other, since it is concerned with the making, has knowledge only of the matter. A helmsman, for instance, knows and prescribes a rudder's form, while the manufacturer knows what wood to make it from and what changes the wood has to undergo. In short, in the case of designed objects, *we* make the matter (and do so because of the work the object has to do), whereas in natural objects the matter is already present.*

ᵇ8   Moreover, matter is relative in the sense that a different form requires a different matter.

ᵇ9   How much knowledge, then, does a natural scientist have to have about form and what a thing is? Just as much as a doctor has about sinews or a metal-worker has about metal, which is to say as much as it takes to know what the purpose of a given thing is? And only about those forms which are found in matter, although they may be separable in form?* After all, a man is created by a man, and by the sun as well. Questions remain—in what sense is anything separable? What is it that is separable?—but it is the job of first philosophy* to answer them.

3. *The four types of cause*

ᵇ16 With these distinctions in place, we should look into the question of how many causes there are, and what they are like. For the point of our investigation is to acquire knowledge, and a prerequisite for knowing anything is understanding *why* it is as it is—in other words, grasping its primary cause. Obviously, then, this is what we have to do in the case of coming to be and ceasing to be, and natural change in general. Then, once we know the principles of these things, we can try to analyse anything we are looking into in terms of these principles.

One way in which the word 'cause' is used is for that from ᵇ23 which a thing is made and continues to be made—for example, the bronze of a statue, the silver of a bowl, and the genera of which bronze and silver are species.

A second way in which the word is used is for the form or ᵇ26 pattern (i.e. the formula for what a thing is, both specifically and generically, and the terms which play a part in the formula). For example, the ratio 2 : 1, and number in general, cause the octave.

A third way in which the word is used is for the original ᵇ29 source of change or rest. For example, a deviser of a plan is a cause, a father causes a child, and in general a producer causes a product and a changer causes a change.

A fourth way in which the word is used is for the end. This ᵇ32 is what something is for, as health, for example, may be what walking is for. If asked, 'Why is he walking?', we reply, 'To get healthy', and in saying this we mean to explain the cause of his walking. And then there is everything which happens during the process of change (initiated by something else) that leads up to the end: for example, the end of health may involve slimming or purging or drugs or surgical implements; they are all for the same end, but they are different in that some are actions and some are implements.

These are more or less all the ways in which we use the 195ᵃ3 word 'cause'. The upshot is that there are a number of ways in which the word is used and also that a single thing has a number of causes, even without considering coincidence. For instance, both sculpturing and bronze are causally responsible for a statue, and are so for the statue in its own right, *qua* statue, although they are dissimilar *kinds* of causes, since one is a cause in the sense that matter is a cause, while the other is a cause in the sense that the source of change is a cause.

Some things mutually cause each other. Physical effort, for ᵃ8 instance, causes physical fitness, and physical fitness causes physical effort. However, they are not causes of each other in the same way: one is a cause in the sense that the end is a cause, while the other is a cause in the sense that the source of change is a cause.

ᵃ11 Moreover, a single thing can be the cause of opposites. Something whose presence is the cause of one thing may also, by being absent, be taken to be the cause of its opposite. For instance, the absence of a helmsman might cause the loss of a ship, and his presence its preservation.

ᵃ15 It is very easy to see that all so-called causes fall under four headings. Typical examples of causes in the sense of that from which things come are letters (from which syllables come), matter (from which artefacts come), fire and so on (from which material bodies come), parts (from which a whole

ᵃ19 comes), premisses (from which a conclusion comes).* Some of these (parts, for instance) are causes in the sense that they underlie a thing; others are causes* in the sense that they constitute what a thing is (for instance, the whole, the com-

ᵃ21 pound, and the form). Then consider seed, a doctor, a planner, and any other kind of agent: they are all causes in the

ᵃ23 sense that they initiate change or stability. Then there are things which are causes in the sense that they are the ends of the other things, and are the good for which they are done. Without quibbling about whether it is an actual good or an apparent good, that at which other things are aimed—that is, their end—tends to be what is best.

ᵃ26 That is how many types of cause there are, and what they are. There are all sorts of ways in which things may be causes, but they too can be brought under a smaller number of headings. 'Cause' is a very equivocal term, and even causes of the same type can differ from one another in terms of priority and posteriority: both a doctor and a professional cause health, for instance, and both the ratio of double and number cause the octave. In every case, there are particular things and the

ᵃ32 broader things that include them. It is also possible for both specific and generic coincidentals to be causes: from one point of view the cause of a statue is Polyclitus* and from another point of view it is a sculptor, because being Polyclitus is a coincidental attribute of the sculptor. And then there are the genera which include the coincidental: we might say that a man was the cause of the statue, or even more generally that an animal was. Also, some of the coincidentals are nearer, and others more remote: think of saying that someone pale, or educated, was the cause of the statue. And again, all causes, 195ᵇ3 whether they are properly so called or whether they are so called in virtue of some coincidental attribute, may be called causes either because they are potentially causes or because they are actually functioning as causes: for example, the cause of the building of a house may be either a builder or a builder building.

Equivalent distinctions can also be drawn in the case of the ᵇ6 things of which the causes are causes. A cause may be said to cause this particular statue or a statue or more generally a likeness, and it may be said to cause this particular bronze or bronze or more generally matter. And the same goes for coincidental attributes.

Moreover, both the cause and what it causes can be ex- ᵇ10 pressed as a compound, as when we speak not of Polyclitus or of a sculptor, but of Polyclitus the sculptor.

Nevertheless, despite this variety, there are six kinds of ᵇ12 cause* in all, each of which may be spoken of in two ways. There is the particular, and its genus; there is the coinciden-tal, and *its* genus; and these may be combined or they may be expressed simply. And all six can be either actual or po-tential. The difference between being actual and potential is ᵇ16 that causes which are actual and particular exist and cease to exist at the same time* as the things of which they are causes. For example, a particular person curing coexists with a par-ticular person being cured, and a particular person building coexists with a particular object being built. However, this is not necessarily the case with potential causes, since a house and a builder do not cease to exist at the same time.

We should always look for the most basic cause in every ᵇ21 case; this is as important in this sphere as elsewhere. What I mean is that a man builds because he is a builder, and a builder builds in virtue of the fact that he possesses skill at building. So skill at building is the prior cause here, and any instance should be analysed in an equivalent fashion. We should also look for general causes for general things (a sculptor for a statue, for instance) and particular causes for particular things (a specific sculptor for a specific statue). And we should hold capacities responsible for potential results, and actually functioning causes responsible for actually happening results.

ᵇ28   So much for sorting out how many causes there are, and in what sense they are causes.

4. *Is chance also a cause? Some opinions on this*

ᵇ31   Chance and spontaneity are also counted as causes: people often attribute the existence and occurrence of things to chance and spontaneity. So we had better try to see how they can be among the causes we have distinguished, whether chance and spontaneity are the same as or different from each other, and in general what they are.

ᵇ36   Some people* find even the question of whether or not they exist difficult to answer. According to them, there is no such thing as a chance event; they claim that there is always a determinate cause for everything which is said to be a chance or a spontaneous event. Consider, for example, the case of someone who chanced to come into the city square and met someone he wanted to meet but had not expected to find there; they say that the cause here was his wanting to go and 196ᵃ5 do business in the square. The same goes, in their opinion, for every other so-called chance event: it is always possible to find some cause other than chance. For, they say, if there really were such a thing as chance, it would be a truly extraordinary phenomenon, and so it would be odd that none of the experts from times past who discussed the causes of coming to be and ceasing to be distinguished chance as such a cause, which makes it look as though they too thought that nothing happens as a result of chance.

ᵃ11   It is also surprising, however, that although people are aware of this long-standing argument in favour of doing away with chance, and realize that it is always possible to attribute the numerous events whose existence or occurrence is due to chance and spontaneity to some determinate cause or other, they all still persist in attributing some things—and only some things—to chance. That is precisely why those predecessors of ours should have made some sort of mention of it. In fact, however, they did not even identify it with one of their causes (love, hatred, intelligence, fire, and so on and so forth). Both ᵃ19 alternatives are strange, then: either they did not think there was such a thing as chance, or they did recognize its existence but ignored it. And this is especially strange since they do sometimes rely on it, as Empedocles does when he says that air is not always separated off towards the highest region, but as chance would have it. At any rate, in the cosmogonical section of his work, he says, 'So chanced it then to run, but often otherwise.' And he also says that the parts of animals mostly came about by chance.

Then there are others* who even attribute this world of ᵃ24 ours and all the worlds* to spontaneity. They say that the rotation is a spontaneous event—that the motion which separated things out and established the orderly nature of the world began spontaneously. But *this* certainly should occa- ᵃ28 sion surprise: at the same time as holding nature or intelligence or something (i.e. something other than chance) responsible for the existence and generation of animals and plants—since things do not come from particular seeds by chance, but an olive-tree comes from one kind of seed and a man from another—they claim that the universe and the most divine aspects of the visible world are spontaneous events, and lack any causation remotely similar to that which animals and plants have. And yet, if they were right, that very ᵃ35 fact should have given them pause for thought and they really ought to have addressed it. I mean, leaving aside any other odd features of their view, it is especially strange for them to make this claim when they could see that there are no spontaneous events in heaven, and could see coincidences happening as a result of chance all around them in the realm which, according to them, is free from the influence of chance. As a matter of fact, however, one would expect things to be the other way around.

196^b5 On the other hand, there are those* who do think that chance is a cause, but one which is opaque to the human mind, because it is divine and too supernatural for us to understand.

b7 So we had better investigate spontaneity and chance, and try to see what each of them is, whether they are the same or different, and how they fit into the causes we have already distinguished.

5. *Explanation of how chance is a cause*

b10 First, then, there are obviously things which *always* happen in the same way, and things which *usually* do. It is evident that no one says that either of these two classes of events—either things which happen always and of necessity, or things which are usual—have chance as a cause, or happen by chance. However, since there are also events which are not covered by these descriptions, and since everyone says that such events happen by chance, it is evident that there is such a thing as chance and as spontaneity. After all, we do acknowledge that these kinds of event are the result of chance, and that the results of chance are these kinds of event.

b17 Now, some events serve a purpose and some do not. Of those that do, some are chosen and some are not,* but both kinds still serve a purpose. So it is obvious that some of the events which cannot be classified as either necessary or usual may also serve some purpose or other. Now, all thought-out and natural events do serve a purpose. So it is when such things happen by coincidence that we ascribe them to chance.

b24 The point is that just as something can either exist in its own right or coincidentally, so a cause may be a cause either in its own right or coincidentally. For example, the cause in its own right of a house is house-building ability, but a house may coincidentally be caused by something pale or educated. While things which are causes in their own right are determinate, coincidental causes are indeterminate, because a single

b29 event could turn out to have an infinite number of them. To repeat, then: when this happens in the case of something which serves a purpose, it is ascribed to spontaneity and chance. We will have to sort out the difference between these two later;* for the time being, I assume it is clear that the province of both of them equally is things which serve some purpose.

Here is an example.* *A* would have come for the purpose ᵇ33 of getting his money from *B*, if he had known that *B* was there and was in the process of reclaiming a loan. In fact, though, that is not why he came; it was just a coincidence that he came—he did not† do so in order to collect his money. Besides, he did not usually go there nor was it inevitable that he would: that is, the result—his collecting his money—was not one of the causes operating in him, but *is* the kind of thing which happens as a result of choice and thought. These 197ᵃ2 are the circumstances under which we say that he chanced to come, but if he had chosen to be there and had come for this purpose, or if he always or usually went there when he was in the process of collecting money,† we would not ascribe his coming to chance. Clearly, then, chance is a coincidental cause in the sphere of events which have some purpose and are subject to choice. So thought and chance have the same province, since thought is a prerequisite for choice.

The things which might act as causes of chance events are ᵃ8 bound, therefore, to be indeterminate. That is why chance too is taken to be indeterminate and opaque to people, and why it does make a kind of sense to think that nothing comes about by chance. All these views are, not surprisingly, correct. There *is* a sense in which things happen by chance: they happen *coincidentally*, and chance is a coincidental cause. But in an unqualified sense chance causes nothing. The cause of a house is a builder, for instance, and the coincidental cause of the house is a pipe-player; in the case of the man who came and collected his money, although he was not there for that reason, there is an infinite number of possible coincidental causes. He could have come because he wanted to see someone, or was on his way to court as a prosecutor or defendant, or was on his way to the theatre. It is also ᵃ18 correct to say that chance is inexplicable, because explanations can only be given for things that happen either always or usually, but the province of chance is things which do not happen always or usually. Since these kinds of causes are indeterminate, chance is indeterminate as well. Nevertheless, in some cases one might wonder whether in fact any event whatever can cause a chance event. For instance, if we can say that the wind or the sun's warmth causes health, can we say that getting one's hair cut does as well? After all, some coincidental causes are less remote than others.

^a25 We call it good luck* when the result is something beneficial, and bad luck when the outcome is bad. 'Good fortune' and 'bad fortune' are used when the results are of some importance. That is also why* just missing a harmful or beneficial outcome counts as good or bad fortune: it is because, as far as the mind is concerned, a near miss is no distance at all away, so it treats the harm or benefit as already present. It is also not surprising that good fortune is taken to be unstable: chance is unstable, since it is impossible for anything which is an effect of chance to happen always or even usually.

^a32 Both chance and spontaneity are, then, coincidental causes, as I have said. Their sphere of operation is events which do not have to happen, either in any case or usually, and they apply to just those cases which might have occurred for some purpose.

6. The distinction between chance and spontaneity

^a36 The difference between chance and spontaneity is that 'spontaneity' is the more general term, in the sense that every chance event is a spontaneous event, but not every spontaneous event is a chance event. Chance and happening by chance are relevant only in cases where there is a possibility of good fortune and, in general, of action. That is also why the province of chance is bound to be things that are done; this is borne out by the fact that people take good fortune to be more or less identical with faring well, and since faring well consists in doing well, it is a kind of doing.

197^b5 It follows that anything which is incapable of action is equally incapable of doing anything by chance. The reason, then, why no inanimate object, beast, or child does anything by chance is because it cannot exercise choice.* They do not experience good or bad fortune either, except in the metaphorical sense Protarchus* was using when he said that the stones from which altars are made are fortunate because they receive respect, while their fellows are trodden underfoot. Even these things can in a way be affected by chance, but only when someone who is doing something which involves them acts by chance.

Spontaneity, however, does apply to non-human animals ᵇ13 and often to inanimate objects as well. We say, for instance, that the horse came spontaneously, in that it was saved by coming, but it did not come for the purpose of being saved. And we say that the three-legged stool happened spontaneously to fall upright; in that position it served the purpose of acting as a seat, but it did not fall for the purpose of acting as a seat. Clearly, then, we say that an event happens spon- ᵇ18 taneously when it is the kind of event which, broadly speaking, serves some purpose, when what actually happened did not happen for the sake of that purpose, and when the cause of the event is external.* However, these events are said to be chance events if they are choice-worthy and happen spontaneously to agents who are capable of exercising choice.

The term 'pointless'* confirms this view. We use the word ᵇ22 when an event which has some purpose other than itself fails to achieve that purpose. Suppose, for instance, that someone takes a walk in order to loosen his bowels, but the walk fails to have this effect: we say that there was no point in his walking and that the walk was pointless. We assume, then, that something is pointless when its nature is to serve some purpose other than itself, but it fails to achieve the purpose for which it exists and which it is its nature to serve. After all, it would be ridiculous for someone to say that it was pointless for him to have washed because there was no eclipse of the sun! The one thing is not the point of the other. As the word implies, then, 'spontaneity' is when something happens which is 'in itself pointless'. The stone fell on him, but the purpose of its falling was not to hit him; however, it *could* have had this purpose, and there *could* have been human intention behind its falling; and so the stone's falling was spontaneous.

b32   Chance and spontaneity are furthest apart in the case of naturally occurring events.* When something unnatural happens, we do not attribute it to chance, but to spontaneity. Actually, though, there is a difference between this kind of event and a spontaneous event: the cause of a spontaneous event is external, whereas the cause of this kind of event is internal.

198a1   I have now explained what spontaneity and chance are, and how they differ from each other. As for what kind of cause they are, they are both to be counted as sources of change. The point is that in every such case, the cause is either nature or thought. There are, however, infinitely many such causes. Spontaneity and chance cause outcomes which either intelligence or nature *could* have caused, but which on

a7   this occasion have a coincidental cause. Now, since anything that is in its own right is prior to things that are coincidental, it obviously follows that the same goes for causes too: something that is a cause in its own right is prior to a coincidental cause. Therefore, spontaneity and chance are posterior to intelligence and nature. The upshot of this is that however much spontaneity is the cause of the universe, intelligence and nature are bound to be more primary causes;* and this applies not only to the universe, but also to plenty of other things as well.

7. *The natural scientist should study all four types of cause (but they often coincide)*

a14   It is clear, then, that there are causes and that there are as many of them as we have been saying, since there are just as many different kinds of question covered by the question 'Why?' To ask 'Why?' is ultimately equivalent *either* to asking 'What is it?' (this is what the question comes to in the case of unchanging entities: in mathematics, for instance, it is ultimately equivalent to asking for a definition of 'straight' or 'commensurate' or whatever), *or* to asking 'What initiated the change?' (as in: 'Why did they go to war?'—'Because they had been raided'), *or* to asking 'What is the purpose?' (as in: 'To gain control'), *or* to asking, in the case of things that come to be, 'What matter is involved?'

So it is clear that there are these causes and that there are ᵃ21 this many of them. It is the job of the natural scientist, then, to understand all four of these causes; if he refers the question 'Why?' to this set of four causes—matter, form, source of change, purpose—he will be explaining things in the way a natural scientist should.

In many cases, the last three of these causes come to the ᵃ24 same thing.* What a thing is and its purpose are the same, and the original source of change is, in terms of form, the same as these two: after all, it is a man who generates a man. This applies universally to everything which is changed itself when initiating change*—and things which are *not* like this are not the province of natural science, because they do not initiate change by having change or a source of change within themselves, but do so without changing. In fact, there are three areas of study:* things which are not subject to change, things which are subject to change but not to destruction, and things which are subject to destruction.

In short, then, the question 'Why?' is resolved by answer- ᵃ31 ing it in terms of a thing's matter, what it is and its original source of change. This last is the normal practice of investigating the causes of events by asking, 'What comes after what?'—that is, 'What was it that first caused some effect, and what was it that was first affected?', and so on through-out a whole sequence. But there are two kinds of sources of ᵃ35 natural change, and in one kind the source is not itself a natural object,* in the sense that it does not contain its own source of change. In this latter category comes anything which causes change without itself changing (for example, that which is absolutely unchanging and is the primary entity in the whole universe) and what a thing is, or its form (since that is its end or purpose).

Since a thing's nature involves purpose, then, we also have 198ᵇ4 to understand this cause. A complete elucidation of the question 'Why?' involves explaining that from *this* there neces-sarily comes *that** (that that either comes from this in any case, or usually), and what must be present if the thing is to exist* (analogous to the way the conclusion follows from the premisses), and that this is what it is to be the thing,* and that the thing is as it is because it is better that way*—not better in any absolute sense, but better given what that particular thing actually is.

8. Final causes are crucially important in nature

ᵇ10 So I had better begin by explaining why a thing's nature is a cause in the sense that it is a purpose, and then go on to discuss necessity and its role in natural objects. The point is that people constantly refer things back to necessity as a cause: it is because the nature of heat (or cold or whatever) is so-and-so, they say, that such-and-such a state of affairs must necessarily exist or happen. Even if they do introduce some other cause (love and hatred, for instance, or intelligence), they merely touch on it and then leave it aside.

ᵇ16 The problem is this. What is wrong with the idea that nature does not act purposively and does not do things because they are better? The proper analogy might be that Zeus does not send rain *so that* the crops will grow: it is just a matter of necessity. The vapour drawn up from the earth is bound to cool down; once it has cooled down it is bound to turn to rain and fall back to earth; and it is sheer coincidence that crops grow when this happens. By the same token, if someone's grain is ruined on the threshing-floor, this does not mean that the rain fell for the purpose of ruining the grain:

ᵇ23 it is just a coincidence. So what is wrong with the idea that the parts of natural things are like that too? Take teeth, for instance: what is wrong with the idea that the front teeth necessarily come through sharp and suitable for biting, and the back teeth flat and good for crushing food? Why should there be purpose behind this? Why should it not just be an accident? And the same question could be asked about any other part of the body which seems to have some purpose. So where every part turned out to be just as it would have been if it had had some purpose, the creatures survived because, spontaneously, they happened to be put together in a useful way. But everything else has been destroyed and continues to be destroyed, as Empedocles says* of his 'cow-like creatures with the heads of men'.

This, or this kind of, argument might be used to pose the $^{b}32$ problem. But it is impossible for this to be the way things are. The point is that the things mentioned turn out as they do either always or usually, and so does every other natural object, whereas no chance or spontaneous event does. Frequent rain in winter is not taken to be a chance accident, but it is during the dog-days;* a heatwave during the dog-days is not taken to be a chance accident, but it is in winter. So if $199^{a}3$ we assume that these things are either accidents or have some purpose, then, given that they cannot be either accidents or spontaneous events, they must have some purpose. But the things I have mentioned and everything else which is like them are natural things, as even exponents of the view I have outlined would admit. It follows that purposes are to be found in natural events and natural objects.

Moreover, whenever there is an end, the whole prior se- $^{a}8$ quence of actions is performed with this end as its purpose. Now, unless something intervenes, how an action is done corresponds to how things are in nature, and vice versa. But actions have a purpose, and so therefore do things in nature. For example, a naturally occurring house*—supposing such a thing were possible—would happen in exactly the same way that a skilfully made house does; conversely, if naturally occurring things were made by skill as well as by nature, they would still happen in exactly the same way as when they occurred naturally. It follows that one thing happens for the $^{a}15$ sake of another. And in general human skill either completes what nature is incapable of completing or imitates nature. If artificial products have some purpose, then, natural things obviously do too, since in both cases the relation between the later stages and the earlier stages is the same.

This is particularly clear in the case of non-human animals, $^{a}20$ whose products are not the result of skill, enquiry, or planning. Some people are puzzled by how spiders, ants, and so on make what they make—do they use intelligence, or what? If you gradually follow this line of reasoning through, it looks

as though things happen at the plant level too which serve
$^a26$ some purpose: leaves, for instance, develop to protect fruit. In short, if it is natural for a swallow to make its nest and it also serves some purpose, if it is natural for a spider to make its web and it also serves some purpose, if their fruit is the reason that plants grow leaves, and nourishment is the reason they grow their roots downwards rather than upwards, then it is clear that this type of causation is present in naturally occurring events and objects. Now, 'nature' is ambiguous in the sense that it can refer either to matter or to form; but since the end is form, and everything else takes place for the sake of the end, it is this form that is the cause, since it is that for which everything happens.

$^a33$ Even the province of human skill is not free from error: scribes can write incorrectly, doctors can prescribe the wrong medicine. Evidently, then, mistakes can happen in the province of nature too. Now, if it is possible for there to be products of human skill which correctly serve some purpose, and mistakes in this province constitute failed attempts at some purpose, then the same should go for natural things too, and monstrosities would constitute failures to achieve that natural purpose. In the beginning, then, any combinations like those 'cow-like creatures', which were incapable of achieving some definite end, must have arisen because of some defect in their source, just as defective seed is responsible for the birth of any such creatures nowadays.

199^b7 Besides, seed must come first; creatures cannot just spring straight into existence. The words 'And first whole-natured . . .'* must mean seed.

$^b9$ Moreover, even plants exhibit purpose, although matters are less clear-cut in their case. So were there or were there not in the domain of plants 'vine-like plants with the heads of olives', equivalent to the 'cow-like creatures with the heads of men'? The idea seems absurd, but there must have been, if there were in the domain of animals.

$^b13$ Moreover, chance would also have had to have been a factor in their seeds. But this idea totally subverts any notion of nature and natural things. The point is that those things are natural which undergo continuous change, starting from an intrinsic source of change and concluding at a particular end. Starting from a given source of change does not result in the same end in every case, but it is not just any chance end either; there is in fact always a tendency towards the same end, unless something intervenes. It is true that a thing's ^{b}18 purpose, and what happens for that purpose, can also happen by chance—as when we say that it was by chance that the visitor came and paid the ransom before leaving (the point being that he acted *as if* he had come for that purpose, when he actually had not)—but this is a coincidence, because chance is a coincidental cause, as I have already explained.* However, when this happens always or usually, it is not a coincidental or chance event; and unless something intervenes, this *is* always the case with natural objects.

It is ridiculous for people to deny that there is purpose if ^{b}26 they cannot see the agent of change doing any planning. After all, skill does not make plans.* If ship-building were intrinsic to wood, then wood would naturally produce the same results that ship-building does. If skill is purposive, then, so is nature. The clearest illustration is to think of someone curing himself: that is what a thing's nature is like.

Anyway, it is clear that a thing's nature is a cause, and that ^{b}32 it is the kind of cause I have been saying—namely, purpose.

9. *The role of necessity in nature*

Now, what about necessity? Is necessity present in nature ^{b}34 conditionally, or also unconditionally? As things are, people think of necessity as present in events. It is as if they supposed, for instance, that a wall had developed as a result of necessity, because heavy things naturally tend downwards and light things come to the surface, and that is why the rubble and foundation-stones are at the bottom, and the earth* is above the stonework because it is lighter, and the wood is on top because it is the lightest of the materials. Now, it is 200^a5 true that there could not have been a wall without these materials, but they are not the reason for the wall (except in the sense that they constitute its matter): the wall exists for the purpose of concealing and protecting certain things. The same goes for anything else which has a purpose: its existence is not separable from things whose nature is subject to necessity, but these things are not the reason for the thing (except

^a10 in the sense that they constitute its matter). Its reason is its purpose. In answer to the question 'Why is a saw as it is?', for instance, we say 'So that it can do such-and-such' and 'For such-and-such a purpose.' However, this purpose is unattainable unless the saw is made out of iron. So it has to be made out of iron if it is to be a saw and if its job is to be done. So what is necessary is so conditionally, not as an end, because the necessity is in the matter,* but the end is in the definition.

^a15   The way in which necessity plays a part in mathematics is in a way similar to the way in which it plays a part in natural events. It is because a straight line is as it is that a triangle necessarily has angles equal to two right angles, and the converse does not obtain (although it is true that if the angles of a triangle were *not* equal to two right angles, then the

^a19 straight line would not be as it is). It is the other way round, however, for purposive things: if the end is to be realized or has been realized, the earlier phase must be realized or must have been realized. If this does not happen, then the end or purpose will not be realized, which corresponds to the fact that in mathematics if the conclusion does not obtain, then

^a22 the original principle does not obtain either. The point is that the end is also an originating principle—not the originating principle of a sequence of actual events, but of a chain of reasoning.* (In the mathematical example too no actual events are involved and the originating principle is the originating principle of a chain of reasoning.) So if there is to be a house, certain things must necessarily have come into being or be present, or in general matter which serves the purpose must necessarily exist—for instance, bricks and stones, in the case of a house. But these things are not and will not be the reason for the end's existence (except in the sense that they constitute its matter). Nevertheless, generally speaking, if they did not exist, there would not be a house or a saw—that is, if stones did not exist in the one case, and if iron did not exist in the other. Likewise, in the mathematical example, if triangles did

not contain angles equal to two right angles, the original principles would not exist.

It is clear, then, that in natural phenomena what is neces- ᵃ30 sary is the matter and the changes it undergoes. A natural scientist should discuss both causes, but especially a thing's purpose, because it is the cause of the matter, rather than the matter being the cause of the end. The end is the purpose, and the principle from which everything begins is the definition and specification of what the thing is, as in the case of artificial products. It is because a house is such-and-such that certain objects must necessarily have come into existence and be present, and it is because health is so-and-so that certain other things must necessarily have come into existence and be present. By the same token, if a man is such-and-such, then there must be so-and-so, and if there is to be so-and-so, then also so-and-so.

It is possible that what is necessary* should also be a part 200ᵇ4 of the definition. If we define the job of sawing as a certain kind of division—well, this kind of division will not exist unless the saw has teeth of a certain kind, and these teeth will not exist unless they are made out of iron. Even in a definition there are parts which are, so to speak, the matter of the definition.

III

A. CHANGE

1. *Introduction to Books III–IV**

200^{b}12 Nature is the subject of our enquiry, and nature is a principle of change, so if we do not understand the process of change, we will not understand nature either; we must devote some attention to change, then. Once we have decided what change is, we will have to try to do the same as we tackle the next

^{b}16 issues. The process of change appears to be continuous, and continuity seems to be the primary context of infinity. That is why in defining continuity one is almost bound to rely on the notion of infinity: it is because the continuous is what is

^{b}20 infinitely divisible. Moreover, change seems to be impossible without place and void and time, and in any case place, void, and time are pervasive and common to all kinds of change, so for both these reasons we shall obviously have to look into each of them as we get to grips with our subject. The point here is that the study of what is special to this or that kind of change is subsequent to the study of what is common to them all. Anyway, let us start, as I said, with change.

The definition of change

^{b}26 Things exist either only actually,* or both potentially and actually; and things are either 'such-and-such a particular object' or 'of such-and-such a quantity' or 'of such-and-such a quality', and so on for all the other categories of being. Things are described as 'relative' either because of excess and deficiency, or because of their ability to act and be acted on, or in general because of their ability to cause change and be changed. These are relative in the sense that anything which can cause change must cause *something* to change and it must be something that can be changed. Similarly, what can be changed must be changed *by something* and it must be something that has the ability to cause change.

There is no change over and above the circumstances of ᵇ32 change. For when something changes, it inevitably does so in respect of substance or quantity or quality or place,* and, as I say, it is impossible to conceive of anything which these categories all share which is not itself either a substance or a quantity or a quality or a member of one of the other categories. So there is no change of anything over and above the kinds of change mentioned, because there *is* nothing over and above what has been mentioned. However, there are always 201ᵃ3 two ways in which an item in any category can be the property of something. Where a substantial object is involved, for instance, there is either its form or the privation of that form; in respect of its quality, it may be either pale or dark; in respect of its quantity, it may be either complete or incomplete. The same goes for motion as well: there is upward motion and downward motion, lightness and heaviness. The upshot is that there are as many kinds of change as there are categories of being.*

Now that we have distinguished between potentiality and ᵃ9 actuality in each category, we can see that change is the actuality of that which exists potentially, in so far as it is potentially this actuality. For example, the actuality of a thing's capacity for alteration, in so far as it is a capacity for alteration, is alteration; the actuality of a thing's capacity for increase, and for the opposite decrease, is increase and decrease (there is no single term which covers these two); the actuality of a thing's capacity for being created and destroyed is creation and destruction; the actuality of the capacity for movement is movement. The following case makes it clear that this is what change is: the actuality of something constructable, in so far as it deserves just this description, is when it is being constructed, and this is what the process of construction is. The same goes for the processes of learning, healing, rolling, jumping, maturing, and ageing.

Something can have the same property both potentially ᵃ19 and actually, but not at the same time or in the same respect (consider something which is actually hot and potentially cold). Such things will act on and be acted on by one another in many ways, because all of them are simultaneously capable

57

of acting and of being acted on. Consequently, anything which causes change in a natural way is also capable of being changed, because anything of this kind is itself changed in the process of causing change. (Some people think that *everything* which causes change is changed while doing so, but other considerations will make the facts of the matter clear: there is, in fact, something which causes change without being changed itself.*)

$^a27$ Anyway, change is the actuality of that which is potential, when that which is potential is actually active not as itself but as something which is capable of change. I mean the 'as' in this way. Bronze is potentially a statue, but still the actuality of the bronze, as bronze, is not a change. This is because what it is to be bronze and what it is to be something potentially are different. (If they were, without qualification and in definition, the same thing, then the actuality of the bronze, as bronze, would be a change. But as I have said, they are not $^a34$ the same thing.) The point is clear in the case of opposites.* A capacity for health and a capacity for illness are different (otherwise being ill and being healthy would be the same), but whatever it is that underlies them—the thing which is either healthy or diseased—is one and the same thing (whether it is the moist, or blood, or whatever). Since the underlying thing and what it is to be something potentially are different —just as 'colour' is different from 'visible thing'—it is clear that the process of change is the actuality of what is potential, as potential.

201^b5 It is easy to see that this is what change is, and that change occurs precisely when the actuality is as described† and not before or after (for anything can be active at one time and not at another). Consider something constructable, for example. Its actuality, as constructable, is the process of construction. For the actuality of something constructable must be either the process of construction or the house constructed. But once the house exists, there is no longer anything construct*able*; on the other hand, what is constructable does undergo the process of construction. So the process of construction must be its actuality, and this process is a kind of change. The same account will fit other kinds of change as well.

2. *The definition confirmed*

One can see that this account is correct, both from what ᵇ16
others have said about change and from the fact that it is not
easy to define it in any other way. It is impossible to classify
change in any other genus, and this becomes clear when you
look at some of the alternative classifications* which have
been suggested, by those who call change 'difference', 'inequality' or 'not being'. It is not inevitable that anything will
change just because it is different or unequal or is not; moreover, change is no more into or from these things than it is
into or from their opposites.

The reason they think of change in these ways is that it ᵇ24
seems to be indeterminate, and all the principles in the second
list* are indeterminate because they are privative: none of
them is 'such-and-such a thing' or 'such-and-such a quality',
and none of them belongs to any of the other categories
either. And the reason why the process of change is apparently indeterminate is that it cannot be located either on the
potential or on the actual side of things, since neither something which is potentially such-and-such a quantity nor something which is actually that quantity is necessarily changing.
Also, the process of change does seem to be a kind of actu- ᵇ31
ality, but an incomplete one,* and the reason is that the
potential of which it is the actuality is incomplete. This makes
it hard to grasp what change is. For it has to be assigned
either to privation or to potentiality or to simple actuality,
but none of them seem possible. The only remaining way to
understand it, then, is that it is a special kind of actuality—
the kind I have described, which may be elusive, but is not
impossible.

Further points about change

Everything that causes change is changed, as I stated earlier,* 202ᵃ3
as long as it is capable of change, and as long as it is at rest
when not changing. (It is only things which admit change
that can be at rest—i.e. when they are not undergoing change.)
For to act on something changeable, in so far as it is change-

able, is precisely to change it, and it takes contact to do this,* so the agent of change is also acted on at the same time.

ᵃ7 Change, then, is the actuality of the changeable, *qua* change-able, and this happens as a result of contact with the agent of change, which is therefore also acted on at the same time. The agent of change will always bring with it some form, which will either be 'such-and-such a thing' or 'such-and-such a quality' or 'such-and-such a quantity', and this form will be the principle and cause of any change that the agent of change produces. For instance, it is an actual man who creates a man out of that which is potentially a man.*

3. *Change takes place in the object changed, not in the agent of change*

ᵃ13 We now have a clear answer to a point of difficulty, namely that change takes place *in* the thing that is capable of being changed, since the change is the actuality of this, brought about by what is capable of changing it. In fact, the actuality of what is capable of causing change and the actuality of what is capable of being changed are the same. The one actuality must be the actuality of both, because* it is thanks to the agent's capacity for causing change that it has the potential for causing change, and it is thanks to its actual activity that it actually causes change, but what it actualizes is something that is capable of being changed. So there is only one actuality involved for them both, just as there is the same interval from 1 to 2 as there is from 2 to 1, and just as uphill and downhill* are identical (for these things are the same, although their definitions are different). The same goes for the agent of change and that which is changed.

ᵃ21 At an abstract level,* there is a problem in this. Perhaps both the agent and the subject it acts on necessarily have their own actualities; after all, there are two processes involved— doing and being done to—and the product and end of one is something done, while the product and end of the other is something affected. Both of them are processes of change, then. Now, suppose they are different: if so, where are they located? Either they are both in that which is being affected and changed, or the doing is in the agent and the being done to is in the subject. (If 'doing' is the proper term for 'being done to' as well, then 'doing' is being used in two senses.) But in that case the argument for saying that the change is the being done to, and so is in the subject, will equally show that the change is the doing, and so is in the agent. The upshot* would be that either every agent itself changes, or that without changing it has change in it. $^a28$

If, on the other hand, both doing and being done to are in that which is being changed and acted on (i.e. although there are two processes, both teaching and learning occur in the student), then in the first place the actuality of a given thing will not be in that thing, and in the second place, it is absurd for two processes of change to coincide like that: I mean, what *two* alterations could take place in a *single* subject which is developing towards a *single* form? The idea is impossible. $^a31$

In that case, there must be a single actuality* involved, not two different ones. But it is absurd for two things which are different in form to have a single, identical actuality. If the processes of teaching and learning, doing and being done to, are identical, then teaching will be the same as learning, acting will be the same as being affected, and the result will be that every teacher is bound to be a student and every agent is bound to be a subject. $^a36$

On the other hand, one could argue that it is not absurd for the actuality of one thing to take place in another thing. After all, teaching is the actuality of a capacity to teach, and it is practised *on* someone; it is not practised *in vacuo*, but *by* someone *on* someone else. 202^b5

Secondly, there is nothing to stop two things having a single, identical actuality.* They do not have to be the same in definition, but only in the way that what is potential is related to what is actual. $^b8$

Thirdly, even if acting and being affected are the same, a teacher will not necessarily be a student, as long as the definition of what they are is not one and the same (compare 'mantle' and 'cloak'*). They could still be the same in the sense that the road from Thebes to Athens is the same as the road from Athens to Thebes (as we said before), because $^b10$

things which are in some sense or other the same do not have all the same properties;* that applies only to things which are $^b16$ the same in definition. Indeed, even if the process of teaching is the same as the process of learning, it does not follow that to learn is the same as to teach. Consider the following analogy: even though two separated things have a single interval between them, still the separation from here to there is not one and the same as the separation from there to here. To put the matter generally, then, the processes of teaching and learning are not absolutely identical,* and neither are the processes of doing and being done to; rather, they are properties of the same thing, namely the change. The point is that the actuality of one thing *on* another, and of one thing *by* another,* are different in definition.

$^b23$ We have now said what change is, both in general and in particular, since it is relatively clear how the various species of change are to be defined. For instance, alteration is the actuality of a thing's capacity for alteration, *qua* having a capacity for alteration. To put this more intelligibly: change is the actuality of the potential for acting and being affected, in so far as it is just such a potential. This is the unqualified version, and then it can also be applied to particular cases— the processes of construction or healing, for instance. And every other process of change can be described in the same way.

B. INFINITY

4. *Problems concerning infinity*

$^b30$ Scientific knowledge of nature involves taking magnitudes and change and time into consideration, and each of them is bound to be either infinite or finite. It is not the case that *everything* is either infinite or finite: a quality is not, for instance, and neither is a point, since there is presumably no need for them to be either one or the other. But a student of nature should consider infinity and try to find out whether or not there is such a thing, and if there is, what it is.

$^b36$ A sign that the consideration of infinity is relevant to scientific knowledge of nature is the fact that all those who seem to have made a significant contribution to this branch of philosophy have had something to say about infinity. In fact, they all* make it a principle of things.

Some of them, like the Pythagoreans and Plato,* claim that 203ª4 it is a principle in its own right—that the infinite is itself a substantial entity, not just an attribute of something else. The difference between them is that the Pythagoreans classify it as perceptible (since they do not think of number as an independently existing property) and claim that the region beyond the heavens is infinite, whereas although Plato claims that there is nothing material beyond the heavens (and that the forms are not beyond the heavens either, because they are not anywhere), he finds infinity in perceptible things and in forms as well. Also, the Pythagoreans identify infinity with even number, ª10 on the ground that when an even number is enclosed and limited by an odd number it makes things infinite. As evidence for this they cite what happens with numbers, namely that successive gnomons* placed around unity produce a uniform shape, whereas successive gnomons apart from unity produce a diversity of shapes. Plato, on the other hand, has two infinites,* the great and the small.

Others—the natural scientists without exception*—make ª16 something else (one of the things they identify as elements, such as water or air or something intermediate between them) the subject of which infinity is predicated. But none of those who posit a finite number* of elements makes them infinite in extent, whereas all those who posit an infinite number of elements say that the infinite forms a continuous whole, thanks to contact. Anaxagoras, for instance, relies on his homoeomerous substances for this, while Democritus relies on his 'seed-aggregate'—that is, the atoms with all their different shapes.*

Anaxagoras said that every part is just as much a mixture ª23 as the whole universe is; he based this view on the observation that anything can come from anything. That is also probably why he said that all things were once mixed together. His reasoning was probably as follows: this flesh and this bone are like that,* and so is anything else, so everything must be like that, and must have been like that at one and the same time, because not only is there a beginning of the separating process from which each individual arises, but there must also be a beginning for the universe as a whole. Why? Because anything which comes into being comes from that kind of body, and everything does in fact come into being (although not at the same time), and this process of coming to be must have a source. Moreover, this source must be a single principle, of the kind which Anaxagoras calls intelligence, and there is always a starting-point at which our intellects stop thinking and set to work. And the upshot of all this is that everything must once have been mixed together and must have started changing at some point in time.

a33   Democritus, however, denies that any primary element can come from any other. Nevertheless, for him there is one body, shared by everything, which is a principle of everything, although it comes, part by part, in different shapes and sizes.

203b3   All this makes it clear that infinity is a relevant field of enquiry for natural scientists. It is also plausible for them all to regard it as a principle, because it cannot be ineffective, and its power must be that of a principle. After all, everything either is an original principle or comes from an original principle, and the infinite cannot have an origin, because that

b7   would limit it. Moreover, they take it not to be subject to generation or destruction, on the grounds that it is a kind of principle, because anything generated must have a last part that is generated, and there is also a point at which the destruction of anything ends. That is why, as I say, the infinite is taken not to *have* an origin, but to *be* the origin of everything else—to contain everything and steer everything, as has been said by those thinkers who do not recognize any other causes* (such as love or intelligence) apart from the infinite. They also call it the divine, on the grounds that it is immortal and imperishable; on this Anaximander and the majority of the natural scientists are in agreement.

b15   There are five considerations which particularly lead people* to infer that something infinite does exist. First, there is the fact that time is infinite. Second, there is the division of magnitudes (for mathematicians too rely on infinity). Third, there is the notion that the only possible explanation for the persistence of generation and destruction is that there is an infinite source from which anything which is generated is subtracted. Fourth, there is the idea that there is always something by which the finite is limited, from which it follows that there cannot be an ultimate limit, since one thing must always be limited by another. But above all, the most ^b22 convincing consideration (because no one finds it easy to cope with) is that number and mathematical magnitudes and the region beyond the heavens seem to be infinite because they do not give out in our thought. And if the region beyond the heavens is infinite, then it seems that body must be infinite too (i.e. there must be infinitely many worlds); why, after all, should there be body in one part of the void rather than another? And so, provided that there is body in one place, there must (the argument goes) be body everywhere. Also, as long as the existence of infinite void and place is granted, there is bound to be infinite body too, because for eternal things there is no difference between being possible and being actual.*

Thinking about infinity is not straightforward; there are a ^b30 lot of intractable consequences whether you assume that there is or is not such a thing as infinity. Besides, there are two possible ways in which it might exist: is it a substance or is it in some natural way an attribute in its own right of something else? Or perhaps it is neither, but there is still some thing or things that are infinite in number?* The most relevant question for a natural scientist, however, is whether there is an infinite perceptible magnitude.

The first thing we have to sort out, then, is how many 204ª2 meanings 'infinite' has. One application of the term is for that which cannot be traversed, simply because it is not the kind of thing which can be traversed—just as sound, for instance, cannot be seen. From another point of view, the infinite is that which is capable of being traversed, but the journey would never end. Or it is that which is capable of being traversed, but only with difficulty. Or it is the kind of thing whose nature is such that it might be traversed, but is not in fact traversed or bounded. Moreover, anything infinite is either infinite by addition or by division or both.*

5. *The infinite is not itself a substance**

a8 Now, it is impossible for the infinite to be separable from perceptible things and just itself. In the first place, if the infinite is not a magnitude or a plurality, but is itself a substance rather than a coincidental attribute, it must be indivisible, because divisible things are either magnitudes or pluralities. But if it is not divisible, it is not infinite—except in the sense that sound is invisible. But when people say that the infinite has substantial existence, they do not mean 'infinite' in this sense, and this is not the sense we are interested in either; what they mean, and what we are interested in, is 'infinite' in the sense of 'untraversable'. (On the other hand, if the infinite is not a substance, but a coincidental attribute, then it cannot be, *qua* infinite, an element of things, just as the fact that sound is invisible does not mean that the invisible is an element of speech.)

a17 In the second place, how could there be an independent infinite, if there cannot be independent number and magnitude? Infinity is in its own right a property of number and magnitude, so, of the three, an independently existing infinite is the least necessary.

a20 Also, it is evidently impossible for the infinite to exist in actuality *and* to have substantial existence as a principle as well. The point is that if the infinite is a substance and not a predicate of some underlying subject, then the infinite is nothing but what it is to be infinite, and in that case (assuming that it has parts) any part of it you take will be infinite. So either it will be indivisible or it will be divisible into infinites. But one and the same thing cannot consist of a plurality of infinites.* (Yet if the infinite does have substantial existence as a principle, then just as any part of air is air, so also any part of the infinite will be infinite.) It follows, therefore, that the infinite must be indivisible and must have no parts. But this is out of the question for any actual infinite, since any actual infinite is bound to be a quantity.

a29 So the infinite is not a substance, but a coincidental attribute of things. If so, however, it cannot (as I have already said) be called a principle; it is that of which it is an attribute (the air, for instance, or the even numbers*) which should be called a principle. And so we can see how implausible are those who repeat Pythagorean doctrine, when they maintain that the infinite is not only a substance, but is divisible into parts as well.

There is no infinitely extended body

Now, the issue here might be a very general one, including $^a34$ the question of whether there is a place for infinity among mathematical entities and among things which are intelligible* and which have no magnitude. However, we are conducting an investigation into perceptible things, and what we are trying to find out is whether or not there is a place among the objects we are studying for an infinitely extended body. At an 204^b4 abstract level, the following considerations make it look as though there is not. If we define 'body' as 'that which is bounded by surface', there cannot be an infinite body, whether that body is perceptible or merely intelligible. Nor can any number that exists apart from perceptible things be infinite either, because number, and anything which has a number, is countable; anything countable can be counted, and it follows that it would be possible to traverse the infinite.

If we employ an approach which is more in keeping with $^b10$ natural science, the following considerations lead us to the same conclusion. The point is that an infinite body cannot be composite, but cannot be simple either. Given that it has a finite but plural number of elements,* it cannot be composite, because the opposite elements must always balance each other. So it is impossible for any one of them to be infinite, however much its power falls short of its opposite's power. Suppose, $^b15$ for example, that there was finite fire and infinite air, and suppose a certain amount of fire was n times more potent than the same amount of air (however great n may be, as long as it is a number): even so it is easy to see that the infinite one will outdo and destroy the finite one. Nor can each of the pair of elements be infinite, because a body is what is extended in all dimensions, and an infinite body must have infinite extension, and so an infinite body would be infinitely extended in all dimensions.

b22   However, there equally cannot be one simple infinite body, and this is so not only if, as some say,* it is an extra body over and above the elements, which acts as the source of the elements, but also on a more straightforward view. Those who suggest that the infinite is not air or water, but this extra body, do so because they want to avoid everything else being destroyed by an infinite element. The point is that the elements are related by mutual opposition (air is cold, for instance, while water is moist and fire is hot), and so if any one of them were infinite, the others would have been destroyed by now. So in fact, they say, there is this extra body which

b29   is the source of the elements. But it is impossible for there to be any such extra body. It is not the fact that it is infinite which makes it impossible; to show that, we will have to produce a general argument which applies equally to every infinite body, whether it is air or water or whatever. No, it is the fact that our senses do not show us any such body; all they show us are the familiar elements. The point is that anything with a source is dissolved back into the source it has come from; so such an extra element, if it existed, would have been here, alongside air, fire, earth, and water. But we see no such thing.

b35   Nor can one of the elements—fire, for instance—be infinite: for there is the general consideration, quite apart from any of them being infinite,* that it is impossible for the whole universe (even if it were finite) to be or to become just one of the elements—as Heraclitus says that at some time everything becomes fire. And the same argument applies to any single body, such as the extra body that the natural scientists come up with over and above the elements. This is because everything changes from opposite to opposite—from hot to cold, for example.

205a8   The following argument† shows that it is fundamentally impossible for there to be a perceptible infinite body. It is the nature of perceptible things that they exist somewhere; each of them has some place. This applies to parts as well as to wholes—to a single lump of soil as well as to the whole earth,

a12   to a spark as well as to a fire. What follows? First, consider a homogeneous infinite body: it must be either motionless or in constant motion. But in a homogeneous body, what counts as above rather than below or any direction at all? I mean, for instance, if there were a lump of this body, where would it move, and where would it be at rest? The difficulty is that the place occupied by stuff which is indistinguishable from it is infinite. Are we to say that the lump will occupy the total place? Of course not. So how or where will it be at rest? How or where will it move? Will it be at rest everywhere? If so, it will never be in motion. Will it be in motion everywhere? If so, it will never stop moving. (The reason, then, that the ᵃ25 natural scientists always make their single infinite body either water or air or something intermediate between them, rather than fire or earth, is because fire and earth both have a clear and definite place, whereas water and air vacillate between upward and downward motion.)†

Consider, second, the possibility that the universe is *not* ᵃ19 homogeneous. In that case, there are also differences between one place and another. First, then, the body which constitutes the universe will not be single, except in the sense that its parts are in contact with one another. Second, these parts must be either finitely or infinitely variable in form. But they cannot be finitely variable, because in that case some of them will be infinite and some finite, given that the whole universe is infinite. Suppose that fire or water is infinite; in that case, they will destroy their opposites, as I have already explained.†* If, on the other hand, there are infinitely many simple parts, ᵃ29 there will be infinitely many places too, and then there would have to be infinitely many elements.* If that is impossible and there must be a finite number of places, then the whole too must be finite†—I mean, there has to be an exact correspondence between place and body, since the total place cannot be larger than the body is capable of filling (which implies once more that the body cannot be infinitely large), and the body cannot be larger than the place either. If place was larger than body, there would be void; if body was larger than place, there would be body whose nature was not to exist anywhere.

Anaxagoras' explanation for why the infinite is at rest is a 205ᵇ1 strange one: he says that the infinite itself fixes itself in place, because it is in itself—there is in this case nothing else that contains it. Where a thing is, he assumes, there it is its nature to be. But this is false: a thing might be forced to be somewhere other than its natural place. So however true it may be that the whole is not in motion (granted that what fixes itself in place and is in itself is necessarily motionless), he still

b8 needs to explain why it is not in its nature to be moved. It is not enough just to say what Anaxagoras says and to leave it at that, since it might be motionless because it has nowhere else to move to, while still, quite possibly, being the kind of thing which *can* move. Consider the earth, for example: it does not move either, and would not move even if it were infinite, as long as it was constrained by the centre of the universe; it would be at rest, not because there is nowhere for it to go, but because it is its nature to behave like that. And yet one could say that it fixes itself in place. So if this would not explain the lack of motion of the earth (supposing the earth to be infinite)—that is, if the real reason is that the earth has weight, and anything with weight stays at the centre, and so the earth stays at the centre—then by the same token the infinite too would be motionless within itself not because it is infinite and fixes itself in place, but for some

b18 other reason. It is also clear that any of its parts would have to be motionless as well. Just as the infinite is at rest within itself because it fixes itself in place, so any given one of its parts will also be at rest within itself. The point is that the places of the whole and of the part are identical in form; for example, both earth as a whole and a lump of earth tend downwards, and both fire as a whole and a spark tend upwards. Consequently, if the place of the infinite is 'in itself', the place of the part is the same—it will be motionless within itself.

b24 A general consideration is that it is plainly impossible to maintain simultaneously that there is an infinite body and that there is a particular place for bodies, in the sense that every perceptible body has either weight or lightness, and tends naturally towards the centre (if it has weight) or upwards (if it has lightness). This must apply to an infinite perceptible body as well, but it is impossible either for it as a whole to have one of the two qualities or for half of it to have one quality and half the other. I mean, how will you divide it? Or alternatively, how can there be above and below within the infinite, or an outside and a centre?

Besides, every perceptible body occupies a place, and the forms and varieties of place are above, below, forward, back, right, and left. These are not mere human conventions, but objective divisions of the universe.* But they cannot exist within the infinite. ^b31

In short, if place cannot be infinite,* and if every body occupies place, then there cannot be an infinite body. Now, anything that is somewhere is in a place, and anything that is in a place is somewhere. But just as the infinite cannot be a quantity (because then it would have to be a particular quantity, such as 2 or 3 feet long—that is what 'a quantity' means), so also anything that is in a place is so because it is somewhere, and that is to say that it is above or below or in one of the other six directions. But each of these directions has a limit.† ^b35

These considerations show, then, that there is no actually infinite body. ^206a7

6. *The sense in which infinity does exist*

However, if there is no such thing as infinity in *any* sense, there will clearly be a number of impossible consequences. For instance, there will be a beginning and an end to time, magnitudes will not be divisible into magnitudes, and number will not be infinite. When neither of two possibilities seems feasible in the light of what has already been settled, an arbiter is required; and obviously there is a sense in which there is infinity and a sense in which there is not. ^a9

Now, 'to be' means either 'to be potentially' or 'to be actually', and a thing may be infinite either by addition or by division. I have argued that no actual magnitude can be infinite, but it can still be infinitely divisible (it is not hard to disprove the idea that there are indivisible lines*), and so we are left with things being infinite potentially. We should not take 'being potentially' here as analogous to 'this material is ^a14 potentially a statue', which implies that it will in the future actually be a statue, and so conclude that there will* in the

a21 future also be an actual infinite. Given the ambiguity of 'to be', we use it of the infinite in the same way that we use it of a day or a contest—that is, because one thing happens after another. (The distinction between potential and actual applies to them too: the Olympic Games 'are' both in the sense that there is the potential for the contest to take place and in the sense that it is taking place.) But the way in which the infinite manifests itself is different in the case of time and the human race from what it is in the case of the division of magnitudes. Generally speaking, the infinite exists by one thing being taken after another. What is taken is always finite on its own, but always succeeded by another part which is different from it. But whereas in the case of magnitudes each part persists, in the case of time and the human race the parts cease to be, but in such a way that the process does not fail.

206b3 There is a sense in which the kind of infinity which depends on addition is the same as the kind of infinity which depends on division. Any finite magnitude can include infinity by addition as an inverse process, in the sense that to see division *ad infinitum* going on within it is at the same time to see addition up to a determinate limit going on within it. For if, in a finite magnitude, you take a determinate amount and add to it not by taking the same fraction of the whole, but the same proportion of what remains, you will never traverse the finite magnitude. (However, if you increase the ratio so that you take the same magnitude, whatever it was, you will traverse it, because every finite quantity is exhausted by the repeated subtraction of any definite quantity whatsoever.)

b12 To summarize: the only possible way for the infinite to exist is potentially and as a result of a process of subtraction. It has actual being only in the sense that we ascribe being to a day or a contest. It has potential being in the same way that matter does,* rather than existing in its own right, as finite things do. This is the way in which the kind of infinity which depends on addition has potential being—the kind we are saying is in a sense the same as the kind of infinity which depends on division. In both cases it is possible to take something extra, but in addition the total will not exceed *every* magnitude, as is possible in division, in which the parts become smaller than any determinate magnitude, and there is always something smaller still. However, the kind of infinity which goes $^b20$ beyond every determinate amount by addition cannot exist even potentially, unless there exists something which is actually infinite, but only coincidentally so—in other words, the kind of infinite the natural scientists describe their body beyond the universe as being, whose substance is air or whatever. But if it is out of the question for there to be an actually infinite perceptible body like this, it is obvious that even potentially there cannot be something which is infinite by addition, except (as I have explained) as the inverse of infinity by division. After all, even Plato was led to the notion of two infinites* $^b27$ because it looks as though it is possible to exceed any specified limit and to tend towards infinity both by increase and by subtraction. However, although he comes up with the idea of there being two infinites, he makes no use of them. For there is no infinite subtraction among numbers (since there is a smallest number, namely one), and he does not allow an infinite increase either (since he has number end at ten*).

Infinity turns out to be the opposite of what people say it $^b33$ is. It is not 'that which has nothing beyond itself' that is infinite, but 'that which always has something beyond itself'.* This is the idea which is implicit in people's description of those rings which do not have a bezel as 'endless', because it is always possible to go beyond wherever you are on them. The analogy is suggestive, but ultimately imperfect, because this 'endlessness' is not the only property a thing has to have in order to count as genuinely infinite: in traversing it, you should also never cover the same ground more than once. This does not happen with a circle, however, which only has the property that the next step is different from the previous one.

A thing is infinite, then, if, for any quantity already taken, 207^a7 one can always take some further part. Anything which has no part beyond itself, however, is complete and whole. In fact, that is how we define a whole, such as a whole person or a whole box—we define it as that which has no part missing.* And the definition 'that which has nothing beyond itself' applies just as much to an unqualified sense of 'whole'† as to particular kinds of wholes. However, that which has something missing, whatever it may be—something left outside itself—is not all present. ('Whole' and 'complete' are either utterly identical or very similar in nature.) Nothing is complete unless it has an end, and an end is a limit.

^{a}15   We have to conclude, therefore, that Parmenides' views are superior to Melissus', because according to Melissus the infinite is a whole, whereas according to Parmenides the whole is finite—'poised from the centre all ways equally'. The infinite is not at all the same thing as the all or the whole, and identifying them fails 'to join flax to flax'.* In fact, the attribution of grand properties to the infinite—saying, for instance, that it contains all things and holds everything within itself—is

^{a}21   due to its partial similarity to the whole. The point is that the infinite is the matter* for the completion of a magnitude and is potentially (but not actually) the completed whole. It is divisible (by subtraction and by the inverse addition), and it takes something else to make it whole and finite, which it is not in its own right; and in so far as it is infinite, it does not contain but is contained. That is also why it is unknowable, in so far as it is infinite: it is because matter has no form. It obviously follows that it is better to classify the infinite as a part rather than as a whole, because the matter of anything is a part of the whole—bronze, for instance, is part of a bronze statue. After all, if it does any containing in the perceptible world, then in the intelligible world the great and the small ought to contain intelligible things.* But it is absurd and unthinkable for something unknowable and indefinite to contain and define anything.

7. *Some consequences of this account*

^{a}33   Now, if what we have been saying establishes the kind of infinity which depends on division, but makes it look as though there cannot be an infinite by addition—at least not one that exceeds every definite magnitude—that would be another reasonable outcome of the discussion. The point is that matter and the infinite are contained within things, while form does the containing.

It is also reasonable that whereas in the case of number 207^{b}1 there is a smallest number, which limits its decrease, and yet in increasing it constantly goes beyond every quantity, the opposite is true of magnitudes: they go beyond any determinate magnitude when it comes to decreasing, but, when it comes to increase, there cannot be an infinite magnitude. The reason for this is that the number one is indivisible,* whatever it is that is one—a person, for instance, is one person and not a number of people—and any given number is a plurality of ones, a particular quantity of them. So of course number stops at what is indivisible. (The point is that the terms 'three' and 'two' are derivative,* and the same goes for the names of all the other numbers.) When it comes to in- ^{b}10 crease, however, it is always possible to conceive of a larger number, since any given magnitude can be halved infinitely many times. So this infinite is potential rather than actual, but it is always possible to take more than any specified number. However, this number does not exist apart from the process of halving, and the infinity is not stable but is being generated, as time is, and as the number of time* is. The ^{b}15 opposite is true of magnitudes, however: something continuous is infinitely divisible, but there is no infinity in the direction of increase. Any magnitude of any size that can exist potentially can also exist actually, and so, since there is no infinite perceptible magnitude, there can be no magnitude which exceeds every specified magnitude: that would mean that there was something larger than the universe.

The infinite is not the same in magnitude, in movement, ^{b}21 and in time; it is not a single kind of thing, but it has a primary sense and dependent senses.* For example, you can only describe a movement (or an alteration, or an increase) as infinite if the magnitude traversed is infinite; and you can only describe time as infinite if change is infinite. (For the time being, we can talk of time and change, but later* I will explain what each of them is and why every magnitude is divisible into magnitudes.)

I have argued that there is no such thing as an actual ^{b}27 infinite which is untraversable, but this position does not rob mathematicians of their study. Even as things are, they do not need the infinite, because they make no use of it. All they need is a finite line of any desired length. But any magnitude whatever can be divided in the same ratio as you would divide an enormous magnitude, and so, for the purposes of their proofs, it makes no difference whether the magnitude proposed is one of those which actually exist.

b34   In terms of the fourfold division of causes, it is easy to see that the infinite is a material cause, and that while what it is to be infinite is a privation, what in its own right underlies it is the continuous and perceptible. In fact, all other thinkers seem to treat the infinite as matter as well—so it is odd for them to make it the container rather than the contained.

8. *Response to the arguments for an actual infinite**

208ᵃ5   We still have to deal with those arguments which seem to imply that the infinite can exist not only potentially but also as a distinct thing. Some of them are not compelling; others can be countered by valid responses.

ᵃ8   First, in order for the process of generation to persist, it is not necessary for there to be an actually infinite perceptible body: in a finite universe, it is perfectly possible for the destruction of one thing to be the creation of another thing.

ᵃ11   Second, being finite has to be distinguished from being in contact. Contact is relative and is contact *with* something— anything which is in contact is in contact with something. It is true that contact may be a coincidental attribute of a finite object, but that does not make the finite object relative. Nor is contact possible between just anything and everything.

ᵃ14   Third, it is absurd to rely on what can be thought by the human mind, since then it is only in the mind, not in the real world, that any excess and defect exist. It is possible to think of any one of us as being many times bigger than he is and to make him infinitely large, but a person does not become superhumanly large just because someone thinks he is; he has

to be so in fact, and then it is merely coincidental that someone is thinking it.

Fourth, time (like change and thought) is infinite, but in the ᵃ20
sense that any given part of it does not persist.

Fifth, neither subtraction nor imagined increase make a ᵃ21
magnitude infinite.

This is all I have to say about the senses in which there is ᵃ22
and is not such a thing as infinity, and about what infinity is.

IV

A. PLACE

1. Reasons for supposing that place exists

208ᵃ27 A natural scientist must have the same kind of understanding
of place as he does of infinity—that is, he should know whether
or not it exists, in what sense it exists, and what it is—
because the idea that existing things exist somewhere is universally accepted. I mean, that which does not exist is nowhere.
Where, for example, is a goat-stag or a sphinx? Also, the
most common and most fundamental kind of change is change
of place, which is known as movement.

ᵃ32　　But what is place? The question is beset with difficulties.
Different conclusions seem to follow from considering different factors. Moreover, other thinkers are no help to us on
this topic; they do not contribute even a statement of the
difficulties, let alone a solution.

208ᵇ1　　The phenomenon of replacement seems to make it clear
that there is such a thing as place. There is water here now,
then after it has left (poured from a vessel, perhaps) there is
air here instead, and at another time some other body may
occupy the same place. This makes it look as though the
place is different from all the things that, by replacement,
come to be in it, because the place in which there is air at the
moment previously contained water, and so it is obvious that
the place or space which they alternately leave and enter is
different from both the air and the water.

ᵇ8　　Secondly, the movements of the simple natural bodies (fire,
earth, and so on) show not only that there is such a thing as
place, but also that it has a certain power. For unless prevented from doing so, each of them moves to its own place,
ᵇ12 which may be either above or below where it was. Above and
below and the other four directions are the parts or forms of
place. Directions like above, below, right, and left are not just

78

relative to us.* In the sense in which they *are* relative to us, they are not always the same, but depend on our position— that is, on which way we are facing. That is why the same object might well be, at different times, to our right and to our left, above us and below us, in front of us and behind us. But in themselves each of the six directions is distinct and separate. 'Above' is not just any random direction, but where fire and anything light move towards. Likewise, 'down' is not just any random direction, but where things with weight and earthy things move towards. So their powers as well as their positions make these places different. (Geometrical figures ᵇ22 show this* too: despite the fact that they do not occupy a place, they can still have a right and a left, depending on their position relative to us. It is only their position which allows us to predicate 'right' and 'left' of them; in themselves, they do not have any such property.)

Thirdly, those who claim that void exists are really talking ᵇ25 about place, since what they mean by 'void' is probably place deprived of body.

These facts support the idea that there is such a thing as ᵇ27 place, in addition to body, and that every perceptible body is in place. Hesiod* seems to be on the right track in putting Chasm first in his system. At any rate, the reason he says 'First came the Chasm, and then broad-breasted Earth' is presumably because the first requirement is that there should be space for things. In other words, he shares the common belief that everything is somewhere—that is, in some place. And if place is like that, then it would be truly remarkable and prior to everything, since that which is a prerequisite for other things to exist, but whose existence does not depend on other things, is bound to be primary. The point here is that place is not destroyed when the things it contains are destroyed.

Problems about place

Nevertheless, if there is such a thing as place, it is still diffi- 209ᵃ2 cult to decide *what* it is. Is it in some sense a body's volume, or what? Our first task must be to try to discover its genus.

ᵃ4 First, it is three-dimensional; it does have length, breadth, and depth, by which every body is defined.* But it is impossible for place to be a body, because in that case there would be two coinciding bodies.

ᵃ7 Second, if bodies occupy place and space, obviously surfaces and the other limits of solid bodies do too, since the same argument that applies to bodies applies to them too: where there were previously the surfaces of water, there will be instead the surfaces of air. However, we cannot differentiate* between a point and the place occupied by a point, and so, if there is no difference between a point and its place, there cannot be any difference between the other limits of bodies and their places either, with the result that place is *not* something additional to each of them.

ᵃ13 Third, what on earth could we say place is? Given the kind of thing it is, it cannot be an element and it cannot be made out of elements either, whether they are material or immaterial. It has magnitude, but is not material, whereas the elements of perceptible things are material, and things which are merely intelligible cannot constitute an object with magnitude.*

ᵃ18 Fourth, what aspect of things could we possibly say that place is a cause of? It is not one of the four causes.* It is not a material cause (for nothing is made out of it), nor is it a cause in the sense that it is the form and definition of things, nor is it the end of anything, nor is it an agent of change.

ᵃ23 Fifth, if place is itself an existing thing, then it will exist somewhere. For Zeno's puzzle needs explaining: if every existing thing is in place, an infinite regress occurs, because there will clearly have to be a place for place.

ᵃ26 Sixth, just as every body is in a place, so in every place there is a body. How are we to explain increase, then? It follows* that place must expand along with increasing bodies, since the place of a thing can be neither larger nor smaller than it.

ᵃ29 These difficulties must make us wonder not only what place is, but even whether there is such a thing.

2. *The place of a thing is neither its matter nor its form*

Sometimes we describe things directly, sometimes derivatively.* ᵃ31
'Place' may refer either to the shared place which contains all
bodies or to the particular place which immediately contains
a body. For instance, you are now in the world, because you
are in the air and the air is in the world; and you are in the
air because you are on the earth; and by the same token you
are on the earth because you are in this particular place,
which contains nothing more than you. So if place is what 209ᵇ1
immediately contains a body, it must be a kind of limit, and
the upshot is that a thing's place would seem to be its form
and shape, by which the thing's magnitude is defined and the
matter of its magnitude is determined. For the form of anything is its limit.

If one looks at place from this point of view, then, it is a ᵇ5
thing's form. However, place also seems to be the extension
of the magnitude of a thing, and taken in this way it is the
thing's matter. The extension of the magnitude is not the
same as the magnitude: it is what is contained and determined by form—by a surface and a limit, for instance—and
that description fits the indeterminate nature of matter. The
point is that when a sphere's limit and qualities are removed,
all you are left with is the matter. That is also why in the ᵇ11
*Timaeus** Plato identifies matter and space, because what is
capable of receiving form is the same as space. Actually, in
the *Timaeus* he gives an account of what receives form which
differs from the one he gives in what are called his unwritten
doctrines;* all the same, he did identify place and space.
Everyone assumes that there is such a thing as place, but
Plato is the only one who tried to say *what* it is.

It is not surprising that these considerations seem to make ᵇ17
it hard to understand what place is, if indeed it is one or the
other of these—matter and form. It is not just that matter
and form require the most intense study; it is also that it is
not easy to understand them in isolation from each other. In

fact, however, it is not hard to see that place cannot be either matter or form.

^{b}22 First, the matter and the form are not separate from the object of which they are the matter and the form, but the place *is* separable. For, as we have already said,* the place where air was before now contains water instead, as they replace each other, and the same goes for other bodies too; consequently, the place of any given thing is not a part or a

^{b}28 state of that thing, but is separable from it. In fact, people do think of place as being like a vessel (since a vessel is a movable place) and a vessel is not a part of the object it contains. Anyway, since place is separable from the object, it is not form; and since it is a container, it is different from matter. It also seems as though anything which is somewhere is not only itself, whatever it may be, but also has something else outside itself.

^{b}33 As an aside, we should ask Plato why, according to him, forms and numbers do not occupy place. After all, place is in his view that which is capable of receiving form, whether this is the great and the small or (as he claimed in the *Timaeus*) matter.

210^{a}2 Second, how could a thing move to its own place, if its place was its matter or its form? Nothing can be a place which involves no movement, and which is not either above or below. So we had better look for place among things which have these features.

^{a}5 Third, if place is in the object (as it must be if it is either form or matter), then place will have a place. I mean, both the form and the indefinite matter change and move along with the object, and do not constantly remain in the same place, but go wherever the object goes. Consequently, place will have a place.

^{a}9 Further, when water comes from air, air's place is destroyed,* because the newly generated body occupies a different place. But what is it for place to cease to exist?

^{a}11 We have now reviewed the arguments which force us to conclude that there is such a thing as place, and also those which make it difficult to know what it is.

3. *In the primary sense of 'in', nothing is in itself*

Next, we had better come to understand in how many ways ª14 we use the expression 'One thing is in another.'

First, there is the sense in which we say that a finger is ª15 *incorporated in* a hand and, in general, that a part is incorporated in a whole. Second, we also say that a whole *consists in* its parts, in the sense that there is no such thing as a whole over and above its parts. Third, we say that 'man' *falls within* 'animal' and, in general, that a species falls within a genus. Fourth, we also say that a genus is *included in* a species and, in general, that any part* of the species is included in the definition of the species. Fifth, we say that health *inheres in* hot and cold things and, in general, that form inheres in matter.* Sixth, we say that the affairs of Greece are *in the power of* the Persian king and, in general, that things are in the power of their original agent of change. Seventh, we say that things are *centred in* their good and, in general, their end or purpose. Finally, the most fundamental sense is when we say that something is *contained in* a vessel and, in general, in a place.

It is not easy to decide whether something can be in itself, ª25 or whether nothing can, in which case everything is either nowhere or in something other than itself. But the question is ambiguous: do we mean that something can be in itself directly, in so far as it is just itself, or derivatively, in so far as it is something else? When both contained and container are parts of the same whole, the whole may be said to be in itself, since we can also say of the whole what is true of the parts (as when we say that a person is pale because his skin is pale, or knowledgeable because the thinking part of his mind is knowledgeable). So a jar cannot be in itself and wine cannot be in itself, but a jar of wine can be in itself, because the wine is in the jar, and in this case both contained and container are parts of the same whole, namely the jar of wine.*

So it is possible for something to be in itself—not in the ª33 primary sense of the expression, however, but only in the sense in which we say that pallor is in the body (i.e. because the surface is in the body, being a part of it) or that knowledge is in the mind. So we describe the person as pale or as knowledgeable, but the descriptions apply directly to these other things, which are parts of the person. When the jar and the wine are separated from each other, they are not parts of anything; it is only when they are together that they are parts of the same thing, and that is why, in this case, what they are parts of may be in itself. For instance, pallor is in a person because it is in his body; it is in his body because it is in his surface; and there the sequence of pallor being derivatively in

210^b6 something ends. Now, pallor and surface are two different things, each with its own nature and properties, so if we are arguing from examples we can infer that the same goes for everything: we cannot find a case of anything being in itself, in any of the senses of the expression we have distinguished.

^b9 Moreover, thinking about the matter also shows that it is impossible: it would mean that each of two things would have to be both at once. For instance, the jar would have to be both a vessel and wine, and the wine would have to be both wine and a jar; otherwise it will be impossible for a thing to be in itself. And the upshot is that, however true it may be that one is in the other, the jar will still contain the wine not because it is itself wine, but because it is a jar, and the wine will be in the jar not because it is itself a jar, but

^b16 because it is wine. It is clear that, in terms of what they are, they are different: the definitions of contained and container are different. In fact, it is not even coincidentally possible for a thing to be in itself, because then two things would simultaneously be in the same place.* The jar would be in itself (if a container can be in itself!) and what it contains (i.e. wine, if it is wine it contains) would also be in it.

^b21 We can see, then, that it is impossible for something to be in itself in the primary sense of the expression. Nor is it difficult to find a solution to Zeno's puzzle that if there is such a thing as place, it must be in something. The point is that it is perfectly plausible for the immediate place to be in something else, as long as 'in' is not understood as implying location* within a place, but is taken in the sense in which

health is 'in' hot things (because it is a state of hot things) and in which heat is 'in' the body (because it is an affection of the body). This avoids the infinite regress.

What is also clear is that since the vessel is not a part of ᵇ27 what is in it (because, in the basic sense of 'in', contained and container are different), and since matter and form are parts of what is contained, then place cannot be either matter or form, but must be something else.

That concludes our discussion of the difficulties. ᵇ31

4. *What a place is*

But what actually is place? Here is a way to find an answer. ᵇ32 Let us take as premisses all the properties which seem genuinely to belong to place in its own right. This is what we expect to be true of place: that it is the immediate container of that of which it is the place, that it is not a part of the object it contains, that a thing's immediate place is exactly the same size as it, that it can be left behind by the object and is separable from it, and also that every place admits of the distinction between above and below, and that every body naturally moves up or down to its own proper place and stays there.

So let us take these axioms for granted and proceed with 211ᵃ6 the enquiry. The aims of our enquiry should be to answer the question 'What is place?' in such a way that we are in a position to solve any difficulties, to attribute to it the properties which are supposed to belong to it, and also to explain why any qualms and difficulties arise. Any exposition which achieves all this, on any topic, has succeeded admirably.

The first point* to appreciate is that it would never occur ᵃ12 to us to make place a topic for investigation if there were no such thing as change of place. That is the main reason we think that even the heavens are in place—because they are in constant motion. This kind of change may be either movement or increase and decrease; increase and decrease involve change of place too, in the sense that what was previously in a given place has subsequently been displaced into a larger or smaller one. (Things which move do so either actually, in ᵃ17 their own right, or coincidentally. Some things which move coincidentally can move in their own right (such as the parts of the body and the nails in a ship), while other things are capable only of coincidental movement (such as pallor or knowledge, which change their place only if the object whose properties they are changes its place).)

ᵃ23   We say that we are in the world, meaning that we are in a place, because we are in the air and the air is in the world. And when we say that we are in the air, we do not mean air as a whole, but it is the particular limit of it which contains us that allows us to say that we are in the air. (If air as a whole were our place, the place of a thing and the thing itself would not be the same size; but they are supposed to be the same size—that is what the immediate place of a thing is.)

ᵃ29   Now, when a container is continuous* with what it contains, and not distinguished from it, we do not talk about the container being the place in which the object is contained, but we describe it as the whole of which the object is a part. However, when a container is distinguished from the object, but in contact with it, the object is in the immediate limit of the container, and this limit is not a part of the object it contains, and is not greater in extent than the object it contains, but is the same size, because the limits of any two objects that are in contact with each other coincide. Also, if an object is continuous with its container, we do not say that it moves *in* its container, but *with* its container; it is only if it is distinguished from its container that—whether or not the container itself is moving—we say the object moves in it.

211ᵇ5   It is already clear, as a result of what we have been saying, what place is. We can be pretty certain that it must be one of four things: shape, or matter, or some kind of extension between the limits of the container, or the limits themselves, if the only extension they contain is the magnitude of the contained body. But three of these can evidently be ruled out.

ᵇ10   The idea that it contains the object is what makes it plausible to think of shape as place, because the limits of container and contained coincide. It is true that both shape and place are limits, but they are not limits of the same thing. Form is the limit of the object, but place is the limit of the containing body.

The reason for thinking that between the limits of the ᵇ14
container there is some kind of extension is that the distinct
object contained often moves while the container stays still
(think of water pouring out of a vessel), which makes it look
as though there is something over and above the body which
is being displaced. But this is wrong. What actually happens
is that some other body—it could be anything, as long as it
can be displaced and can fit the container—comes in to replace the original body. If place was some kind of extension ᵇ19
which was capable of independent and permanent existence,
there would be an infinite number of places in the same thing,
because when water (or air) is being displaced bit by bit,
every bit behaves in the same way* within the whole as the
whole mass of water does in the vessel. Also, place will not
in fact be a stable entity,* and so one place will occupy
another place, and there will be a plurality of coincident
places. But when the whole vessel is being displaced, that
does not alter the place inside which any given part is moving:* it remains the same, because air and water (or water's
parts) do not replace each other in the place they are going
to be next; they replace each other in the place where they are
now. The place they are going to be next is a fragment of the
place which is the place of the whole world.*

Focusing on something which is at rest and not separate but ᵇ29
continuous with its surroundings* might incline one to suppose
that matter was place. For just as in alteration there is something which is now pale, but was previously dark, or which
is now hard, but was previously soft—and that is why we say
that there is such a thing as matter—so a similar phenomenon
makes it plausible to say that there is such a thing as place
too. The only difference is that in the former case something
which was air is now water, while in the case of place where
there was air, there is now water. However, as I have already
explained, matter is not separable from the object, and does
not contain it either, whereas place has both properties.

So if place is none of these three—form, matter, and some 212ᵃ2
kind of unchanging extension over and above the extension
of the displaced object—then it must be the remaining one of
the four. It must be the limit of the containing body, by which

the container makes contact with what it contains. By 'what it contains' I mean a body which is capable of movement.

ª7 Now, there are two reasons why place is supposed to be something profound and hard to understand. First, the comparison with matter and shape confuses the issue; second, the displacement of the contents happens within a container which is at rest. This is a cause of confusion because it makes it seem possible for there to be an extension between the limits which is not the same as the magnitude of the moving bodies. The apparent immateriality of air also contributes towards this confusion, by making place seem to be not only the limits of the vessel, but also an apparent void between them.

ª14 Just as a vessel is a movable place, so place is an immovable vessel. That is why when something is in motion inside a *moving* object (imagine a ship on a river), the container functions as a vessel rather than as a place. Given that place is meant to be immovable, the whole river is really the place for the ship, because taken as a whole the river is immovable. And so place is the nearest unmoved limit of the container.

ª21 We can now see the reason why the centre of the world and the inner limit of the heavenly revolution are taken to give us 'above' and 'below' in the most basic sense. It is because of their constancy: the centre is absolutely stable and the limit of the rotation always stays in the same state. So since what is light is what naturally moves upwards, and what is heavy is what naturally moves downwards, the containing limit which lies in the direction of the centre* is 'below' (as is the centre itself) and the containing limit which lies in the direction of the periphery is 'above' (as is the periphery itself). And we can see why place is thought to be a kind of surface, and like a vessel or container. Moreover, the place of an object coincides with it, because limits do coincide with what they limit.

5. The world as a whole is not in a place

ª31 A body is in place, then, if there is a body outside it which contains it, but not if there is none. So, if water were such a body,* then although its *parts* would still move (because they are contained by one another), the whole of it would not

move, except in a particular sense. What I mean is that it
cannot change place all at once, considered as a whole, but,
because its periphery is the place of its parts, it can rotate,
while some of its parts rotate (but do not move up and down),
and others (all those that are liable to condensation and rarefaction) move both up and down.

Now, as I explained earlier,* some things are potentially in 212^{b}3
place, while others are actually in place: when something has
continuous, indistinguishable parts, these parts are potentially
in place, whereas when the parts are distinct but in contact
(as in a heap), they are actually in place. Also, some things ^{b}7
are in place in their own right: for instance, every body which
is capable either of movement or of increase is somewhere in
its own right. (The world, however, as I have said,* is not as
a whole somewhere or in a particular place, since there is no
body which contains it; but it does move in a way, and in
that way its parts also have place, since one is consecutive to
another.) Other things, however, such as a person's mind and
the world, are coincidentally in place. The point is that all the ^{b}12
world's parts are in a sense in place, because one part contains
another in the rotation. That is why its upper part rotates.
But the universe as a whole is not anywhere, because for a
thing to be somewhere it does not just have to be something
itself: there also has to be something else beyond it, which it
is in and which contains it. But there is nothing beyond the
whole sum of all things, and therefore all there is is within the
universe, since the universe is presumably the sum of all things.
However, the place of things is not the same as the world;
their place is a part of the world, a limit, which is in contact
with the movable body. And so earth is in water, water is in
air, air is in fire, and fire is in the heavens; but there the
sequence stops—the heavens are not in anything else.

This account resolves the problems*

It is clear from what I have been saying that on this account ^{b}22
of place all the difficulties can be solved. There is no need for
place to expand along with the object, nor for there to be a
place for a point, nor for two bodies to coincide, nor for
there to be an extension which is itself a kind of body (what

is inside a place is a body—any body at all—but not a body's extension). And place is in fact somewhere—not in the sense that it is in a place, however, but in the sense that the limit is in the thing which is limited* (for only a movable body is in a place, not everything).

b29   It is reasonable, too, that each element should have its own place to move to, because two successive elements, which are in contact naturally (and not by force), are akin to one another.* (If two things have fused together,* they cannot act on each other, but if they are in contact they can act on and be acted on by each other.) It is also reasonable that everything of its own nature stays in its own place. That is what a part does, and the relation between a thing and its place is analogous to that between a separable part and a whole; think of someone moving a portion of water or air.*

213ª1   What is the relation between water and air? It is as if one was the matter, the other the form. Water is the matter of air, and air is like the actuality of water, since water is potentially air (and air is also potentially water, but in a different way). I will draw the relevant distinctions later,* but I had to mention the point now, because the opportunity arose to do so. Anything which remains unclear now will be explained more fully later. Anyway, whenever something is simultaneously matter and actuality (as water is—one potentially and the other actually*), it will conform in a sense to the relation of part to whole. That is also why these things are in contact (as distinct from both forming what is actually one thing, which is what happens when two things fuse together).

ª10   I have now explained that there is such a thing as place, and I have said what it is.

B. VOID

6. *The existing arguments for and against void*

ª12   We are bound to conclude that the same questions—whether or not there is such a thing, in what sense it exists, and what it is—are just as important to a natural scientist where void is concerned as where place is concerned. Indeed, the assumptions people make about void lead to views which are persuasive or not in a very similar way to their assumptions about place. For those who say that there is such a thing as void think of it as a kind of place, a kind of vessel, which may be either full or empty depending on whether or not it contains the body it is capable of receiving. They are assuming, then, that 'a void', 'a plenum', and 'a place' all refer to the same thing, though they are different in definition. Our enquiry should take as its starting-point the views not only of those who say that there is such a thing as void, but of those who say that there is no such thing as well, and thirdly any shared opinions* about it.

Now, those who try to demonstrate the non-existence of ^{a}22 void are refuting their own mistaken conception of void, rather than what people generally mean by 'void'. I am thinking here of Anaxagoras, and others who address the issue in the way that he does. What they do is demonstrate that air is something by torturing wineskins* and showing that the air offers resistance, and by enclosing it inside a water-thief.* However, what people mean by 'void' is an extension in which ^{a}27 there is no perceptible body. Because they think that all there is is body, they claim that anything which contains absolutely nothing is a void—and that is why they think that anything which is full of air is a void. What needs demonstrating is not that air is something, but that the only kind of extension there is is the extension of bodies, and that this cannot be separated from bodies or exist without them in actuality, and cannot break up the material universe so that it is not continuous (which is what Democritus, Leucippus, and plenty of other natural scientists* claim). It also needs to be shown that, even if the universe of bodies is continuous, there is no extension separate from bodies that exists outside it.

So this lot have not even reached the threshold of the issue, 213^b2 whereas those who claim that there is such a thing as void do better. Their arguments are, first, that without void it is inconceivable that there could be such a thing as change of place (i.e. movement and increase), since it is impossible for a plenum to be receptive of anything. If a plenum could receive something, two objects would be in the same place, and then you could have *any* number of bodies coinciding, since it would be impossible to specify a point at which this coincidence would stop. And if this coincidence were possible, then however small a body was, it could hold the largest thing in the world, because anything large consists of a number of small parts. So if many equal objects could coincide, there would be nothing to stop many unequal objects coinciding

b12 too.* Melissus also relies on the same argument to demonstrate that the universe does not move: he says that there has to be void for movement to take place, but there is no such thing as void.

b14   These considerations gave them one way to demonstrate that there is such a thing as void, and a second argument is based on the observation that some things contract and are compressed. For instance, they claim that a wine-cask can hold not only the wine, but also the wineskins* which the wine is in, and they explain this by claiming that a compressed body contracts into the void which is within it. Third, they all use void to explain the phenomenon of growth,* the point being that food is a body, and it is impossible for two bodies to coincide. They also cite as evidence what happens to ash: ash in a vessel can hold as much water as the empty vessel can.*

b22   The Pythagoreans* also claim that there is such a thing as void. According to them, it enters the world from the infinite breath because the world breathes in void as well as breath. What void does, they say, is differentiate things; they think of void as being a kind of separation and distinction when one thing comes after another. This happens first among the numbers, because on their view it is the void that distinguishes one number from another.

b27   So much for a sketch of the kinds of arguments people produce for and against the existence of void; these are more or less all their arguments.

7. What 'a void' means

b30 In order to decide whether or not it exists, we must understand what the term 'void' means. People take the void to be a place in which there is nothing, and they do so because they think that all that exists is body, and every body is in a place, and void is place in which there is no body, and so anywhere there is no body, there is nothing.

Also, they think that all body is tangible, which means that ᵇ34 it must have weight or lightness. It follows from this that void is that in which there is nothing heavy or light. But although it is true, as I said, that this follows from the premisses, it also has the absurd consequence that a point is void. A void must be a place which contains an extension that a tangible body might have.

Anyway, it does look as though one thing people mean by 214ᵃ6 'void' is what is not full of body that is perceptible by touch, and what is perceptible by touch is what has weight or lightness. (One might ask whether or not an extension with colour or noise is void, and wonder what reply to give; but perhaps it is clear that if it could hold body in a tangible form, it is void, and if it could not, it is not.) And another ᵃ11 thing people seem to mean by 'void' is that in which there is no identifiable individual or bodily substance. This is why some people* say about void what they also say about place— that it is the matter of a body. But they are wrong, because matter is not separable from objects, whereas the void is, as they understand it.

Refutation of the arguments for a void

Now, we have already settled the issue of place, and if there ᵃ16 is such a thing as void, it must be place deprived of body. But we have also said in what sense there is and in what sense there is not such a thing as place, and so it is clear that there is no such thing as void in the sense we have just been considering, and that goes whether void is separated or inseparable.* For a void is not a body, but is intended to be the extension of a body, and so the reason why void is supposed to exist is because place does too, and for the same reasons.

The fact of movement and change supports both those who ᵃ22 say that place is something distinct from the bodies that come to occupy it, and those who say that void exists. People take void to be the cause of change in the sense of being that in which change occurs—which is the kind of thing some people say about place. But the reality of change does not mean that ᵃ27 there has to be void. In general, there is no need for void to be the cause of all change, because it is perfectly possible for a plenum to alter qualitatively (which is also something Melissus overlooked*). Nor do we need void to explain change of place either, since it is possible for things to make way for one another without there being any separate extension besides the moving bodies. It is as easy to see this in the case of the rotation of continuous objects as it is in the rotation of liquids.

ᵃ32 Also, a thing might be compressed not into void but because whatever is in it is squeezed out (for example, the compression of water might involve the air inside it being squeezed out); and growth might occur not by something entering the growing body, but by alteration (as when air 214ᵇ3 comes from water). And a general point is that the argument about growth and the argument about pouring water on to ash get in their own way.* For either growth is restricted to only some parts of a body, or it is not caused by the addition of body, or it is possible for two bodies to coincide (in which case they are raising a common problem, which requires a solution, but they are not demonstrating the existence of void), or the whole of the body has to be void, if it is growing all over and if growth is through void. The same argument applies to the ash as well.

ᵇ10 Evidently, then, it is easy to resolve the arguments they use to demonstrate the existence of void.

8. Proof that there is no void separate from bodies

ᵇ12 Let us again argue the point that there is no separated void, as some say. If each of the simple bodies *naturally* has its own proper motion (as fire moves upwards and earth moves downwards and towards the centre of the world), it is easy to see that void is not responsible for their motion. Which motion might void be responsible for? It is supposed to be responsible for change of place, but it is not.

Besides, if void is something like place deprived of body, ᵇ17
then when there is void, where will a body placed in it move
to? I mean, it cannot move to all quarters of the void. The
same argument is also valid against those who take place to
be a separate something* into which body moves: how will
a body placed there move? Or will it be motionless? This
argument applies equally to 'up' and 'down' as to the void,
and it is not surprising that it does, since the champions of
the existence of void do not differentiate between it and place.
Next, in what way will anything be in a place or in the void? ᵇ24
Imagine some body located as a whole in a place—a place
thought of as separate and as persisting. They do not get the
results they want: since no part of this body is located separately from the whole, no part will be in a place, but in the
whole.* Moreover, if place is not separate, void is not separate
either.

The notion that there has to be void for movement to ᵇ28
occur turns out, on reflection, to be quite the opposite of the
truth: void makes it impossible for anything to move. The
idea that the earth is at rest because of the equilibrium of
things* is analogous: by the same token, anything in a void
is bound to be at rest, since there is nowhere for it to move
to more or less than anywhere else, because the void by
definition contains no differentiation.*

Secondly, every movement is either forced or natural. Now, 215ᵃ1
if there is forced movement, there is natural movement as
well, since forced movement is movement which is unnatural,
and unnatural movement is secondary to natural movement.
And the upshot of this is that if the natural bodies lack their
own specific forms of natural movement, they will not move
in any other way either. But how can there be such a thing ᵃ6
as natural movement if there are no distinctions within that
which is void and infinite? Since it is infinite, there is no
above or below or centre; since it is void there is no distinction* between above and below. There is as little differentiation within the void as there is within nothing, since the void
is supposed to be something without being, a kind of privation. But there *are* differences between the various natural
movements, and so there must be natural distinctions. There

are two alternatives, then: either there is no such thing as natural movement anywhere for anything, or, if there is, there is no such thing as void.

^a14   Thirdly, what about the fact that when things are thrown, they continue to move* when the thrower is no longer touching them? This may be due to replacement, as some say, or to the air's being pushed faster, so that it overcomes the natural movement of the pushed object towards its own proper place. But none of these conditions obtain in a void; the only way anything can move is by riding on something else.

^a19   Fourthly, it would be impossible to explain why something which has been set in motion should stop anywhere: why should it stop here rather than there? Either it never moves or it has to go on and on moving for ever, unless something stronger than it impedes it.

^a22   Fifthly, as things are, it is suggested that a thing moves into the void because the void yields to it. But in the void this yielding would happen without any distinction of direction, and so the object would move in all directions at once.

^a24   Also, the following considerations make it easy to see that our claims are correct. It is clear that there are two reasons why one moving body* might move faster than another with the same weight. This is due either to a difference in the medium through which they are moving (for example, one might move through water as opposed to earth, or water as opposed to air), or to a difference in the moving object: while all the other conditions are the same, one object might have greater weight or lightness.

^a29   Now, the medium through which the object is moving makes a difference by impeding the object. It does so especially if it is moving in the opposite direction, but also if it is still. The resistance is greater if the medium is not easy to divide, which

^a31   is to say, if the medium is denser. An object *A* will move through *B* in time *C*, but through the less dense medium *D* in time *E* (assuming that *B* and *D* are equal in extent), and the times will be proportionate to the resistance exerted by the impeding body. Let *B* be water and *D* air; in proportion as air is less dense and less material than water, *A* will move this much faster through *D* than it does through *B*. Let the

speeds have the same ratio to each other, then, as the density of air does to the density of water, so that if air is twice as refined, A will take twice as long to traverse B as it does to traverse D, or in other words the time C will be double the time E. And it will always be the case that in proportion as the medium is less material, less impeding, and easier to divide, A will move that much faster.

However, there is no ratio to measure the extent to which 215^b12 the void is exceeded in density by body, just as what is nothing has no ratio to any number. Four exceeds three by one, and it exceeds two by more than one, and it exceeds one by still more; but there the sequence of ratios ends. The excess of four over nothing cannot be expressed by any ratio,* because it has to be possible to split the greater number into the lesser number and the remainder, which would mean that four would be the sum of the remainder and nothing. That is why a line does not exceed a point (unless a line is made up of points*). By the same token, a void cannot stand in any ratio to a plenum.

The upshot of this is that motion through a void cannot $^b20$ stand in any ratio to motion through a plenum. Suppose an object takes a certain time to travel a certain distance through a very refined medium: the distance it travels through the void in the same time exceeds all proportion. Let F be a void which is equal in extent to B and to D.* So if A traverses F and takes time G for its journey, and given that G is less than E, then the ratio between G and E will be the same as the ratio of the void to the plenum. But in an amount of time equal to G, A will traverse a distance H of D. And A will also traverse in the same amount of time any body F whose density is less than air in the same proportion as G is less than E. For if the body F is less dense than D in the same ratio as $^b30$ E exceeds G, then A (assuming A to be in motion) will traverse F in a time inversely proportional to the speed, in an amount of time equal to G. If, then, there is no body in F, A will traverse it even faster. But we found that it will traverse it in time G. Therefore, it takes just as long for A to traverse a plenum as it does for it to traverse a void; but that is unthinkable. Evidently, then, if there is a time in which it traverses *any* portion of the void, the same unthinkable result will follow: it will be found to take just as long for it to traverse some plenum as it does for it to traverse a void, because there will always be some body which stands in the same ratio to some other body as the time does to the time.

216ᵃ8   In short, the reason for this result is clear: it is because there is always a ratio between one movement and another (because they take time, and there is always a ratio between one stretch of time and another, since they are both limited), but there can be no ratio between the void and a plenum.

ᵃ11   So much for the consequences of differences in the media; the consequences of one moving body exceeding another in weight or lightness are as follows. It is a fact of experience that the greater the impulse of weight or lightness things have, the faster (other things being equal) they complete a given journey, in accordance with the ratio the magnitudes have to one another. So the same should be true of magnitudes travelling through a void as well; but that is impossible. Why would one move faster than another? In a plenum one is bound to move faster than another, because the greater the object the faster it cuts through the medium with its strength, since the moving or projected object cuts through either with its shape or with its impulse. But it follows that in a void everything will travel at the same speed, which is impossible.

ᵃ21   As a result of these arguments we are in a position to see that, if there is void, its consequences contradict the reasons given for its existence by its champions. Some people think that without a void, separated off in its own right, there could be no change of place. But this is no different from claiming that there is such a thing as place in isolation from what occupies it, and I have already argued that this is impossible.

ᵃ26   However, if we do think of what void is in its own right,* it will turn out to deserve its name and to be really void! If you put a cube in water, an amount of water equal to the cube will be displaced. Exactly the same happens to air as well, though it is hard to notice it. And the same holds true universally for every body which is capable of being displaced: assuming that it is not compressed, it is bound to be displaced, always in the direction in which it is its nature to be displaced,* either downwards (if its natural motion, like that of earth, is downward) or upwards (if it is fire), or in both directions; and this is true whatever kind of object is put in it. But this is impossible in a void, since a void is not a ᵃ33 body. What must happen instead is that an already existing extension within the void, of equal dimensions to the cube, must have penetrated the cube—as if the water or air were completely to permeate the wooden cube rather than being displaced by it. But the cube is also exactly the same size as the void it occupies, and whatever qualities it has—it may be hot or cold, heavy or light—it is in itself different from all these qualities (even if it is not separable from them). I am talking about the volume of the wooden cube. Therefore, 216ᵇ6 even if it is isolated from everything else and is not heavy or light, it will still occupy an amount of void equal to itself, and it will still coincide with a bit of place and a bit of void equal to itself. So what will be the difference between the body of the cube and the void and place which are equal to it? And if two things can behave like this, why cannot any number of things coincide?*

This is one absurd and impossible consequence. It is also ᵇ11 clear that the cube will have this volume even while changing place (the same goes for all other bodies too). And so, if there is no difference between volume and place, why should we conceive of place for bodies, as something apart from any-thing's volume,* if the volume cannot be affected by any-thing? It contributes nothing for there to be some other such extension, equal to the volume, but different from it.†

These considerations show, then, that there cannot be a ᵇ20 separated void.

9. The true explanation of compression and expansion

Some people, however, think that the existence of void is ᵇ22 clearly shown by the existence of rarity and density: without rarity and density, compression and contraction are imposs-ible too; but if they are impossible, either change will be eliminated altogether, or the universe will bulge (as Xuthus* said), or there must be no loss or gain during change[†] (for instance, if some air comes from a ladleful of water, then an equal volume of air must at the same time generate the same amount of water), or there has to be void (because without void, the argument goes, it is impossible for compression and expansion to occur).

[b]30   Now, if by 'rare' they mean 'that which has a plurality of separated voids within itself', it is clear—given that there can no more be separable void than there can be place with its
[b]33   own extension—that this sense of 'rare' is ruled out. The alternative—that something rare does contain void, but not separable void*—is more plausible, but it still entails, first, that void is not responsible for all movement, but only for upward movement (since anything rare is light, which is why fire is said to be rare), and, second, that void is not responsible for movement in the sense that it is that in which movement happens. Rather, it carries things upwards in the same way that wineskins,* when lifted upwards, carry with them
217[a]3   what is continuous with them. And yet how can void move or have a place? The place to which it moves would have to become void of void.* Moreover, how will they explain the downward movement of heavy things? Also, if the more a thing is rare and void, the faster it will move upwards, it is clear that nothing would move faster than something which is entirely void. But it is also possible that this absolute void cannot move. The same argument applies: just as everything in a void is immobile, so the void too is immobile, because the speeds are incomparable.*

[a]10   Although we deny that there is such a thing as void, the rest of the difficulties do genuinely arise: if condensation and rarefaction do not happen, then either change will be ruled out, or the universe will bulge, or when air comes from water an equal amount of water will always come from air (since obviously the volume of air produced is greater than the
[a]15   volume of the water it came from). If there is no such thing as compression, then, it necessarily follows either that when things are pushed outwards, one after another, the last thing will make a bulge (this is a consequence of any change of place, unless the motion is circular; but movement goes in

straight lines as well as in circles),† or that somewhere else in the world some air is changing into an equal amount of water so that the total volume of the universe as a whole remains constant, or that nothing is changing.

These are the reasons which lead some people to claim that ᵃ20 there is such a thing as void. Our position, however, based on considerations we have already established,* is that opposites (hot and cold, and the other naturally existing oppositions) have a single underlying matter; that something actual comes to be from a state of potential; that while the matter is not separable, it is different in definition; and that numerically it is the same matter for the hot and for the cold and, if it so happens, for colour too.

Moreover, the matter of a body when it is large is the same ᵃ26 as its matter when it is small. This is obviously so: when water turns into air, the same matter becomes something else. Nothing is added to it; all that happens is that something which formerly existed potentially comes to exist actually. The same goes for when air turns into water as well. In the one case the change is from smallness to largeness, in the other it is the other way round. By the same token, then, ᵃ31 when a large amount of air diminishes in volume, and when a small amount increases, it is the matter, with its potential, which becomes either smaller or larger. Just as the same matter becomes hot instead of cold and cold instead of hot, because it was so potentially, so it changes from being hot to being more hot, without anything in the matter becoming hot which was not already hot when the matter was less hot.

Analogously, if the convex circumference of a larger circle 217ᵇ2 becomes that of a smaller circle (it does not matter whether or not it is the same circumference), convexity has not become a property of something which was straight rather than convex before. Increase or decrease of degree do not depend on the quality failing in some parts; it is impossible to find any part of flame, of any size, which does not possess the properties of heat and brightness. This is an analogy for the relation between the earlier heat and the later heat, and it follows that it is not because the matter of an amount of perceptible stuff is added to that its largeness or smallness is extended, but because it was potentially larger or smaller. It is the same thing, then, that is dense and rare; there is a single matter for them both.

^b11 Anything dense is heavy, anything rare is light. Each of them—rarity and density—is associated with two qualities. Both the heavy and the hard are thought to be dense, and conversely both the light and the soft are thought to be rare. Heaviness and hardness do not coincide, however, in the case of lead and iron.

^b20 These arguments show, first, that there is no such thing as a separate void (whether it is taken to be absolutely separate, or to be contained within rare things*) and, second, that there is no such thing as potential void either, unless one is determined to call the cause of movement 'void', whatever that cause turns out to be. If so, the matter of heavy and light, *qua* their matter, would be the void, because considered as heavy and light, the dense and the rare would give rise to movement, and considered as hard and soft, they would give rise to being affected and not being affected—that is, to alteration rather than to movement.

^b27 We have now decided in what way there is and in what way there is not such a thing as void.*

C. TIME

10. *Problems about time*

^b29 After this discussion, the next thing to look into is time. It makes sense to start by rehearsing the difficulties which the issue generates, and in doing so we will draw on non-specialist ideas as well. The question is, first, whether or not it is a real entity and, second, what its nature is.

^b32 Some suspicion that it either does not exist at all, or at least that its existence is tenuous and faint, arises as a result of the following considerations. Some of it has happened and does not exist, and some of it is in the future and does not yet exist; these constitute both the infinite stretch of all time

and the time that is with us at any moment; but it would appear to be impossible for anything which consists of things that do not exist to exist itself.

Moreover,* for anything which is divisible into parts, if it 218ᵃ3 exists, then when it exists some or all of its parts must exist. But time has parts, and some of them have existed, while others will exist, but none of them currently exist. The now is not a part of time, because a part measures the whole and the whole must consist of its parts; time, however, does not seem to consist of nows.*

Moreover, the now appears to divide past from future, but ᵃ8 it is not easy to see whether it always stays the same or whether it is always different.* Suppose, first, that it is always different. If none of the parts of time, which are successively different, are simultaneous (except that one might contain another, as a longer stretch of time contains a shorter stretch), and if a now which does not exist, but which existed earlier, must have ceased to exist at some time, then, first, nows will not be simultaneous with one another either, and, second, earlier nows must have ceased to exist. An earlier now cannot ᵃ16 have ceased to exist during itself, because that is when it exists; but it is also impossible for the earlier now to have ceased to exist during some other now, given that it is as impossible for nows to be consecutive as it is for points. So since there is no next now during which the earlier now ceased to exist, but it has ceased to exist during some now other than itself, then it must have existed during all the infinitely many nows between itself and that other now. This is impossible, however. But it is also impossible for it to stay ᵃ21 perpetually the same, because nothing that is divisible and finite has only one limit, whether it is continuous in one or in more than one dimension. But the now is a limit, and a finite time can be grasped. Secondly, assuming that to be temporally simultaneous, rather than being earlier or later, is to be in one and the same now, if both earlier and later events are within this present now, then things which happened ten thousand years ago would be simultaneous with today's events, and nothing would be either earlier or later than anything else.

ᵃ30 This will have to do as a statement of the difficulties connected with time's properties. The views that have been handed down to us and our earlier discussion are equally unhelpful in clarifying what time is and what its nature is. Some say that time is the movement of the universe, others* that it is the heavenly sphere itself. And yet even a partial rotation of the heavens is in a sense time, even though it is not a rotation (since we took just a part of the rotation, not the rotation). Besides, if there were a plurality of universes, the movement of any one of them would be time, just as much as the movement of any other one of them, and the upshot would
218ᵇ5 be a plurality of simultaneous times. The idea that the heavenly sphere is time is based on the fact that everything is in time and in the heavenly sphere—but there is no point in considering the impossibilities such a naïve statement entails.

ᵇ9 What is worth considering, however, is the prevalent idea that time is variation and change. Now, the change of anything exists only in the thing that is being changed, or where that changing thing happens to be; time, however, is both
ᵇ13 everywhere and present alike to all things. Moreover, change can be faster and slower, but time cannot, since 'fast' and 'slow' are defined in terms of time: anything which changes a lot in a short stretch of time is fast, and anything which changes little in a long stretch of time is slow. Time, however, is not defined in terms of time: it is not defined as being such-and-such an amount of time, or as being such-and-such a kind of time. So it is easy to see that time is not change. (For the moment let us assume* that it makes no difference whether we say 'variation' or 'change'.)

11. *What time is*

ᵇ21 Nevertheless, it is also true that time is not without change. For without any change (or any noticeable change) in our minds, time does not seem to pass, as in the story* about those who sleep in the sanctuary of the heroes in Sardinia, who wake up and do not think time has passed; what they do is amalgamate the later now into a unit with the earlier now and

eliminate all the time in between because they have not noticed its passage. There would be no time if there were only ᵇ27
a single now, rather than different nows, and by the same
token, if the difference between the nows is not noticed, the
time between them seems not to exist. So if thinking that time
does not exist is something that happens when we do not
distinguish any change and when the mind seems to remain
in a single, undifferentiated condition, and if when we do
notice and discern change, we say that time has passed, then
clearly time does not exist without change.

It is clear, then, that time is not change, but at the same 219ᵃ1
time that it does not exist without change. So in our attempt
to discover what time is we had better start with this fact and
try to see what aspect of change time is. After all, we do
notice change and time simultaneously. If it is dark and our
bodily experience is nil, but some change is happening within
the mind, we immediately suppose that some time has passed
as well; also, whenever some time seems to have passed, we
suppose that some change has occurred. Consequently, time
is either change or an aspect of change. Therefore, since it is
not change, it must be an aspect of change.

Since any change has a starting-point and an end-point, ᵃ10
and since every magnitude is continuous, the change follows
the nature of the magnitude. It is because magnitude is continuous that change is too,* and because change is continuous, time is too. For the amount of change corresponds on
any occasion to the amount of time that seems to have passed.

Now, what is before and after is found primarily in place. ᵃ14
In that context it depends on position, but because it is found
in magnitude, it must also be found, in an analogous fashion,
in change. And since time always follows the nature of change,
what is before and after applies also to time. In the province
of change, what is before and after is a change* (i.e. it is the
actual thing that is the change), but what it is to be before
and after is different from what it is to be a change.

However, we know time too when we distinguish change ᵃ22
by distinguishing its limits as before and after; and we say
that time has passed when we have received an impression of
the before and after in a process of change. We distinguish

time by taking the before and the after of the change to be different and by supposing there to be something which comes between them; the point is that in order for us to say that there is time, we have to think of the extremes as different from the middle, and the mind has to say that there are two nows, an earlier one before and a later one after. For we can take for granted the notion that what is limited by a now is a stretch of time.

ᵃ30　So when the impression we receive of the now is that it is single (i.e. when there is no impression of it as being before and after in change, or when the now is perceived as identical but is not perceived as a limit of something before and something after), no time seems to have passed, because no change seems to have happened either. But when we notice before and after, then we say that there is time. For this is what time is: a number of change in respect of before and after.

219ᵇ2　Time is not change, then, but it is that feature of change that makes number applicable to it. Evidence for this may be found as follows. We assess 'more' and 'less' by number, but we assess more and less change in terms of time. So time is a kind of number. But 'number' is ambiguous:* we describe not only that which is numbered and numerable as number, but also that by which we number. So time is number in the sense of that which is numbered, not in the sense of that by which we number. That by which we number is not the same as that which is numbered.

ᵇ9　Just as change is perpetually different from what it was before, so time is too. But all simultaneous time is the same, since the actual thing that is the now is the same (but what it is to be the one now is different* from what it is to be the other now), and the now determines time, in respect of before and after.

ᵇ12　In a sense, the now is something single and identical, but in a sense it is not. In so far as it is to be found at successively different points, it is different—this is what it is to be 'now'—but the actual thing that is the now is the same. For (to repeat) change follows magnitude, and time follows change,

ᵇ16　as we claim. By the same token, then, a moving object, by which we know change, and what is before and after in change, follows a point. The actual thing that is the moving object is the same (for the point† is a stone or something like that), but in definition it is different, just as the sophists take Coriscus in the Lyceum to be something different from Coriscus in the city square. A moving object, then, is different by being successively in different locations. And a now follows a moving object, just as time follows change; for it is the moving object that enables us to know before and after in change, but the now exists in so far as the before and after are numerable.* So in the case of before and after too, what- ᵇ26 ever it is that the now is is the same (since it is what is before and after in change), but what it is to be the now is different (since the now exists in so far as the before and after are numerable). And this is what is especially knowable, because change too is known through a changing object and move- ment is known through a moving object, since a moving object is a particular identifiable thing, whereas movement is not. So in a sense the now is always the same, and in a sense it is not, because the same goes for a moving object.

It is also clear that if there were no such thing as time, ᵇ33 there would be no such thing as the now, and that if there were no such thing as the now, there would be no such thing as time. For just as the moving object and the movement exist simultaneously, so also do the number of the moving object and the number of the movement. Time is the number of movement; the now is equivalent to the moving object and is, as it were, a unit of number.*

So time is not only continuous thanks to the now, but is 220ᵃ4 also divided at the now, because this too follows the nature of the movement and the moving object. The point is that the change and the movement are unities* because the moving object is a unity (not the actual thing that is the moving object, because that might stop moving; I mean what is by definition the moving object). Also, the moving object distin- ᵃ8 guishes earlier and later stages of the movement. This too in a way follows the nature of the point; for the point both constitutes the continuity of a length by holding it together* and distinguishes it, since it is the beginning of one part of the length and the end of the other. When you look at the point

in this latter fashion, however, and treat it as two despite its singleness, there must be a pause,* because otherwise the same point could not be both beginning and end. The now, however, is never the same, because of the motion of the
ᵃ14 moving object. So time is a number,* not by being the number of a single point (treated as both beginning and end), but more as the ends of a line form its number; and it is not a number as the parts of the line are, both for the reason already stated (for one will treat the middle point as two, so there will be a lack of movement), and because it is clear that the now is not a part of time, nor is the division of a movement a part of the movement, any more than a point is a part of a line. (It is the two lines that are the parts of the single
ᵃ21 line.) So in so far as the now is a limit, it is not time (except coincidentally), but it is in so far as it numbers.†* For limits belong only to that of which they are the limits, but the number of these horses—that is, ten—can be the number of other things as well.

ᵃ24   Evidently, then, time is a number of change in respect of before and after; and because it is a number of something continuous, it is continuous itself.

12. *Notes on the above account*

ᵃ27 The smallest number, without qualification, is two. But where a particular kind of number is concerned, in a sense there is, and in a sense there is not, a smallest number: for instance, in terms of plurality the smallest number of a line is two lines (or perhaps one), but in terms of magnitude there is no smallest number,* because every line is always divisible. The same goes for time, then: in terms of plurality, the smallest number is one period of time, or two; in terms of magnitude, however, there is no smallest number.

ᵃ32   As we all know, time is not described as fast and slow, but as plenty and little (in so far as it is a number), and as long and short (in so far as it is continuous). But it is not fast and slow. No number is fast and slow either—none of the numbers by which we number,* I mean.

Also, time is the same everywhere at once, but time before 220^b5
is not the same as time after. The reason is that the same goes
for change too: the change that is occurring at present is one
and the same, but a past change is different from a future
one, and time is not a number in the sense of that by which
we number, but in the sense of that which is numbered.
Because the nows are different, this kind of number turns out
to be always different before from what it is after. (However,
the number of a hundred horses and the number of a hundred people is the same, but the objects whose number it is—
the horses and the people—are different.) Moreover, just as ^b12
it is possible for one and the same change to recur again and
again, so too can the same time* (for instance, a year or a
spring or an autumn).

Not only do we measure change by time, but we also ^b14
measure time by change, because they are determined by each
other; time determines change in the sense that it is a number
of change, and change does the same for time. We talk about
'plenty of time' and 'little time' by measuring the time in
terms of a change, just as we count the number of anything
in terms of a numbered thing. For we understand how many
horses there are by assigning them a number, but that number
itself is understood as a number of *horses* by using the one
horse as a measure. The same goes for time and change as
well: we measure change by time and time by change.

It is not surprising that we should find this to be so, given ^b24
that change follows magnitude, and time follows change in
respect of being quantifiable, continuous, and divisible. It is
because magnitude is like this that change has these attributes,
and it is because change has these attributes that time has
them too. Moreover, we also measure magnitude by change
and change by magnitude. For we say that the road is long
if the journey is long, and also that the journey is long if the
road is long; and we say that the time is long if the change
is long, and that the change is if the time is.

What it is to be 'in time'

Time is a measure of change and of being changed, and it ^b32
measures change by defining some change which will exactly

measure out the whole process of change (just as a foot measures a length by defining a magnitude of which the whole length is a multiple). Now, to say that a change is 'in time' is to say that both the change itself, and its existence, are measured by time (for time simultaneously measures both the change and its existence, and this—that its existence is measured—is what it is for the change to be in time). Therefore, the same obviously goes for anything else as well: what it is for it to be in time is for its existence to be measured by time.

221ᵃ9 Being in time is one of two things. It is either being in existence when time is in existence, or it is similar to being in number (as we say some things are). When we say that something is 'in number' we mean either that it is a part and attribute of number (and in general that it is an aspect of number), or that it has a number.

ᵃ13 Since time is number, the now and before and so on are in time in the same way that a unit and odd and even are in number: the latter are aspects of number, while the former are aspects of time. Objects, however, are in time† in the same way that they are in number. If so, they are contained by time in the same way that things which are in number are contained by number and things which are in place are contained by place.

ᵃ19 It is also clear that being in time is no more being in existence when time is in existence than being in the process of change is being in existence when change is in existence, and being in place is being in existence when place is in existence. After all, if this is what it is to be in something, everything will be in anything: the universe will be in a millet-seed, because the universe and the millet-seed are both in existence at the same time. This is a coincidental fact, whereas the other is a necessary outcome: it is a necessary consequence of something's being in time that a time should exist when it does, and it is a necessary consequence of something's being in a process of change that the process of change should exist then.

ᵃ26 Since the way in which anything is in time is equivalent to the way in which anything is in number, for anything which is in time there will be a time greater than its time. That is why everything in time is bound to be contained by time, just as anything which is in anything is; anything in place, for instance, is bound to be contained by place. They are also affected by time, then, as is suggested, in fact, by familiar expressions such as 'Time wears away', and 'All things are aged by time', and 'Time has made him forgetful'—but note that we do not say 'Time has made him learn' or 'Time has made him young' or 'Time has made him good-looking.' In 221ᵇ1 its own right, time is responsible for destruction rather than for generation, because it is a number of change, and change removes present properties. Evidently, then, anything eternal, in so far as it is eternal, is not in time: it is not contained by time, nor is its existence measured by time. This is indicated by the fact that it is not affected at all by time either, which suggests that it is not in time.

Since time is a measure of change, it will also be a measure ᵇ7 of rest; after all, all rest is in time. For although anything which is in the process of change is necessarily changing, the same does not necessarily go for something that is in time, since time is not change; it is a number of change, and something at rest can be 'in a number of change' just as much as something changing. The point is that if something is unchanging, it does not follow that it is at rest; as I explained earlier,* for a thing to be at rest, it has to be naturally capable of change, but to have been deprived of change. Now, what it is to be in number is for the object to have some number and for the object's existence to be measured by the number in which it is. It follows that if a thing is in time, its existence will be measured by time. Time will measure a ᵇ16 changing object in so far as it is a changing object, and an object at rest in so far as it is an object at rest; it will measure the extent of the change of the one, and the extent of the rest of the other. A changing object, then, is not measured by time just in so far as it has some quantity or other, but in so far as its change has a quantity. So anything which does not change, and does not rest either, is not in time. The point is that to be in time is to be measured by time, and time is a measure of change and rest.

Clearly, then, not everything that does not exist is in time ᵇ23

either; I am thinking, for example, of things which cannot be otherwise, such as the diagonal of a square being commensurate with the side. The general point is that if time is in its own right a measure of change, and is coincidentally a measure of other things, it obviously follows that all those things whose existence time measures will exist in a state of rest or a state of change. So all those things which are liable to destruction and generation (which, in general, sometimes exist and sometimes do not) are necessarily in time, because a stretch of time exists which lasts longer than them—which will exceed both their existence and also the time which
b31 measures their existence. On the other hand, all those things which do not exist, and which are contained by time, either used to exist (as Homer once did) or will exist (as any future event will), depending on the direction in which time contains them; and if time contains them in both directions, then they have both modes of existence. However, all those things which time does not contain in any manner neither were nor are nor will be, and this category includes all those things that do not exist and whose opposites do exist—as, for instance, the incommensurability of the diagonal always exists, and so it is not in time. Its commensurability, therefore, is not in time either; it never exists because it is the opposite of something that always exists. All those things whose opposites do not always exist, however, can both exist and not exist, and are subject to generation and destruction.

13. *Definitions of various temporal terms*

222ª10 The now is what holds time together, as I have said,* since it makes past and future time a continuous whole; and it is a limit of time, in the sense that it is the beginning of one time and the end of another. However, this is less easy to see in its case than it is in the case of a stationary point. But it does divide time potentially, and in so far as it does so, it is always different; but in so far as it joins one time to another, it is always the same. In this respect the now is equivalent to the point in mathematical lines, because the dividing point is not always the same in thought: when we divide with it, we must think of it as now one thing and now another, but in so far as it is a single thing, it is the same all the way along the line. Likewise, the now too is in one way a division of time, but only potentially, and in another a limit of both past and future, unifying the two. The division and the unification are the same, and they involve the same thing, but in terms of what they are, they are different.

This, then, is one of the meanings of 'now', but it is also ᵃ20 used when the time of what is called 'now' is close: 'He will come now', because he will come today; 'He came just now', because he came today. However, the events in Troy did not happen 'just now', and neither will the Flood,* because although the time from now until these events is continuous, they are not close.

'At some time' means a time which is defined by its relation ᵃ24 to the first kind of now. We say, for example, 'Troy fell at some time' and 'There will be a flood at some time', the point being that it is limited by its relation to the now. There will, therefore, be a definite quantity of time between now and then, and there was a definite quantity between now and the past event.

But if there is no time which is not 'at some time', then all ᵃ28 time will be finite.* Will time fail, then? Presumably not, since change is everlasting. Will time, then, always be different, or does the same time recur again and again? It is clear that whatever obtains for change will also obtain for time: the recurrence of the identical time depends on whether or not the identical change happens at some time. Now, the now is an end and a beginning of time, but not of the same time: it is the end of past time and the beginning of future time. It follows that just as a circle is in a sense simultaneously convex and concave, so time too is always at a beginning and at an end. This explains why time always seems to be different; it is because the now is not a beginning and an end of the same time. If it were, that would be a case of opposites occurring at the same time and in the same respect. And so time will not fail, because it is always at a beginning.

'Soon' refers to that part of future time which is near the 222ᵇ7 present indivisible now ('When are you going for a walk?'—

'Soon', because the time when I intend to walk is near), and 'already' to that part of past time which is not far from the now ('When are you going for a walk?'—'I have already been.') However, the expression 'Troy has already fallen' is not one we use, because the event took place too far from the now.

b12   'Recently' also refers to that part of past time which is near the present now: in answer to the question 'When did you come?', we say 'Recently', provided that the time is near the present now. 'Long ago' refers to that part of past time which is far from the present now. 'Suddenly' refers to a shift which takes an imperceptibly small time to happen.†

b16   Everything which comes into being and ceases to be does so in time, which led some to say that there is nothing wiser than time; Paron the Pythagorean,* however, was closer to the mark when he said that there was nothing more stupid than time, because in time people also forget. It is obvious, anyway, and this is to repeat a point I made earlier,* that time is in its own right responsible more for ceasing to be than for coming to be (because in its own right change involves a change of state), and it is only coincidentally respons-

b22 ible for coming to be and for stability of being. Adequate evidence for this is provided by the fact that nothing comes to be without being itself changed in some way and without being acted on,† but it ceases to be without changing even in the slightest.* In fact, if there is one kind of cessation we usually attribute to the agency of time, it is this kind. Nevertheless, time is not actually responsible for it; even this change merely happens to occur in time.

b27   I have now explained that time exists and have stated what it is, how many senses 'now' has, and what 'at some time', 'recently', 'already', 'long ago', and 'suddenly' mean.

14. *Further notes on time*

b30 Now that we have settled these matters in this fashion, it is clear that every change and every changing object are in time. After all, 'faster' and 'slower' apply to every change, as is obvious in every case. I say that something changes faster when of two changing bodies, both of which have the same interval to cross and both of which are changing at a uniform pace, one changes into a given state before the other (in the case of movement, for instance, suppose that both are moving along a circumference or along a straight line, and similarly for other kinds of change). But 'before' is in time. We use the terms 'before' and 'after' to refer to distance in relation to the now, and the now is the limit between the past and the future. So since nows are in time, before and after will be in time as well, because distance from the now will be in the same thing the now is in. (When the point of reference is the   223ᵃ8 past, the way 'before' is used is the opposite of the way it is used when the point of reference is the future. When we are referring to the past, we describe that which is further from the now as 'before' and that which is nearer as 'after', but when we are referring to the future, we describe that which is nearer as 'before' and that which is further away as 'after'.) Since what is before is in time, then, and since every change involves one thing being before another, it is obvious that every change is in time.

The relation between time and the mind also deserves our   ᵃ16 attention, as does the question of why time seems to be everywhere—on the earth, in the sea, and in the heavens. Presumably it is because time is a property or state of change (since it is the number of change), and all these things are subject to change (since they are all in place), and time and change go together both potentially and actually.

It might be wondered whether or not there would be time   ᵃ21 if there were not mind: if the existence of anything to do the numbering is ruled out, the existence of anything numerable is also ruled out, with the consequence that there would be no such thing as number either (since number is either that which has been numbered or that which is numerable*). If nothing else except mind (and in particular the part of the mind which is intelligence) is such that it can number, it is impossible for there to be time if there is no mind—except that there might still be whatever it is that time is. For example, it might be possible for there to be change without mind, before and after are in change, and time is what is before and after in so far as they are numerable.

a29　It might also be wondered what kind of change time is a number of. Could it be the number of any kind of change? And in fact things come to be and cease to be in time, increase in time, alter in time, and move in time. So in so far as there is such a thing as change, time is a number of any and every change. And so, speaking generally, it is a number of continuous change, rather than a number of a particular kind of change.

223b1　Suppose, however, that two things undergo change now, with the result that† time would be the number of both changes. Then there is another time and there are two equal times at once. Or perhaps this is not so, because any time which is equal and simultaneous with another is in fact one and the same time; even those which are not simultaneous are

b4　specifically the same. If there are seven dogs and seven horses, there is the same number of each of them; by the same token, any changes whose limits are simultaneous have the same time, even if one change is fast, say, while the other is slow, and one is a movement while the other is an alteration. The time of the alteration is still the same, provided it is equal and simultaneous, as the time of the movement. And this explains why, although changes differ from one another and occur in different places, time is everywhere the same; it is because the number of things which are equal and simultaneous is also everywhere one and the same.

b12　Now, there is such a thing as movement, and one kind of movement is circular movement. Also, every kind of thing is numbered in terms of some one thing of that kind—units in terms of a unit, horses in terms of a horse, and so time too is numbered in terms of some determinate time. Moreover, time, as we said,* is measured by change and change is measured by time (and this is because the quantity of the change and of the time is measured by a change determined

b18　in time). It follows from all this that if that which is primary is the measure of everything akin to itself, then uniform circular movement is a measure *par excellence*, because its number is the most intelligible number there is. (There is no

116

uniform alteration or increase or coming into being, but there is uniform movement.*) The reason, then, why people think $^{b}21$ of time as the change of the heavenly sphere is because all other changes are measured by this change and time too is measured by this change. This has also led to the commonly expressed idea that human affairs, and the affairs of all other things which, by their nature, change and are generated and destroyed, are cyclical. This is due to the fact that they are all assessed by time and to the fact that their beginnings and endings seem to conform to a cycle. In fact, people think of $^{b}28$ time itself as a kind of cycle, and this, in turn, is because time measures that kind of movement and is itself measured by that kind of movement. And so to say that things which are generated form a cycle is to say that time is a kind of cycle, which is due to the fact that it is measured by a circular movement. For, apart from the measure, one does not notice anything in the thing measured, except that the whole of it is more measures than one.

Also, it is correct to say that the number of sheep and of $224^{a}2$ dogs is the same number (assuming that the two numbers are equal), but this no more makes the ten of them, or the ten objects, the same ten than the fact that an equilateral and a scalene triangle are the same shape, because they are both triangles, makes them the same triangles. Things are said to $^{a}6$ be the same if they do not differ by a specific difference, and not otherwise. For example, what makes one triangle differ from another is a difference in triangularity—that is why they are different triangles. They are not different in shape, however, but in fact belong in one and the same subdivision of shape. One kind of shape is a circle and another is a triangle, and under 'triangle' fall both isosceles and scalene triangles. So their shape (i.e. triangularity) is the same, but they are not the same triangle. Likewise, then, the number is the same (because it does not differ by a numerical difference), but it is not the same ten (because the objects it is predicated of are different—dogs in one instance, horses in the other).

We have now discussed time—time itself and those matters $^{a}15$ related to time which are relevant to our enquiry.

V

CHANGE

1. Coincidental changes

224ᵃ21 One kind of change is coincidental change* (as, for example, when we say that it is an educated person who is walking, because what is walking is by coincidence someone educated).

ᵃ23 Secondly, we do say that something is changing *tout court* when some aspect of it is changing, as when we describe things as changing because their parts are changing (for example, health is restored to the body because it is restored to

ᵃ26 the eye or the chest—i.e. to parts of the whole body). Thirdly, however, there is the kind of change where something is not changing coincidentally or because something else, one of its parts, is changing, but because it itself is immediately changing. Such a thing is something which is in its own right capable of change, but it is differently described depending on the kind of change it is capable of. For instance, it might be capable of alteration, and more precisely of becoming healthy or hot.

ᵃ30 The same goes for the agent of change as well. It can impart change either coincidentally, or by a part (when some aspect of it is causing the change), or immediately and in its own right (as, for instance, when a doctor heals or a hand hits).

ᵃ34 There is, then, an immediate agent of change and an object which is being changed. Then there is the time when the change is happening, and since every change is from something and to something else, we also have to take into consideration the starting-point and the end-point of the change.

224ᵇ1 The object which is immediately changing, the starting-point of the change and its end-point are all different. Consider wood, heat, and cold; they might be respectively the object, the end-point, and the starting-point. The change obviously takes place in the wood, not in its form, because form (like place and quantity) does not cause change* and is not changed

118

either. No, there is that which causes change, that which is changed, and the end-point of the change. I say this because a change is described by its end-point rather than its starting-point. That is why destruction is a change to a state of non-existence, despite the fact that anything which is being destroyed is changing from a state of existence, and generation is change to a state of existence, despite the fact that it is from a state of non-existence as well.

I have already* stated what change is. Now, the end-points ᵇ10 (which may be forms or affections or place*) are not subject to change. How could knowledge or heat, for instance, be subject to change? (All the same, one might wonder whether affections* might not be processes of change, and whether pallor might not be such an affection. If so, we would have a case of changing to a change. But presumably it is paling, not pallor, which is a process of change.) Again, these end- ᵇ16 points may be the end-points of change either coincidentally, or because of a part (i.e. depending on something other than just themselves), or immediately (i.e. not depending on something other than themselves). For instance, something which is becoming white might coincidentally change to an object of thought* (since by coincidence its colour is being thought about), or it might change to a colour in the sense that white is a part of colour (and Europe might be the end-point of a change because Athens is a part of Europe), or it might in its own right change to the colour white.

It is clear, then, what it means to say that something is ᵇ22 changing or causing change in its own right, what it means to say that it is changing or causing change coincidentally, and what the difference is between its changing or causing change thanks to something other than itself and its changing or causing change immediately, because of itself. It is also clear that change does not take place in the form, but in that which is being changed—which is to say, something capable of change when it is actually changing.

Now, I propose to ignore coincidental change, because it is ᵇ26 a constant aspect of everything in every respect. However, non-coincidental change is restricted to things that are opposites or intermediates, or in contradiction.* A survey of examples would convince one of this. Things can change from an intermediate state because it acts as the opposite of either of the two extremes, since it *is* in a sense the extremes. This also explains why we talk as if there were opposition between an intermediate and the extremes, and between the extremes and an intermediate; for example, the middle string makes a high note relative to the lowest string, and a low note relative to the highest string, and grey is pale relative to black and dark relative to white.

The distinction between change and variation

$^b35$ All change is from something to something. The word itself* shows this: one thing comes 'after' another thing—that is, there is an earlier phase and a later phase. Since all change is from something to something, there are four possible ways in which it might occur. There is either change from an entity to an entity, or from an entity to a non-entity, or from a non-entity to an entity, or from a non-entity to a non-entity. By 'an entity' I mean something signified by an affirmative term. It necessarily follows from this that there are three kinds of change: from an entity to an entity, from an entity to a non-entity, and from a non-entity to an entity. Change from a non-entity to a non-entity is impossible* because there is no opposition involved: they are neither opposites nor contradictories.

225^a12 Change from a non-entity to an entity, where contradiction is involved, is coming to be; it is coming to be in an unqualified sense when the change is unqualified, and it is coming to be of a particular kind when the change is of a particular kind. For example, when something is changed from not being pale to being pale, this is the coming to be of this particular quality, whereas when something is changed from simply not being to being a substance, this is simple coming to be; hence we say that it simply comes to be, not that it comes to be something.

$^a17$ Change from an entity to a non-entity is ceasing to be; it is simple ceasing to be when it is change from being a substance to not being, and a particular kind of ceasing to be when it is change to an opposite negation, as I have already explained in the case of coming to be.

'Not being' is ambiguous.* Something which 'is not' in the ᵃ20
sense that it is a false combination or separation of subject
and predicate cannot vary, and nor can anything which 'is
not' in the sense that it is only potentially (i.e. in the sense
that it is opposed to that which actually is, *tout court*). For
although something not pale or not good can still vary coincidentally (since 'something not pale' might be a person),
nevertheless something which just is not an individual thing
cannot vary at all. It is impossible, then, for that which does
not exist to vary. It follows from this that it is also impossible ᵃ26
for coming to be to be a kind of variation, because it is something that does not exist that comes to be. I mean, however
true it may be that it is coincidentally coming to be,* it still
remains true to say that not being does belong to that which
simply comes to be. And by the same token, that which does
not exist cannot be at rest either.* Apart from these awkward
consequences, everything that undergoes variation is in some
place or other, but something non-existent is not in place,
because then it would be somewhere. So ceasing to be is not
a kind of variation either, because the opposite of a variation
is either a variation or a state of rest, but ceasing to be is the
opposite of coming to be.

Since every variation is a kind of change, and there are ᵃ34
three kinds of change (as already mentioned), and of these
the ones involving coming to be and ceasing to be are not
variations—these are the ones which involve contradiction—
it necessarily follows that the kind of change which is from
an entity to an entity is the only one that is a variation. And
entities are either opposites or intermediates (assuming that
we may take a privation* as an opposite) and are signified by
affirmative terms such as 'naked', 'toothless', and 'dark'.

2. *The different kinds of variation**

Now, predications are divided into the various categories— 225ᵇ5
that is, into predications of substance, quality, place, relation,
quantity, and action or affection.* So it necessarily follows
that there are three kinds of variation—qualitative variation,
quantitative variation, and variation of place. There is no ᵇ10

variation in respect of substance, because nothing that exists is opposite to a substance. Nor is there any variation in respect of relation, because when one of two things which are relative to each other changes, it is possible for the relation to cease to obtain† even though the other thing is not changing at all, with the result that their variation is coincidental. Nor is there variation in respect of the action of what is acting (i.e. the agent of variation) or the affection of what is acted on (i.e. the thing that varies), because a variation cannot vary and a coming to be cannot come to be; to put it generally, a change cannot be changed.

^b16   The first point to notice here is that there are two ways in which variation of a variation might be possible. The variation which is varying might be an underlying thing; this is how a person varies when he changes to being dark instead of pale. But can a variation too become hot or cold in this way, or alter its place, or increase or decrease? No, this is impossible, ^b21 because change is not an underlying thing. Alternatively, the variation might vary because some underlying thing (which is not to be identified with the variation itself) changes from a process of change to a different species of existence. But this is impossible too, except coincidentally, because variation itself is change from one form to another, as when a person changes from illness to health. (The same goes for coming to be and ceasing to be as well, except that they involve opposites of one ^b27 kind, whereas variation involves opposites of another kind.) It follows that the person of our example, at the same time as changing from health to illness, would also be changing from this particular change to some other kind of change. So obviously, by the time he is ill, he will have changed to whatever other change it may be (it could even be a state of rest), and moreover the change he changes to will not be just any kind of change;† for that too—the process of changing from one change to another—must be a change from something to something opposite, and so what he changes to will be the opposite process—namely, becoming healthy.* No, a change can only change coincidentally, as when there is a change from remembering to forgetting* because the subject involved changes at one time to knowing and at another time to ignorance.

Secondly, the idea that a change could change and a coming ᵇ33
to be could come to be generates an infinite regress. If the later
change is to happen, then the one that comes before it must
happen too. For instance, if a simple coming to be was itself
coming to be at some time, then the thing that would come to
be was also becoming a thing that comes to be, and so there
was not yet anything which was simply coming to be, but
only a thing becoming something (i.e. becoming a thing that
comes to be). And if this too was also coming to be at some
time, then at that time it was not yet even becoming a thing
that comes to be. And since there can be no first term in an
infinite series, then this sequence cannot start anywhere, and
so it cannot continue either. Consequently, on this hypothesis
coming to be, variation, and change are completely impossible.

Thirdly,* a single changing thing also has the capacity for 226ᵃ6
the opposite kind of change (and also for the state of rest
which is opposite to its change); and similarly coming to be
and ceasing to be are properties of the same thing. Now,
anything which is coming to be a thing that comes to be
cannot be ceasing to be that thing just as it is coming to be
it, because in order for anything to cease to be, it first has to
be, and it cannot do so after it has come to be either; therefore
it is ceasing to be a thing that comes to be at the very time
that it has become it!

Fourthly,* anything which is coming to be and anything ᵃ10
which is changing have to have an underlying matter. On the
present hypothesis, what will this be? Just as it is either a
body or a mind which is capable of alteration, what will it
be that becomes a change or becomes a coming to be? And
again, what will it be that they change to? After all, the
change or the coming to be of any specific thing has to be a
change from one state to another. Moreover, what kind of
changes will these be? The coming to be of learning cannot
be learning, and so also the coming to be of coming to be
cannot be coming to be, and the coming to be of anything
cannot be that thing either. Also, both the underlying thing
and the end-point of the change must be one of our three
forms of variation, and so, for instance, movement must itself
alter or move.

^a19   In short, then, since everything that changes changes in one of three ways—either coincidentally or by a part changing or in its own right—the only way in which change can change is coincidentally (as when someone who is recovering from illness runs or learns); but we dismissed coincidental change a long time ago.

^a23   Since there can be no variation of substance or relation or action and affection, it is only in respect of quality and quantity and place that there can be variation. For each of these

^a26   categories admits opposition.* Let variation of quality be 'alteration', since this is a general term which links both the opposites together. By 'quality' I do not mean here the kind of quality which is an essential property of substance (for the differentia of a substance is a quality), but an affective quality,* which allows us to describe something as affected or as incap-

^a29   able of being affected. There is no general term for variation of quantity, but it is described by reference to each opposite separately as 'increase' or as 'decrease', increase being change towards the completion of a thing's size and decrease being

^a32   change away from this. As for variation of place, there is no general term which covers both the opposites together or each of them separately, but let us use 'movement' as the general term, despite the fact that,* strictly speaking, only those things are said to be moving which are such that, once they are changing place, they do not have the power to stop by themselves, and which do not initiate their own change of place.

226^b1   Change within a single form—that is, change to a different degree of that form—is alteration, since alteration is, in a qualified or unqualified sense, change from one opposite or to the other. When the direction of the change is towards a lesser degree of the form in question, we tend to say that the end-point of the change is the opposite of that form, and when the direction of the change is towards a greater degree, we tend to say that the starting-point of the change is the opposite of the form and that the end-point is the form itself. The point is that it does not make any difference whether the change is qualified or unqualified, except that in the qualified version the opposites will have to be present in a qualified sense. The form is present to a greater or lesser degree, depending on whether or not there is more or less of the opposite present in it.

These arguments show that there are only the three kinds ᵇ8 of change. Now, something is 'unchanging' not only if it is completely incapable of change (compare the invisibility of sound), but also if it takes a lot of time and effort to get it changing or if its change is initially sluggish (when it is described as 'hard to change'). Then there is also that which is by its nature capable of change, but which is not changing when, where, and how it is its nature to change. This is the only unchanging thing I describe as being at rest, since rest is the opposite of change and so will be the privation of change in that which is capable of admitting change.

I have now explained what change is, what rest is, how ᵇ16 many kinds of change there are, and what kinds of change there are.

3. *Definitions of various terms*

Next, we had better define 'together' and 'apart', 'in contact', ᵇ18 'between', 'successive', 'consecutive', and 'continuous', and explain the kinds of situation to which each of them is by its nature applicable. I say that things are *together* in respect of ᵇ21 place when they coincide in a single immediate place,* *apart* when they are in different immediate places, and *in contact* when their extremes are together.

Now, every change involves opposites, and opposites can 227ᵃ7 be either contraries or contradictories; but since there is nothing intermediate between contradictories, it obviously follows that there must be contraries for there to be something in between. So *between* involves at least three terms. For the final stage of a change is the opposite, and that which is in between is what the changing thing naturally reaches before it reaches the final stage of its changing, assuming that the process of change is continuous and in accordance with the thing's nature. The change is continuous* if none of the pro- 226ᵇ27 cess is left out, or only a very little. It is what happens to the process in which the change occurs, not to the time, which is relevant, because there is nothing to prevent some time being

left out, and moreover it is perfectly possible to sound the
b31 highest note immediately after the lowest note. Change of
place makes clear what I mean, but it is equally evident in
other kinds of change as well. In change of place, 'opposite'
refers to that which is furthest away in a straight line, because a straight line is the shortest distance between two
points and is therefore limited; so it acts as a measure, just
as anything limited does.

b34 A thing is *successive* when it comes after a beginning, either
in position or in form or by some other distinguishing criterion,
and when there is nothing of the same kind as itself between
it and that to which it is successive. Think of a line (or a
number of lines) succeeding a line, or a unit (or a number of
units) succeeding a unit, or a house succeeding a house: it
does not matter if something of a different nature comes in
between. The point is that for anything to be successive, it
has to succeed something and it has to come later than that
thing. After all, the number one is not successive to the number
two, and first day of a month is not successive to the second
day either; it is the other way round.

227a6 A thing is *consecutive* if it is both successive and in contact.
Something continuous is consecutive in a sense, but I say that
something is *continuous* rather than consecutive when the
limits by which the two objects are in contact have become
identical and, as the word implies, enable one object to continue into the other. This is impossible where there are two
separate limits. It is clear from this definition that continuity
is a property of things which naturally form a unity by their
contact with one another. And the way in which what makes
them continuous* is single determines the way in which the
whole is single too; this may involve pinning, for instance, or
gluing, or contact, or grafting.

a17 It is also clear that successiveness is primary. For contact
inevitably implies successiveness,* but successiveness does not
necessarily imply contact. (That is why successiveness is a
property of things which are prior by definition—numbers,
for instance—but contact is not.) Also, wherever there is
continuity there is contact, but not vice versa, because the
fact that the extremes of the things involved are together does

126

not necessarily mean that they are one. If they are one, however, they are bound to be together as well. And so growing
together is the last of the series, because any extremes that
grow together are bound to be in contact, but the fact that
things are in contact does not necessarily mean that they are
growing together. Where there is no contact, however, there
is obviously no growing together either. It follows from this ᵃ27
that even if, as some say, a point and a unit both have independent existence, they cannot be identified, because contact
is a property of points,* whereas successiveness is a property
of units, and also because there can be something between
points (every line is between points), whereas there cannot be
anything between units (nothing comes between one and two).

I have now defined 'together' and 'apart', 'contact', 'between', ᵃ32
'successive', 'consecutive' and 'continuous', and I have explained
the kinds of situation to which each of them is applicable.

4. *What counts as a single change*

To say that a change is one is ambiguous, because 'one' is 227ᵇ3
ambiguous. Generic unity of change depends on the categories
of predication; any movement is generically the same as any
other movement, but alteration is generically different from
movement. A change is specifically the same as another change,
however, when it is not only generically identical, but also is
a change in the same indivisible species. For example, there
are specific differences within the genus colour, and that is
why becoming black is specifically different from becoming
white, but there are no specific differences within whiteness.
So any instance of becoming white is specifically the same as
any other.

(If there are things which are species as well as genera, ᵇ11
there is obviously a sense in which any change which falls
within such a genus is specifically identical to any other, but
not unqualifiedly so. Consider learning, for example, given
that knowledge is a species of apprehension as well as a
genus consisting of the various branches of knowledge.)

Someone might wonder whether what it takes for the change ᵇ14
to be specifically one is that the same thing changes from the

same starting-point to the same end-point (think, for instance, of a single point moving from this place to that over and over again). However, if this is so, circular movement will be the same as movement in a straight line and rolling will be the same as walking. Or have we already decided* that where what the change is in is specifically different, the change is specifically different, and that a circular route is specifically different from a straight route?

b20   That is what it is for change to be generically and specifically one, and it is simply one, without any qualification, when it is one in definition and in number. The following distinctions will show what kind of change this is. There are three factors we take into account when talking about change—what, in what respect, and when. What I mean is that there has to be something which is changing (a person, say, or gold), and it has to be changing in a certain respect (in place, perhaps, or in some quality), and it has to be changing at a certain time, because every change happens in time.

b27   Of these, the respect is responsible for the generic or specific unity of the change, and time is responsible for the consecutiveness of the change; but all of them together are responsible for unqualified unity. For this, the respect (the species of the change) must be one and indivisible, the when (the time of the change) has to be single with nothing left out, and the changing thing has to be single, and not coincidentally so. Think of a pale object turning dark, and of Coriscus walking: the pale thing may be the same thing as Coriscus, but that is coincidental. Moreover the changing thing must be single, not just in the sense that a single common character is involved; two people might simultaneously be recovering from the same ailment (an eye infection, say), but this process of recovery is not absolutely single, only specifically so.

228a3   But what if Socrates alters, with an alteration which is specifically the same, and does so first at one time and then again at another time? If it is possible for something which has ceased to be to come into being again as numerically the same one thing, then this alteration of Socrates' could be one as well; if it is impossible, however, it is the same alteration, but not a single alteration.*

There is a problem which is closely related to this one, ᵃ6
namely whether health and bodily states and affections in
general are one in substance.* The problem arises because
their seats are evidently changing and in flux. If the health I
had in the morning and the health I have now are one and
the same, why should health which is restored after an interval not be the same health too? Why should that health and
this health not be numerically one? After all, the argument is
the same in both cases. There is this much difference between ᵃ12
them,* however: if there are two states, because they are the
states of individuals which are in this way numerically two,†
there must be two activities as well, since an activity is numerically one only if it is the activity of something numerically one. But if there is one state, that still might not lead us
to count the activity as one as well, because when a person
stops walking, the walking no longer exists, but it will exist
again if he walks again. So if the walking is one and the
same, it would have to be possible for something which is
one and the same to cease to exist and then to exist again
time after time. But these problems lie outside our present
enquiry.

Every change is continuous,* since every change is divisible, ᵃ20
and so change which is unqualifiedly one must be continuous
too, and if it is continuous, it must be one. For instance, it
is not the case that every change is continuous with every
other change, just as it is not the case that any two objects
taken at random are continuous; two things are continuous
only if their limits are one. Now, some things do not have
limits, and although other things do, their limits are different
in form from one another and share only a name. How could
the end of a line and the end of a walk be in contact or
become one? Changes which are specifically or generically ᵃ26
different could be consecutive to one another: someone could
go for a run and then immediately come down with a fever.
And there could be movements which are consecutive, such
as the race where torches are passed on in relay. They are not
continuous, however. After all, we have established that continuity involves unity of limits. So continuity of time makes
changes consecutive and successive, but it is continuity of the

changes themselves that makes them continuous—that is, when
228^b1 their two limits are the same. That is why for a change to be
continuous and one in an unqualified sense it has to be specifically the same, there has to be only one changing thing
involved, and the change has to happen in a single period of
time. It has to happen in a single period of time, because otherwise there will be a period of non-change in between, since
when change is interrupted, there is bound to be rest. Any
change which is interrupted by rest is a number of changes,
not a single change, and so a change which involves intervals
of rest is neither single nor continuous, and these intervals
b7 occur if there are intervening periods of time. But even if the
time involved is continuous, as long as the change is not specifically the same, it makes no difference that only a single stretch
of time is involved: the change remains specifically different.
The point is that although specific identity is a prerequisite
for a change to be single, it is not the case that a change that
is specifically single is necessarily unqualifiedly one.

b11   I have now explained what it is for a change to be unqualifiedly one. Moreover, a change is also described as one—
generically, specifically, or in substance*—if it is complete,
just as in other cases completeness and wholeness are properties
of something that is one. But sometimes a change is described
as one even if it is incomplete, as long as it is continuous.

b15   A uniform change is also described as one, though not in
the same way as the changes we have already discussed. What
I mean is that there is a sense in which a non-uniform change
gives the impression of not being single, whereas this description seems more suited to a uniform change (as it is to a
straight line, for instance). After all, a non-uniform change is
divisible. But non-uniform changes seem to differ in degree,
some being more uniform than others. Every kind of change
can be either uniform or non-uniform. An object can alter uniformly, it can move uniformly (around a circle, for instance, or
in a straight line), and the same goes for increase and decrease.
Non-uniformity is an inconstancy which may occur in the
path the change is taking, since it is impossible for a change
to be uniform if it is moving over a non-uniform magnitude,
such as an angled line or a spiral or any other magnitude

which is not such that any two parts, chosen at random, will fit on to one another. Alternatively, non-uniformity may occur ᵇ25 not where the change takes place,† nor in the time, nor in the end-point, but in the way the change happens. For instance, there may be inconstancy of speed; any change which happens at the same speed is uniform, whereas any change where the speed differs is non-uniform. That is why quickness and slowness are neither species nor differentiae of change: it is because they are found in all the various species of change. It follows* that greater or lesser heaviness and lightness, considered as tendencies in a single direction (i.e. when one compares the relative weight of one piece of earth with that of another, or the relative weight of two bits of fire), do not constitute different species of change either.

Although non-uniform change is single because it is con- 229ᵃ1 tinuous, then, it is less single than uniform change, as we can see in the case of angled movement; and 'less' always implies the infiltration of the opposite. Now, every change that is single can be either uniform or non-uniform, and therefore* changes which are not specifically the same but are consecutive with one another cannot form a single, continuous unity. How could a change which is a combination of alteration and movement be uniform? In order to be uniform, its parts would have to fit on to one another.

5. How a change is opposite to a change

We should also decide which changes are opposite to which, ᵃ7 and do the same for rest as well. First, what is it for one change to be the opposite of another? Is it that the starting-point of one is the end-point of the other (one might be a change from health, for instance, while the other is a change to health)? This is what seems to apply to coming to be and ceasing to be. Or is it that they have opposite starting-points (one being a change from health, say, while the other is from illness)? Or is it that they have opposite end-points (one being a change to health, for instance, while the other is a change to illness)? Or is it that the starting-point of one is the opposite of the end-point of the other (one being a change from health, for instance, while the other is a change to illness)? Or is it that both their end-points and starting-points are opposed to one another (one being a change from health to illness, for instance, while the other is a change from illness to health)? One or more of these kinds of opposition must apply, since the list is exhaustive.

^a16   Now, changes where the starting-point of one is the opposite of the end-point of the other (where one, for instance, is a change from health, while the other is a change to illness) are not opposites: they are identical, in fact (except that what it is to be one is different from what it is to be the other, in the sense that it is not the same to change from health and to change to illness).

^a20   Also, changes with opposite starting-points are not opposites either, since a change from an opposite is at the same time a change to the opposite of that opposite (or to something in between the two opposites)—but I will discuss this in a moment.*

^a22   Opposite end-points would seem to account for opposition between changes better than opposite starting-points, since in the latter case there is a loss of opposition, whereas in the former case opposition is gained. Besides, we do describe any given change in terms of its end-point rather than its starting-point: the change to health, for instance, is called 'recovering health', and the change to illness is called 'falling ill'.

^a27   We are left, then, with changes to opposite end-points, and those with both opposite end-points and opposite starting-points. Now, it may be that changes to opposite end-points are also changes from opposite starting-points, although what it is to be the one is perhaps different from what it is to be the other. What I mean is that the change to health is different from the change from illness, and the change from health is different from the change to illness. But since change and variation are different (in the sense that variation is change from an entity to an entity), then variations whose end-points and starting-points are both opposed are opposites; for example, the variation from health to illness is the opposite of the 229^b2 variation from illness to health. A survey of particular examples shows the kinds of cases which are taken to be opposites. For

instance, falling ill is taken to be the opposite of recovering health, and being taught the facts is taken to be the opposite of being misled by someone else, since the end-points are opposed (after all, just as one can acquire the truth from someone else, as well as from oneself, so one can also be misled by someone else or by oneself). Or again, movement upwards is taken to be the opposite of movement downwards (since the end-points are opposed on the dimension of length), movement to the right is taken to be the opposite of movement to the left (since the end-points are opposed on the dimension of breadth), and movement to the front is taken to be the opposite of movement to the back (since here too there are opposite end-points).

However, processes in which only the end-point is an op- ^{b}10 posite (as when pallor comes into being, but not from anything) are changes, but not variations.* And in cases where there is no opposite, one change is the opposite of the other if its starting-point is the end-point of the other; that is why coming to be is the opposite of ceasing to be, and loss is the opposite of gain. But these are changes, not variations. In ^{b}14 cases where the opposites have an intermediate, variations to the intermediate point should be counted as variations to opposites, in a sense, because as far as the variation is concerned, the intermediate point acts as an opposite, in whichever direction the change is proceeding. For instance, grey acts as black in the change from grey to white, and as black again in the change from white to grey, but it acts as white in the change from black to grey. For, as I have already said, compared to either of the two extremes, the mid-point is described as in a sense being either of them.

So the way in which changes are opposed to one another ^{b}21 is if both their starting-points and their end-points are opposed to one another.

6. How change and rest are opposites

However, change is not only opposed by change: rest seems ^{b}23 to be the opposite of change too. So here is another issue we had better settle. A change is opposed in an unqualified sense by a change, but rest is also opposed to it in that it is the privation of change, and there is a sense in which we describe the privation of anything as its opposite. Now, the opposite of change of a particular kind is rest of the same kind; for instance, the opposite of change of place is staying in one place. But now this statement needs qualifying. Which is the opposite of staying in a given place? Is it movement to that place or movement from that place? Clearly (remembering that two entities* are involved in movement), staying in place *A* is opposed by moving from there to place *B*, and staying in place *B* is opposed by moving from there to place *A*.

$^{b}31$ At the same time, staying in place *A* and staying in place *B* are also the opposites of each other. After all, there could hardly be opposition between changes and not between states of rest; and rest in an opposite state is an opposite state of rest. For instance, remaining in a state of health is the opposite of remaining in a state of illness (as well as being the opposite of the change from health to illness. It would be ridiculous for it to be opposed by the change from illness to health, because the change whose end-point is the state of remaining healthy is a process of coming to rest rather than the opposite of the rest, or rather the process of coming to rest happens to coincide with the change. But it must be opposed by either the change from health to illness or the change from illness to health.) After all, it cannot be remaining in the state of pallor that is the opposite of remaining in the state of health.

$230^{a}7$ In cases where there is no opposite, the change where something is the starting-point is the opposite of the change where that thing is the end-point; this is not a variation, however. Change from being, for instance, is the opposite of change to being. Also, in these cases there is no such thing as rest,* but only changelessness. And if some entity was involved, its changelessness in existence would be the opposite of its change-

$^{a}12$ lessness in non-existence. But one might object that there is no such thing as a non-existent thing, and ask what changelessness in existence is opposed to, and whether changelessness in existence is in fact a state of rest. But if it is, then either it is possible for the opposite of a state of rest not to be a variation, or else coming to be and ceasing to be are variations.

Clearly, then, since coming to be and ceasing to be are not in fact variations, we should not describe changelessness in existence as a state of rest; we should acknowledge its similarity to a state of rest, but call it a state of changelessness. And it will be the opposite either of nothing or of changelessness in non-existence or of ceasing to exist, because ceasing to exist is a change which has this state of changelessness as its starting-point, while coming to be is a change which has it as its end-point.

The opposition between natural and unnatural change and rest

Why, someone might ask, should both rest and change be ᵃ18 either natural or unnatural in the case of change of place, while this does not obtain for other kinds of change? For instance, the contrast between natural and unnatural does not apply to alterations: recovering health is no more natural or unnatural than falling ill, and paling is no more natural or unnatural than darkening. The same goes for increase and ᵃ23 decrease: they are not opposed to each other in the sense that while one is natural, the other is unnatural, and one increase is not opposed to another increase in that way either. The same argument applies to coming to be and ceasing to be: it is not the case that coming to be is natural, while ceasing to be is unnatural (after all, growing old is natural), and it is also evident that the contrast between natural and unnatural does not apply to instances of coming to be.

However, if what happens by force is unnatural, then forced ᵃ29 ceasing to be is unnatural, and is opposed to natural ceasing to be. There are also some instances of coming to be that are forced and are not as destined (and are therefore opposed to natural instances). And are there not forced instances of increase and decrease, such as the rapid growth to maturity of those who have been given a luxurious diet, or the ripening† of grain even when it is not compressed?* And what about alteration? Is it not the same for it too? If so, some alterations would be forced, while others are natural. An example of the difference between the two kinds of alteration might be people recovering from fever on non-critical days, as distinct from those who do so on the critical days: the second kind of alteration would be natural, the first unnatural.

230^b6 So there will also be instances of ceasing to be which are opposed to one another, not to instances of coming to be. Why not? Why should there not be a sense in which this is so? After all, one instance of ceasing to be may be enjoyable, while another is distressing. In other words, it is not the case that instances of ceasing to be are opposed to each other in an unqualified sense, but in the sense that one of them has one quality, while the other has a different quality.

^b10 Generally speaking, then, changes are opposed to changes, and states of rest to states of rest, in the way I have described. For example, since up and down are opposites in the category of place, upward movement is the opposite of downward movement. Fire naturally moves upwards, and this is the opposite of the natural movement of earth, which is downwards; the natural upward movement of fire is the opposite of its ^b15 unnatural downward movement. The same goes for states of rest: rest high up (unnatural for earth) is the opposite of movement down from high up (which is natural for earth); so unnatural rest is the opposite of a thing's natural movement. After all, a thing's movements are opposed on the same principle: one of its movements—either upward or downward—is natural, while the other is unnatural.

^b21 There is a problem here: if a state of rest is not everlasting, does it come into being and is its coming into being the same as coming to rest? Coming into being would then be something that happened to unnatural states of rest (such as earth being at rest high up), and it would follow that when earth is forced to move upwards, it is coming to rest. But every process of coming to rest seems to involve the object accelerating, whereas the opposite is the case when a thing has been forced to move. So it would be at a standstill without having come to a standstill. Also, a thing's coming to rest is generally taken to be either the same as its moving to its proper place, or at least to be something that coincidentally happens at the same time.

^b28 It might be objected that rest in a given place might not be the opposite of movement from that place, on the grounds

that when something is moving away from a place and losing touch with it, it seems to retain what is being left behind;* therefore, it might be said, if the state of rest in question is the opposite of movement from there to somewhere which is opposed to there, the object will have two opposites simultaneously. However, surely it is in *some* sense at rest, provided that it remains in its former state; and, to put the matter generally, whenever something is changing, part of it is at the starting-point of the change and part of it is going towards the end-point. That is also why the opposite of a change is better regarded as another change, rather than a state of rest.

I have now explained in what sense change and rest are 231ᵃ2 each single, and what kinds of opposition exist among them.

VI

CONTINUITY

1. Proof that no continuum is made up of indivisible parts

231ª21 If our earlier definitions* of 'continuous', 'in contact', and 'successive' were correct (we defined as 'continuous' things whose limits formed a unity, as 'in contact' things whose limits are together, and as 'successive' things which have nothing of the same kind as themselves between them), it is impossible for a continuum to consist of indivisible things. For instance, a line, which is continuous, cannot consist of points, which are indivisible, first because in the case of points there are no limits to form a unity (since nothing indivisible has a limit which is distinct from any other part of it), and second because in their case there are no limits to be together (since anything which lacks parts lacks limits too, because a limit is distinct from that of which it is a limit).

ª29   Moreover, in order for points (or any indivisible things, for that matter) to form a continuum, they must be either continuous or in contact. Now, they cannot be continuous (for the reason already stated), and contact is always between wholes* or between parts or between a part and a whole. However, since anything indivisible has no parts, points must be in contact with one another as wholes. But the contact between a whole and a whole does not constitute a continuum, because any continuum has distinct parts and is divisible into parts which are distinct in the sense that they occupy different places.

231ᵇ6   Also, a point cannot be successive to a point, nor can a now be successive to a now, in such a way that they form a length or a stretch of time. I mean, things are successive if there is nothing of the same kind as themselves between them, but there is always a line between points* and a stretch of time between nows.

Furthermore, anything can be divided into its components, ᵇ10
and so on this hypothesis a length or a stretch of time could
be divided into indivisible things. But we found that no continuum is divisible into things which lack parts. And it is
impossible for there to be something else of a different kind
between the parts of a continuum,* because any such thing
must be either indivisible or divisible, and if it is divisible, it
must be divisible either into indivisible parts or into infinitely
divisible parts, and anything divisible into parts that are infinitely divisible is a continuum. It is also clear that every
continuum is divisible into infinitely divisible parts. For if a
continuum were divisible into indivisible parts, that would be
a case of indivisible things being in contact, because the limits
of continuous things form a unity and are in contact.

Proof that distance, time, and movement are all continua

Magnitude, time, and movement are all liable to the same ᵇ18
reasoning. Either they all consist of indivisible components
and are divisible into indivisibles, or none of them does. The
following considerations will make this plain. If a magnitude
consists of indivisible components, movement over that magnitude will consist of the same number of indivisible movements. For example, if the magnitude *ABC* consists of the
indivisible components *A*, *B*, and *C*, each of the parts of the
movement *DEF* of an object *X* over *ABC* is indivisible. Now,
where there is movement, there has to be something in motion, and where there is something in motion, there has to be
movement; therefore, being in motion will also consist of
indivisible parts. So *X* moved the distance *A* when it was in
motion with the movement *D*, it moved the distance *B* when
it was in motion with the movement *E*, and by the same
token it moved the distance *C* when it was in motion with the
movement *F*.

Now, a thing which is in motion from one place to an- ᵇ28
other cannot simultaneously be moving and have moved over
the distance over which it was moving when it was moving.
For example, if something is walking to Thebes, it cannot

simultaneously be walking to Thebes and have finished walking to Thebes. But now consider X's movement D over a distance A which is not divisible into parts:* given that X completes its traversal of A *after* it was in the process of traversing A, the movement must be divisible, because at the time when it was in the process of traversing A, it was not at rest and it had not completed its journey. The alternative is for its journey to be in the process of happening and at the same time to have finished, in which case the walking thing, at the time when it is walking, will have walked to its destination and will have moved over the stretch it is moving over.

232ᵃ6 Suppose something is moving over ABC as a whole (where the movement involved is DEF); given that nothing is ever moving over A (because A has no parts)—it only has moved over A—then a movement will not be made up of movements but of discrete changes of place; it will be the result of something having moved without moving, because it will have completed its traversal of A without ever having been in the process of traversing A. Consequently, it will be possible for something to have walked without ever having been walking, because it will have walked this indivisible distance ᵃ12 without walking that distance. Since, then, everything is inevitably either at rest or in motion, X is at rest in each of A, B, and C, and so it will be possible for a thing to be continually at rest and moving at the same time. For X was in motion over ABC as a whole and was also at rest in any of its parts, and was therefore at rest in ABC as a whole as well. Also, if the indivisible parts of DEF are movements, it will be possible for something in motion to be not moving but at rest; and if its indivisible parts are not movements, it will be possible for the components of movement not to be movements.

ᵃ18 The same would necessarily go for time as for distance and movement: time too would be indivisible and its components would be indivisible nows. For if every distance is divisible, and if something which moves at a constant speed traverses a shorter distance in less time, then time too will be divisible; and if the time in which something moves over A is divisible, the distance A will also be divisible.

2. *Further proofs that distance and time are continua**

Since every magnitude is divisible into magnitudes (for it has ᵃ23
been shown that no continuum can consist of indivisible
components, and every magnitude is a continuum), it necessarily follows* that if one thing is faster than another, it will
cover a greater distance in an equal amount of time, and it
will take less time to traverse an equal distance, and it will
take less time to traverse a greater distance. Some people take
these properties to define 'faster'.

Let *A* be faster than *B*. Since it is what changes sooner that ᵃ27
is faster, then in the time *MN* in which *A* has changed from
C to *D*, *B* will not yet have reached *D*, but will fall short of
it, and so in an equal amount of time the faster one traverses
a greater distance.

It will also take less time to traverse a greater distance: by ᵃ31
the time *A* has reached *D*, the slower *B* will have reached *E*,
let us say. So since *A* has taken all of *MN* to reach *D*, it will
take less time than *MN* to reach *F*; let us call this lesser amount
of time *MO*. So *CF*, which *A* has traversed, is a greater distance than *CE*; and the time *MO* is less than *MN* as a whole,
and so it has taken less time to traverse a greater distance.

It is also clear from this that a faster object takes less time 232ᵇ5
to traverse an equal distance. Since it takes less time to traverse
a greater distance than a slower object, and since taken on its
own it takes more time to traverse a greater distance (*GH*, say)
than it does to traverse a shorter distance (*GI*, say*), then the
time *PQ* in which it traverses *GH* will be greater than the
time *PR* in which it traverses *GI*. Therefore, if *PQ* is less than
the time *X* which the slower object takes to traverse *GI*, then
PR will also be less than *X*; for *PR* is less than *PQ*, and if
one thing is less than another thing which is less than a third
thing, then the first thing is less than the third thing. So a
faster object takes less time to move over an equal distance.

Besides, any object must take less or more or an equal ᵇ14
amount of time over a given change, compared to any other
object. One which takes more time is slower, and one which
takes the same amount of time has the same speed; but since

the one which is faster is changing neither at the same speed nor more slowly, it cannot take an equal amount of time or more time to change. The only remaining possibility, then, is that it takes less time, and it necessarily follows that a faster object also* takes less time to traverse an equal distance.

b20 Since every change is in time and there is no time in which change cannot occur, and since any changing object can be faster or slower in its changing, there is no time in which either a faster or a slower change cannot occur. It necessarily follows from these facts that time too must be continuous. By a continuum I mean that which is divisible into parts which are always further divisible. If we accept this as a definition of continuity, it necessarily follows that time is continuous. For, as we have demonstrated,* a faster object takes less time

b27 to traverse an equal distance. So, let A be the faster object and B the slower one; and let the slower object take time FG to move over a magnitude CD. Obviously, then, the faster object will take less time than this (FH, say) to move over the same magnitude. Again, since the faster object has taken FH to traverse CD as a whole, the slower object will traverse a lesser distance (CE, say) in the same time. Since the slower object B has taken FH to traverse CE, the faster object will take less time to traverse it, with the result that the stretch of time FH will again be divided. And the division of the time allows us to divide the magnitude CE in the same proportion. But the division of the magnitude allows us to divide the time

233a5 as well. And we can carry on doing this for ever, taking first the faster object, then the slower one, then the faster one again, and so on, each time using the point demonstrated; for the faster object will allow us to divide the time and the slower object will allow us to divide the distance. So if this correlation is always valid and allows us to make a division each time it occurs, it is clear that any stretch of time is a continuum. For both the time and the magnitude are subject to the same divisions and to the same number of divisions.

a13 Further, we can show that the continuity of magnitude follows from that of time by considering the things we normally say about them, since it does take half the time to traverse half the distance, and in general less time to traverse a shorter distance; both time and magnitude are liable to the same divisions. And if either of them is infinite, the other one will be too; and the way in which one of them is infinite will be the way in which the other one is too. For example, if time is infinite in extent, distance will be too;* if time is infinitely divisible, distance will be too; and if time is infinite in both respects, magnitude will be infinite in both respects as well.

That is why Zeno's argument makes a false assumption, ^a21 that it is impossible to traverse what is infinite or make contact with infinitely many things one by one in a finite time. The point is that there are two ways in which distance and time (and, in general, any continuum) are described as infinite: they can be infinitely divisible or infinite in extent. So although it is impossible to make contact in a finite time with things which are infinite in quantity, it is possible to do so with things which are infinitely divisible, since the time itself is also infinite in this way. And so the upshot is that it takes an infinite rather than a finite time to traverse an infinite distance, and it takes infinitely many rather than finitely many nows to make contact with infinitely many things.

It is impossible, then, to traverse an infinite extent in a ^a31 finite time, and it is also impossible to traverse a finite extent in an infinite time. If the time is infinite, the magnitude is infinite too, and if the magnitude is, so is the time. Let *AB* be a finite magnitude, and let *C* be an infinite stretch of time. Let us take a finite portion of the time, and call it *CD*. In this amount of time, then, a moving object will traverse a portion of the magnitude—let us call the traversed portion *BE* (and it does not make any difference whether *AB* is an exact multiple of *BE*, or whether it is larger or smaller than the nearest multiple of *BE*). If the moving object traverses any magnitude which is equal to *BE* in the same amount of time as it takes to traverse *BE* (and let us assume that *AB* is an exact multiple of *BE*), then the total time it takes over the journey will be finite, because it will be divided into the same number of portions as the magnitude has been divided into by *BE*.

Again, if the moving object does not take infinite time to 233^b7 traverse every magnitude, but can traverse a magnitude such as *BE* in a finite time, and if *AB* is a multiple of *BE*, and if a magnitude equal to *BE* is traversed in the same amount of time as it takes to traverse *BE*, then the time will be finite too. And it becomes clear that it does not take infinite time to traverse *BE* if we begin with a time which is limited in one of its two directions; for since it takes a shorter time to traverse a part than a whole, and since one of the limits is already given, the shorter time is bound to be finite. The same proof applies also to the case of an infinite distance and a finite time.

^b15   These arguments* show, then, that no continuum such as a line or a plane is indivisible. This is clear not only for the reasons already given, but also because otherwise the indivisible will end up divided. This follows from the premisses that objects may be faster or slower than one another within any time-period, that a faster object traverses a greater distance in an equal time, and that it may traverse a distance which is double, or one and a half times as long (since their relative
^b22 speeds may stand in this ratio). Suppose, then, that the faster object moves over one and a half times the distance in an equal time, and that the distance covered by the faster object is divided into three indivisible parts, and that covered by the slower object into two. Let the three parts of the distance traversed by the faster object be *AB*, *BC*, and *CD*, and the two parts of the distance traversed by the slower object be *EF* and *FG*. The time taken by the faster object, then, will also be divided into three indivisible parts, since equal distances are traversed in equal times. So let the time of the faster object be divided into *KL*, *LM*, and *MN*. But on the other hand, since the slower object has moved over *EFG*, the same amount of time will also be divided into *two* segments. It follows that the supposed indivisible will be divided, and that it will take not an indivisible amount of time, but more time, to traverse a distance which was supposed not to have any parts.

^b31   It is clear, then, that there is no such thing as a continuum which is not divisible into parts.

3. *A* now *is indivisible; hence nothing moves in a* now

^b33 The now—in its own right and in the primary sense* of the word, not in any secondary sense—must also be indivisible.

This is the kind of 'now' which occurs in any and every stretch of time, since it is a limit of the past (because there is nothing of the future on one side of it) and also of the future (because there is nothing of the past on the other side of it). We call it, then, a limit of both at once. And the proof that it is such a limit—that the limit of the past is the same thing as the limit of the future—would simultaneously be the proof of its indivisibility.

It is bound to be the same now that is the limit of both 234ª5 these times. If there were two different limits, one could not succeed the other, because no continuum consists of things without parts, and if they were separate, there would be a stretch of time between them. This is because what is between limits in a continuum has the same name as the continuum itself. And if what comes between the limits is time, it will be divisible, because we have demonstrated that all time is divisible.

The now, then, would on this hypothesis be divisible. But ª11 if the now is divisible, there will be something of the past in the future* and something of the future in the past, because the point at which the division occurs will form a boundary between past and future time. Also, such a now will not be the now in its own right, but in some secondary sense, because the division in it is the now in its own right.† Furthermore, part of the now will be past and part of it will be future; and since it will not always be the same parts that are past and future (because any stretch of time is divisible at many points), it will not always be the same now.*

Since these consequences are impossible, the now that be- ª19 longs to the past must be the same as the now that belongs to the future. But if it is the same now, it is evidently indivisible as well, because if it were divisible, the same impossibilities would arise as before. So these arguments make it clear that there is something indivisible in time, and this is what we call the now.

The following considerations will show that nothing moves ª24 in the now.* If it were possible for something to move in the now, there could be both faster and slower motion in it. So let N be the now, and let AB be the distance the faster object has travelled. In the same now, then, the slower object will have traversed a distance less than *AB*, and let us call this shorter distance *AC*. But since the slower object takes the whole of the now to move over *AC*, the faster object will take less than this to move over *AC*, and the consequence will be that the now will be divided. But we found that the now is indivisible. So it is impossible for there to be movement in the now.

ᵃ31 It is also impossible for there to be rest in the now. For we speak of rest only in the case of something whose nature is to move, but which is not moving when, where, and how it is its nature to move. Consequently, since there is nothing whose nature is to move in the now, obviously there is nothing whose nature is to rest in the now either.

ᵃ34 Moreover, it is the same now which belongs to both past and future time, and it is possible for something to be in motion throughout the whole of one of these periods and at rest throughout the whole of the other. Now, something which is moving throughout the whole of a period of time will be moving in any part of it in which it is naturally capable of moving; and something which is at rest throughout a whole period will likewise be at rest in any part of it in which it is naturally capable of being at rest. The upshot of all this is that the same thing will simultaneously be at rest and in motion,* because both periods of time have the same limit, namely the now.

234ᵇ5 Also, we say that something is at rest when both it and its parts are in the same state both now and earlier. But there is no 'earlier' in the now, and so nothing can be at rest in the now.

ᵇ8 It necessarily follows, then, that anything in motion and anything at rest are in motion and at rest in time.

4. The changing object, the time, the change, and the respect of the change are all divisible

ᵇ10 Everything that changes is necessarily divisible. For every change has a starting-point and an end-point, and when something—the thing itself and all its parts—is at the end-point of its change, it is no longer changing, and when it is at the starting-point of its change, it is not yet changing, because anything which remains the same in itself and in its parts is not changing. It necessarily follows, therefore, that part of ᵇ15 the changing object is at the one point and part is at the other point. After all, it cannot be at both points or at neither point. (By the end-point of a change, I mean in this context the first end-point* which occurs during the change. In the case of change from white, for instance, I mean grey rather than black. The changing object does not have to be at one of the two extremes.) So it is clear that every changing thing will be divisible.

There are two ways in which change is divisible.* First, it ᵇ21 is divisible because the time involved is divisible; second, it is divisible into the changes of the parts of the changing object. For example, if the whole of *AC* is changing, then both *AB* and *BC* will change. So let *DE* be the change of the part *AB* and *EF* be the change of the part *BC*. Then the whole change *DF* is bound to be the change of *AC*; *DF* will be the change of *AC* precisely because *DE* and *EF* are the changes of its two parts. The change of a thing is the change of just that thing, not of anything else, and so the whole change *DF* is the change of the whole magnitude *AC*.

Besides, every change is the change of something, and the ᵇ29 whole change *DF* is not the change of either of the parts (since they each have their own change) and is not the change of any other whole either (for the partial changes are changes of the parts of the whole whose change is the change as a whole; and the partial changes are the changes of the parts *AB* and *BC* and of no other parts; for we found* that a single change cannot be the change of a number of things). Therefore, the whole change *DEF* is the change of the magnitude *ABC*.

Also, suppose that the whole magnitude was liable to an- ᵇ34 other change, which we can call *HI*. Now, the changes of each of the parts can be subtracted from *HI*, and these partial changes will (on the principle that a single thing cannot have more than one change) be equal to *DE* and *EF*. The consequence is that if the whole change *HI* is divided in this way into the changes of the parts, *HI* will be equal to *DF*. If, on 235ᵃ4 the other hand, there is some remainder (call it *JI*), it will be

a change which is the change of nothing, because it will not
be the change of the whole or of the parts (given the principle
that a single thing cannot have more than one change) or of
anything else (because a continuous change is the change of
continuous things); and the same goes also if *DE* and *EF* add
up to more than *HI*. So since *HI* cannot be greater or smaller
than *DF*, it must be equal and identical to it.

ₐ9   So much, then, for the division of change which corresponds to the changes of the parts of the changing object.
The changes of anything divisible are necessarily subject to
this kind of division. The other kind of division of change
depends on time. Since every change is in time, since time is
always divisible, and since in less time there is less change, it
necessarily follows that any change is divided as the time it
takes is divided.

ₐ13   Since every changing object changes in a certain respect
and takes a certain amount of time, and since there is a
process of change for every changing object, there must be
identical divisions of the time, the change, the changing,
the changing object, and the respect* (although not all the
respects of change are divisible in the same way: place is
divisible in its own right, while quality is only coincidentally

ₐ18 divisible). Let us call the time in which the change occurs *A*
and the change *B*. If the whole change occupies all the time,
in half the time there will be less than the whole change, in
a smaller portion of the time there will be still less, and so on
ad infinitum. Conversely, if the change is divisible, the time
is also equivalently divisible. For if it takes the whole of the
time for the whole change to happen, it takes half the time
for half the change to happen, and still less time for still less
of the change to happen.

ₐ25   The changing will also be divisible in the same way. Let *C*
be the changing. Corresponding to half the change, then,
there will be a changing which is less than the whole, and
corresponding to half of half the change there will be even
less of a changing, and so on *ad infinitum*. Also, if one sets
out the changing which corresponds to each of the two phases
of the change—*DC*, say, and *CE*—one can claim that the
changing as a whole will correspond to the change as a whole

(for if it were some other changing that corresponded to the complete change, more than one changing would correspond to a single change). In order to make this claim, one could use the same arguments we used to prove that the change too is divisible into the changes of its parts, because once we have a changing corresponding to each of the two phases of the change, the whole changing will turn out to be formed into a continuous whole by them.

The same argument will also show the divisibility of the ᵃ34 distance and in general of every respect in which change takes place (although some respects are only coincidentally divisible—i.e. they are divisible only because the changing object is divisible); for the divisibility of any one of them implies the divisibility of them all. Likewise, on the issue of whether they are finite or infinite, what goes for one of them will go for them all. The fact that all of them are divisible and are infinite is a consequence above all, we have found,* of the fact that the changing object is divisible and infinite. For the properties of divisibility and infinity belong initially to the changing object. We have already established the position on divisibility, and now we turn to explaining infinity.

5. *There is always a first instant at which a change has been completed*

Since everything that changes changes from a starting-point 235ᵇ6 to an end-point, it necessarily follows that what has changed, as soon as it has changed, is in the state to which it has changed. For the changing object leaves or departs from the starting-point of its change, and change and departure are either identical or departure is a consequence of change. If departure is a consequence of change, then having departed is a consequence of having changed, since the relation between the two is the same in either case.

Now, one kind of change involves contradiction; so at the ᵇ13 moment when a thing has changed from non-existence to existence, it has departed from non-existence. Therefore, it will be in a state of existence, because everything must either exist or not exist. In the case of the kind of change that involves contradiction, then, it is clear that once the changing object has changed it will be in that to which it has changed. And if this is true in this case, it is true for the other kinds of change too, because what goes for one goes equally for the rest as well.

b19   This is also clear if we take each kind of change separately. In each case, once the changing object has changed, it must be somewhere or in some state. Since it has departed from the starting-point of its change and since it must be somewhere, then it must be either in the end-point or somewhere else. If something that has changed to its end-point *B* is somewhere else—in *C*, say—then *C* becomes the starting-point of a further change to *B*; for *B* is not consecutive to *C* and the change is continuous. And so the object which has changed, when it has changed, is changing to that to which it has changed. But since this is impossible, the object which has changed must be in the state to which it has changed.

b27   It is also clear, then, that when an object which has come into existence has come into existence, it will exist, and an object which has ceased to exist will not exist. For what we have been saying applies generally to every kind of change, and is particularly clear in the case of the kind of change that involves contradiction.

b30   We can see, then, that as soon as a changing object has changed, it is at its end-point. Now, the immediate occasion when it has changed is bound to be indivisible. (By 'immediate' here, I mean that which is what it is directly, and not because some part of it is so.) We can see this as follows. Let *AC* be divisible, and suppose it is divided at *B*. If the object has finished changing in *AB* (or in *BC*), *AC* cannot be the immediate occasion on which it has finished changing. And if it was changing in both *AB* and *BC* (and it must either have finished changing or have been changing in each of them), then it would have been changing in *AC* as a whole, when on 236a2 our hypothesis it has finished changing in *AC*. The same reasoning is also valid if it is in the process of changing in one of these parts, but has finished changing in the other, because there would then be something prior to what was supposed to be immediate. The upshot is that the occasion on which it has finished changing must be indivisible. It is also clear, then, that something which has ceased to exist has ceased to exist at an indivisible moment, and that something that has come into existence has done so at an indivisible moment.

There is not a first instant at which a change has been begun

'The immediate occasion when it has changed' is an ambigu- [a7] ous phrase. On the one hand, it could mean the immediate occasion when the change has been completed, because that is the time when it is true to say that the changing object has changed; on the other hand, it could mean the immediate occasion when it has begun to change. Now, when the expression is used for the end of the change, it refers to an immediate occasion that exists and is real, because it is possible for change to be completed and there is an end to change—and in fact this end is what we have shown to be the limit of change and therefore indivisible. However, there is [a13] no such thing as the immediate occasion when a change has begun, because change has no beginning and there is no immediate occasion in time when a changing object began to change. Let *AD* be such an immediate occasion. This cannot be indivisible, because it would follow that the nows are consecutive.* Besides, if the object is at rest (we can assume that it is) throughout the whole period of time *CA* which leads up to *AD*, then it is at rest at *A*;* and so, if *AD* is indivisible into parts, the object will be at rest and will have changed simultaneously, because it is at rest at *A*, but has changed at *D*. Since *AD* does have parts, then, it must be [a20] divisible—and then there are necessarily no occasions within *AD* when the object has not changed. Suppose *AD* is divisible into two parts: if the object has not changed in either part, it has not changed in *AD* as a whole either, and if it is changing in both parts, it is changing in *AD* as a whole as well. But if† it has changed in one of the two parts, then *AD* as a whole is not the immediate occasion when it has changed. It necessarily follows that it must have changed in each part. Obviously, then, there is no immediate occasion when the object has changed, because there is no end to the divisions of any proposed candidate.

Two corollaries

^a27 Also, nothing which has changed has a part which is the first part to have changed. Let *DF* be the putative first part of *DE* that has changed; after all, we have proved* that every changing object is divisible. Let *HI* be the period of time in which *DF* has changed. So if *DF* has changed in the whole period of time, in half the time a part smaller than *DF* has changed, and it does so before *DF* does; and then there will be another that has changed before this one, and then yet another one, and so on *ad infinitum*. And so nothing which has changed has a part which is the first to have changed.

^a35   These arguments show, then, that there is no part of the object that is the first to change, nor any part of the time that is the first in which change has occurred. However, matters are different with the end-point or respect of the change.†

236^b2   There are three factors to take into consideration in change —the thing which changes (a person, for instance), that in which the change takes place (time), and the end-point of the change (pallor, for example). The person and the time are divisible, but the pallor is another matter—except that it and everything like it are all coincidentally divisible, because the subject of which pallor or some quality is a coincidental attri-

^b8 bute is divisible. The point is that there is in fact no first part to have changed in anything which is said to be divisible in its own right rather than coincidentally. Consider magnitudes, for example. Let *AB* be a magnitude, and let it move in the first instance from *B* to *C*. Now, if *BC* is indivisible, something indivisible into parts will be consecutive with something else which is indivisible* into parts; if, on the other hand, *BC* is divisible, there will be a point to which *AB* has changed before it has changed to *C*, and then another point before that one, and so on *ad infinitum*, because there is no end to the process of division. And so there can be no first point to

^b16 which the object has changed. The same goes also for the case of change of quantity, since this change too is change in respect of something continuous. We can see, then, that the only kind of change which can be indivisible in its own right is change of quality.

6. *Whatever is changing has changed already*

Every changing thing changes in time, but 'time' in this ex- ᵇ19 pression may refer either to the immediate time of the change or to the time in a derivative sense. For instance, something changes in a given year because it changes on a given day. Therefore, there is bound to be no part of the immediate time of a change when the thing is not changing. Our definition makes this clear, because that is how we have been defining 'immediate';* but the following considerations also make it clear. Let *AB* be the immediate time in which the changing ᵇ25 object is changing, and let it be divided at *C* (after all, time is always divisible). In the time *AC* the object is either changing or it is not changing, and the same goes for the time *BC*. If it is not changing in either time, it would be at rest throughout, since it is impossible for it to change in *AB* as a whole when it is not changing in any of its parts. If it is changing in one of the two parts, but not the other, *AB* cannot be the immediate time in which it is changing, because *AB* is only derivatively the time of the change. There must, then, be no part of *AB* in which it is not changing.

Now that we have established this point, it is clear that ᵇ32 every changing object must have changed earlier.* For if it takes the immediate time *AB* for it to have moved over a magnitude *KL*, it will take half the time for something which moves at the same speed and which started moving at the same time to have moved over a magnitude half the extent of *KL*. But given that it takes this object which moves at the same speed a certain time to have moved a certain distance, it necessarily follows that the first object too must have moved over the same magnitude in the same time. Consequently, an object that is moving has already moved.

Secondly, if it is our grasp of the limiting now of *AB* that 237ᵃ3 allows us to say that the object has moved in the time *AB* as a whole, or in general in any part of *AB* (because it is the limiting now that defines a period of time, and time is what comes between nows), by the same token we could also say that it has moved in the other parts of *AB* as well. But by dividing *AB* we have provided half of it with a limiting now, and so the object will have moved in half of *AB* and in general in any given part of *AB*. For to make a division is simultaneously to provide a time defined by those nows. So if any and every time is divisible, and if time is what lies between nows, then every changing object has completed an infinite number of changes.

ᵃ11 Thirdly, something which is continuously changing and has not ceased to exist or stopped changing must always either be changing or have changed. But since it is impossible for it to be changing in the now, then at each now it must have changed. Therefore, since there are infinitely many nows, every changing object must have completed an infinite number of changes.

Whatever has changed was changing earlier

ᵃ17 As well as being necessary that what is changing has changed before, it is also necessary that what has changed was changing before, because everything that has changed from a starting-point to an end-point has taken time to complete the change. Suppose that a thing has changed, from state *A* to state *B*, in a now. Then it has not done so in the same now in which it is in state *A* (which would mean that it was both in state *A* and in state *B* simultaneously), because we have already shown* that when a thing has changed from a given state it
ᵃ24 is no longer in that state. Alternatively, it is a different now, in which case there will be a stretch of time between the two nows,* because nows are not consecutive. So since it takes time to have changed, and since all time is divisible, then in half the time it will have completed a different change, and in half that time a different one again, and so on *ad infinitum*. Therefore, that which has changed must have been changing earlier.

ᵃ28 Moreover, where the change is a change in some magnitude, the fact that this magnitude is continuous makes the point even more clearly. Suppose something has changed from *C* to *D*. Then if *CD* is indivisible, something without parts is consecutive* with something else which lacks parts. But since this is impossible, there must be something between C and D and it must be a magnitude, and so infinitely divisible. If so, there are infinitely many dividing points it changes to before it reaches D.

Everything that has changed must have been changing ᵃ34 earlier, then. This is universally true, because the same proof also holds good for changes involving non-continuous things, such as changes involving opposites and contradictories. All we have to do is take the time* in which the thing has changed, and repeat the same argument. So that which has changed must have been changing, and a changing object must have changed; the condition of having changed will precede changing, and the condition of changing will precede having changed. So there cannot be a first change for us to take. The reason for this is that something which is indivisible into parts cannot be consecutive to something else which is indivisible into parts. Division can go on *ad infinitum*, just as it can in the case of lines being increased and diminished.

Obviously, then, the same goes for coming into being too: in 237ᵇ9 the case of things which are divisible and continuous,* what has come to be must earlier have been coming to be, and what is coming to be must earlier have come to be; however, what has come to be is not always what is coming to be, but it may be one of its parts (for instance, the foundation-stone of a house). The same goes for that which is ceasing to be and that which has ceased to be as well, because anything which comes to be and anything which ceases to be is continuous and therefore is infinite in a way. So nothing can be ᵇ15 coming to be unless something has come to be and nothing can have come to be unless something has been coming to be, and likewise for ceasing to be and having ceased to be. Having ceased to be will always precede ceasing to be, and ceasing to be will always precede having ceased to be. Obviously, then, because every magnitude and every period of time is divisible, that which has come to be must earlier have been coming to be, and that which is coming to be must earlier have come to be. And so whatever stage such a thing may be at cannot be a first stage.

7. *If any of the time, the distance,*
and the moving object is infinite in extent,
so are the others

ᵇ23 Since every moving object takes time to move, and since in more time it will move over a larger magnitude, it cannot move over a finite magnitude in an infinite time. I exclude cases of perpetual repetition of the same motion or of some part of that motion; I am talking about cases where a whole magnitude is covered in a whole stretch of time. But if the object is moving at a constant speed, it is plainly bound to move over a finite distance in a finite amount of time. After all, we may take some part of which the whole distance is a multiple, and then completion of the movement over the whole distance takes as many equal stretches of time as there are parts equal in extent to this given part. And so, since these parts are finite (both individually in extent and together in number), the time too must be finite. That is, the total time will be equal to the time taken over the given part multiplied by the number of parts.

ᵇ34 In fact it does not make any difference even if the object is not moving at a constant speed. Let *AB* be a finite distance which an object has traversed in an infinite time (which we can call *CD*). Now, because the distance covered in a shorter time is different from the distance covered in a longer time (and it does not make any difference whether or not the movement happens at a constant speed, and whether the rate increases or decreases or remains steady), then the distance covered earlier is different from the distance covered later, and therefore one part of the distance must have been traversed before another part. So let us take a part of the distance *AB* (call it *AE*), and let *AB* be a multiple of *AE*. Now, traversing *AE* takes a certain part of the infinite time; it cannot take an infinite time, because *ex hypothesi* traversing the whole distance takes infinite time. Now suppose I take another part of *AB*, equal in size to *AE*: again, this must be traversed in a finite amount of time, because *ex hypothesi* 238ᵃ11 traversing the whole distance takes infinite time.* Suppose, then, we take parts of AB like this; now, infinity is not a multiple of any finite part of itself, because it is impossible for infinity to consist of finite components* whether they are equal or unequal to one another. This is because it is things which are finite in number or magnitude which are multiples of some unit, and this remains the case whether their parts are equal or unequal to one another, just so long as they are of a determinate size. Therefore, and since *AE* taken a certain number of times does measure out the finite distance *AB*, *AB* will be traversed in a finite time. And the same goes also for the process of coming to rest. Consequently, it is impossible for one and the same thing to be perpetually coming into existence or ceasing to exist.

The same argument also demonstrates the impossibility of ᵃ20 something taking a finite time to move or come to rest over an infinite distance, whether or not it is moving uniformly. For if we take a part of which the whole stretch of time is a multiple, a moving object will take this much time to traverse a certain amount of the magnitude, but not the whole magnitude (because that would take the whole of the time), and it would traverse another portion of the magnitude in the same amount of time again, and so on for each stretch of time, whether or not it is equal to the original stretch of time we took. It does not make any difference whether or not each ᵃ26 stretch is equal to the original stretch, as long as it is finite,* because it is clear that when the time is exhausted the infinite magnitude will not be exhausted, since the process of subtraction involves finitely many steps and finite quantities at each step. So infinity cannot be traversed in a finite time. Nor does whether the magnitude is infinite in one direction or in both directions make any difference to the validity of the argument.

Now that we have established these points, it is also clear ᵃ32 that a finite magnitude cannot traverse an infinite magnitude in a finite time and that the reason is the same: it will take part of the time to traverse a finite part of the infinite magnitude, and so on for each part of the time, so that it will have traversed only a finite magnitude in the whole stretch of time. And since a finite magnitude will not traverse an infin- ᵃ36 ite magnitude in a finite time, an infinite magnitude will obviously not traverse a finite magnitude either in a finite time. The point is that if an infinite magnitude could traverse a finite magnitude, a finite magnitude could traverse an infinite magnitude as well. It makes no difference which of the two is the moving object; both possibilities involve something finite traversing something infinite. I mean, when the infinite magnitude *A* is moving, some part of it (*CD*, say) will occupy the finite magnitude *B*, and then another part of it will occupy B,

238^b7 and then another, and so on *ad infinitum*. And the end result of this is that not only will the infinite have moved over the finite, but the finite will also have traversed the infinite. For it is presumably impossible for the infinite to move over the finite except by the finite traversing the infinite, whether the finite is in motion over the infinite or is measuring out the infinite length by length. Since this is impossible, then, something infinite cannot traverse something finite.

^b13 Moreover, something infinite will not traverse something infinite* in a finite time. If it could, something finite could too, because the infinite includes the finite. The same point can also be demonstrated in terms of the time involved.

^b17 Since in a finite time a finite magnitude cannot traverse an infinite magnitude, and an infinite magnitude cannot traverse a finite magnitude, and an infinite magnitude cannot traverse an infinite magnitude, there obviously cannot be an infinite movement in a finite time. I mean, what difference does it make whether we take the movement or the magnitude to be infinite? If either of the two is infinite, the other one is bound to be infinite as well, because all movement is in place.

8. *There is no last time of coming to rest, and no first time of being at rest**

^b23 Since everything whose nature it is to move or rest is either in motion or at rest when, where, and how it is its nature to be so, something which is coming to a standstill is in motion at the time when it is coming to a standstill. For if it is not in motion, it is at rest, but something at rest cannot be coming to rest. Now that we have established this point, it is clear that it must also take time to come to a standstill, because

anything moving is moving in time and we have shown that
something which is coming to a standstill is moving, and therefore it must be coming to a standstill in time. Moreover, we
apply the terms 'faster' and 'slower' to things which take time,
and it is possible to come to a standstill faster and slower.

An object which is coming to a standstill must be in the ᵇ31
process of doing so in any part of the immediate time in
which it is coming to a standstill. Suppose we divide the time
into two parts. If it is coming to a standstill in neither of the
two parts, it is not doing so in the whole either, which would
mean that an object that is coming to a standstill is not
coming to a standstill. If, on the other hand, it is coming to
a standstill in just one of the two parts, then the whole time
is not the immediate time of its coming to a standstill; it
would be coming to a standstill in the time as a whole only
in a derivative sense, as I also explained earlier when discussing
changing objects.

Just as there is no immediate time in which a changing ᵇ36
object is changing, so also there is no immediate time when
an object that is coming to a standstill is coming to a standstill. For there is no first stage* of either changing or coming
to a standstill. Let *AB* be the immediate time when an object
is coming to a standstill. Then *AB* must have parts, since
there cannot be movement in what has no parts; for anything
that is in motion has moved already in some earlier part of
the time,† and we have shown* that an object which is coming to a standstill is in motion. But if *AB* is divisible, then, the
object is coming to a standstill during any part of *AB*. For we
have already seen* that it is coming to a standstill in any part
of the immediate time when it is coming to a standstill. So
because the immediate time in which an object is coming to
a standstill is a period of time, not something indivisible, and
because every stretch of time is infinitely divisible into parts,
there can be no immediate time when an object is coming to
a standstill.

Nor, then, is there an immediate time when something at 239ᵃ10
rest was at rest. It cannot have been at rest in an indivisible
moment of time, because* there can be no movement in what
is indivisible; and whenever rest is possible, movement is

possible too, because for something to be at rest, we said, it has to be something whose nature it is to move, but which is not moving when—that is, in the time in which—it is in its
ᵃ14 nature to do so. Secondly, we also say that something is at rest when it is in the same state now as it was earlier; in other words, we need not one but at least two moments to assess a state of rest, and so that in which something is at rest is not indivisible into parts. If it is divisible, then, it must be a stretch of time, and an object at rest will be at rest during any part of the time. We can use the same arguments to prove this as we used in earlier cases too. And the upshot is that there is no immediate first part of the time, because everything at rest and everything in motion is at rest or in motion in time, and there is no first time, no first magnitude, and in general no first part of any continuum. For every continuum is infinitely divisible.

ᵃ23 Every changing object changes in time and there is both a starting-point and an end-point to the change. Therefore, in the time when a changing object is changing—that is, the precise time when it is changing, and not just some time of which this is a part—it is impossible for it to be immediately opposite to anything. After all, for a thing to be at rest—for it itself and each of its parts to be at rest—it has to be in the same state for a certain period of time. I mean, we say that something is at rest when it is true to say that it and its parts
ᵃ29 are in the same state for one now after another. And if this is what it is for something to be at rest, it is impossible for any changing object to be as a whole opposite to anything in the immediate time of its change. For time is always divisible, and so it will be true to say that the object and its parts are in the same state for one after another of the parts of time. If this is not so, and it is in the same state only for a single now, it will not be opposite to anything for any period of
ᵃ35 time, but only at the limit of a period of time. It may be that at any given now it is opposite to something, but it is not at rest, because there is no such thing as being in motion or at rest in the now. And although it is true to say that in the now it is not in motion and is opposite to something, it is not possible for it to be opposite to something at rest for any

period of time, because that would mean that something in motion was at rest.

9. *Zeno's arguments on motion present no difficulty for us*

Zeno's reasoning is invalid. He claims that if it is always true $239^{b}5$ that a thing is at rest when it is opposite to something equal to itself, and if a moving object is always in the now, then a moving arrow is motionless. But this is false, because time is not composed of indivisible nows, and neither is any other magnitude.

Zeno came up with four arguments about motion which $^{b}9$ have proved troublesome for people to solve. The first is the one about a moving object not moving because of its having to reach the half-way point before it reaches the end. We have discussed this argument earlier.*

The second is the so-called Achilles. This claims that the $^{b}14$ slowest runner will never be caught by the fastest runner, because the one behind has first to reach the point from which the one in front started, and so the slower one is bound always to be in front. This is in fact the same argument as the $^{b}18$ dichotomy, with the difference that the magnitude remaining is not divided in half. Now, we have seen that the argument entails that the slower runner is not caught, but this depends on the same point as the dichotomy; in both cases the conclusion that it is impossible to reach a limit is a result of dividing the magnitude in a certain way. (However, the present argument includes the extra feature that not even that which is, in the story, the fastest thing in the world can succeed in its pursuit of the slowest thing in the world.) The solution, then, must be the same in both cases. It is the claim that the one in front cannot be caught that is false. It is not caught as long as it is in front, but it still is caught if Zeno grants that a moving object can traverse a finite distance.

So much for two of his arguments. The third is the one I $^{b}29$ mentioned a short while ago,* which claims that a moving arrow is still. Here the conclusion depends on assuming that time is composed of nows; if this assumption is not granted, the argument fails.

$^b33$ His fourth argument* is the one about equal bodies in a stadium moving from opposite directions past one another; one set starts from the end of the stadium, another (moving at the same speed) from the middle. The result, according to Zeno, is that half a given time is equal to double that time. The mistake in his reasoning lies in supposing that it takes the same time for one moving body to move past a body in motion as it does for another to move past a body at rest, where both are the same size as each other and are moving 240^a4 at the same speed. This is false. For example, let AA ... be the stationary bodies, all the same size as one another; let BB ... be the bodies, equal in number and in size to AA ..., which move from the middle of the stadium; and let CC ... be the bodies, equal in number and in size to the others, which start from the end of the stadium and move at the same speed as BB ... Now, it follows that the first B and the first C, as the two rows move past each other, will reach the end of each $^a10$ other's rows at the same time. And from this it follows that although the first C has passed all the Bs, the first B has passed half the number of As; and so (he claims) the time taken by the first B is half the time taken by the first C, because in each case we have equal bodies passing equal bodies. And it also follows that the first B has passed all the Cs, because the first C and the first B will be at opposite ends of the As at the same time, since (according to Zeno) the first C spends the same amount of time alongside each B as it does alongside each A,[†] because both the Cs and the Bs spend the same amount of time passing the As. Anyway, that is Zeno's argument, but his conclusions depend on the fallacy I mentioned.

Other alleged difficulties resolved

$^a19$ We will also find no impossibilities arising from the kind of change which involves contradictories—if, for example, something is changing from not pale to pale and is currently in neither state, so that it is not pale, but is not not-pale either. Even if it is not *entirely* in either state, that will not stop us calling it either pale or not pale. After all, in order for us to describe things as pale or not pale, we do not require them to be entirely pale or not pale; it is enough that most of their parts are pale, or their most important parts are. It is one thing not to be in a certain state, and it is another thing not to be entirely in that state. The same goes for being and not being and all other contradictories: the changing object is bound to be in one or the other of the two opposed states, without being entirely in either.*

We can also cope with the argument that circles and spheres ᵃ29 and all those things which move within their own dimensions are actually at rest, because both the objects themselves and their parts will occupy the same place for a certain stretch of time, and so they will simultaneously be at rest and in motion. First, the parts do not stay in the same place for any time at ᵃ33 all. Second, the whole is in fact constantly changing place, because the circumference taken from a point *A* is different from that taken from *B*, or from *C*, or from any other point; the only way in which they are the same is coincidentally, as an educated person is the same as a person. And so, as it moves, one circumference becomes another, and it is never at rest. The same goes for a sphere as well, and for everything else which moves within its own dimension.

10. *What has no parts cannot change in its own right*

Now that we have established these points, the next thing 240ᵇ8 to prove* is that a thing which lacks parts cannot change, except coincidentally—if (for instance) it is contained within a moving body or magnitude. Think of something in a ship being moved by the motion of the ship, or a part being moved by the motion of the whole.

(By 'something which lacks parts' I mean something which ᵇ12 is quantitatively indivisible. Indeed, wherever parts are subject to change, there are differences not only between the changes the parts themselves undergo, but also between the changes of the parts and the changes of the whole. The difference is especially obvious in the case of a sphere, since the speed at which the parts at the centre are moving differs from the speed at which the parts at the outside are moving, and differs from the movement of the whole sphere, which shows that no single movement is involved.)

^b17   As I said, then, something which has no parts can change as a man sitting in a ship is moving while the ship proceeds on its way, but it cannot change in its own right. Suppose it is changing from *AB* to *BC** (from one magnitude to another, or from one form to another, or from one state to its contradictory), and let *D* be the immediate time of its change. At the time when it is changing, it must be either in *AB* or in *BC* or partly in the one and partly in the other; for we found*

^b26   that this is so for every changing object. Now, it cannot be the case that part of it is in each of *AB* and *BC*, because then it would be divisible into parts. But it cannot be in *BC* either, because then it would have completed its change, when *ex hypothesi* it is in the process of changing. The only remaining possibility, then, is that at the time when it is changing it is in *AB*. But then it must be at rest, because we found that to

^b30   be in the same state for a period of time is to be at rest. The upshot is that it is impossible for something which has no parts to move, to vary, or to change in any way at all. For the only condition which would make it possible for it to change would be if time consisted of nows,* because then at any given now it would have completed a change; it would never be in the process of changing, but would always have finished changing. But we have already proved* the impossibility of this: time does not consist of nows, a line does not consist of points, and change does not consist of discrete changes. The idea that it does is no more or less than the idea that change consists of indivisible parts, and is equivalent to claiming that time consists of nows or that a length consists of points.

241^a6   There is also another way to show that it is impossible for a point* or for any other indivisible thing to move. No moving object can move over a magnitude larger than itself until it has first moved over a magnitude equal to or smaller than itself. Therefore, it is clear that, if a point can move, it too will move over a magnitude smaller than or equal to itself. But since it is indivisible, it cannot first move over a magnitude smaller than itself, so it must first move over a magnitude equal to itself. Therefore, the line over which it moves will consist of points, because since the point always moves over a magnitude equal to itself, it must eventually measure out the whole line. But since this is impossible, it is also impossible for something indivisible to move.

Also, since everything that changes changes in time and ^a15 nothing changes in the now, and since all time is divisible, then for anything that moves there would be some time smaller than the time in which it moves over a magnitude as large as itself. There will be a time for this move, since everything moves in time, and we have already demonstrated* the divisibility of every stretch of time. So if it is a point that is ^a19 moving, there will be a stretch of time smaller than the time in which it has moved over itself. But that is impossible, because what it moves over in a smaller stretch of time must be a smaller magnitude, and so the indivisible point will be divisible into something smaller, just as the time too will be divisible into the smaller stretch of time. For the only condition which would make it possible for something without parts, something indivisible, to change would be if it were possible to change in an indivisible now. The same argument covers both change in a now and the possibility of something indivisible changing.

Can change be infinite?*

Now, no process of change is infinite, because (as we have ^a26 seen*) every change, whether it involves contradictories or opposites, has a starting-point and an end-point. So change between contradictories is limited by the affirmation and its negation (for example, existence is the limit of the process of coming into existence, and non-existence is the limit of the process of ceasing to exist), and change between opposites is limited by the opposites, since they are the extremes of the process of change. Opposites, then, are the extremes of every process of alteration too, since alteration depends on certain opposites; they are likewise the extremes of increase and decrease as well, since the extreme in increase is the limit consisting in the completion of a thing's natural size, and the extreme in decrease is the limit consisting in the removal of 241^{b}2 the thing's natural size.* However, this reasoning will not prove that movement is finite, because not all movements are between opposites. But since something which is incapable of having been divided (in the sense—'incapable' being ambiguous—that it is impossible for it to have been divided) cannot be in the process of being divided, and since in general something which is incapable of having come to be cannot be in the process of coming to be, then the same goes for something which is incapable of having changed too: it could not possibly be in the process of changing to that which it cannot have changed to. So if a moving object is in the process of changing to something, it follows that it will also be capable of completing the change. Consequently, movement cannot be infinite and a moving object cannot be in motion over an infinite magnitude (which it cannot traverse).

^{b}11   Anyway, it is clear that there cannot be an infinite change in the sense of a change that is not defined by limits. But we should also consider whether change can be infinite in the sense that a process of change may be infinite because of the time taken, while remaining one and the same process of change. Now, if no single process of change is involved, there is presumably nothing to stop its being infinite in this sense; imagine, for instance, that a process of alteration comes after a movement, and then a process of increase comes after the alteration, and then a process of coming to be after that. Although this would mean that there was always change going on throughout the time, it would not be a single process of change because there is no single kind of change which consists of all these different kinds. As long as a single process of change is involved, there is only one kind of change which can be infinite because of the time taken, and that is circular motion.

VII

VARIOUS POINTS ABOUT CHANGE

1. *Everything that changes is changed by something*

Everything that changes must be changed by something.* For 241^b34 if the source of the change is not to be found within the changing object itself, it must obviously be changed by something other than itself, because the agent of change must under these circumstances be something different. However, ^b37 what if the source of the change is within the changing object itself? Let us take an object *AB* that is changed in its own right, rather than because any of its parts is changing. First, to suppose that *AB* is changed by itself, just because it is changing as a whole and is not changed by any external object, is equivalent to thinking that if *KL* causes *LM* to change and is itself changing too, *KM* is not changed by anything,‡ simply because it is not obvious which of the two parts is the agent of change and which is the one that is being changed. Second, there is no need for a changing object which is *not* changed by anything‡ to stop changing because something else is at rest; however, if the fact that something else has stopped changing causes an object to rest, then it must be that* this object was being changed by something.‡ But if this 242^a37 is accepted, it will follow that everything is changed by something. Let *AB* be a changing object, as we have assumed. It is bound to be divisible, because every changing object is divisible.* So let it be divided at *C*. Now, if *CB* is not changing, *AB* will not be changing, because if *AB* is changing, then obviously *AC* will be changing while *CB* is at rest, with the consequence that *AB* will not be changing in its own right and in the first instance. But *ex hypothesi AB* is changing in its own right and in the first instance. So if *CB* is not changing,

---

‡ See Explanatory Notes, p. 281, for explanation of these symbols.

AB must be at rest. But we have agreed that anything which is at rest because something[†] is not changing is changed by something.[†] It follows that everything that changes must be changed by something, because the changing object will always be divisible, and if one of its parts is not changing, the whole must be at rest too.

There cannot be an infinite regress of movements caused by other movements

[a]49 Since everything that changes must be changed by something, if something is changing place and is being moved by something else, and this agent of change in its turn is being moved by some other moving object, which is being moved by another moving object, and so on, there must be some first mover—the sequence cannot go on *ad infinitum*. Suppose this was not the case and the sequence was infinite. Then let *A* be moved by *B*, *B* by *C*, *C* by *D*, and so on, with each consecutive member of the sequence being moved by the next object in the sequence. Since *ex hypothesi* the mover imparts movement by itself being moved, it necessarily follows that the movement of the moved object and that of the mover take place at the same time, because the mover moves and the moved object is moved at the same time. It is clear, then, that *A*'s movement and *B*'s and *C*'s and that of each of the movers and the moved objects will all be simultaneous. Let us take each movement separately, and let that of *A* be *E*, that of *B* be *F*, and those of *C* and *D* be *G* and *H*.

[a]64   It is true that there has to be a second object to move any given object, but it is still possible to take the movement of a given object to be numerically single, because every movement has a starting-point and an end-point, and these limits guarantee that it is not infinite.[*] I say that movements are 'numerically the same' if they take a numerically single stretch of time to move from a numerically single starting-point to a [a]69 numerically single end-point. For changes can be generically or specifically or numerically the same: changes are generically the same if they fall within the same category (for instance, substance or quality); they are specifically the same if their starting-points and end-points belong to the same species (for

instance, if they change from white to black or from a good state to a bad state, where these are indistinguishable in terms of species); they are numerically the same if they take a single stretch of time to change from a numerically single starting-point to a numerically single end-point (for instance, if they change from this particular whiteness to this particular blackness, or from this particular place to that, and take this particular stretch of time to do so; the point is that if the time is different, the changes will not be numerically the same, although they would still be specifically the same; but we have discussed this issue earlier*).

Now, let us also take the time in which A has completed 242ᵇ42 its movement, and let us call it K. Since A's movement is finite, the time it takes over the movement will also be finite. But since on our hypothesis there are infinitely many movers and infinitely many moved objects, the movement EFGH . . . which consists of all the movements will also be infinite. For there are two possibilities: either all the movements—A's movement, B's movement, and so on—are equal to one another, or the size of the movements progressively increases; in either case, whether they are equal or increasing in size, the upshot is that the total movement is infinite. For we are assuming what is possible.* Now, since A and all the other objects move at the same time, the total movement takes the same time as A's movement; but A's movement takes a finite period of time; and that would mean that an infinite movement happens in a finite time, which is impossible.

It might seem as though we have now proved our original ᵇ53 proposition, but the argument falls short of being such a proof because it fails to prove that the opposite proposition entails an impossibility. After all, there can be infinite movement in a finite time, if the movement belongs to a plurality of things rather than to a single thing. And this is so in the present case, in fact. For each thing moves with its own motion, and there is nothing impossible in a plurality of things moving at the same time. However, the immediate agent of bodily ᵇ59 change of place must be either in contact with or continuous with* the moved object, as we always observe to be the case. So it necessarily follows that the moved objects and the movers

are either continuous or in contact with one another, so that all of them together form a single thing. Now, it does not make any difference in the present context whether this single thing is finite or infinite, because the movement will be infinite in any case,* since it consists of infinitely many parts—at any rate, it will be infinite if its component movements are either equal to one another or increasing in size; either of these alternatives is possible, and we can assume that what is pos-
b67 sible is in fact the case. Anyway, if the objects *ABCD* . . . form either a finite or infinite magnitude, which takes the finite time *K* to perform its movement EFGH . . . , then something traverses an infinite magnitude in a finite time; and it does not matter whether this something is finite or infinite—in either case, the conclusion is impossible.* And so the sequence must come to an end: there must be a first thing which moves others by being moved itself. It does not make any difference that the impossibility we have just noted is a consequence of a hypothesis, because the hypothesis we assumed is a possibility and nothing impossible should follow from assuming a possibility.

2. *The agent of change and the object changed must be in contact*

243a32 Any immediate agent of change—not in the sense that it is the purpose of the change, but in the sense that it is the original source of the change—is contiguous with what is changed (by 'contiguous' I mean that there is nothing between them). This is common to every agent of change and changed object. Now, there are three kinds of change—change of place, change of quality, and change of quantity—and so there must also be three kinds of agents of change, one which causes movement, one which causes alteration, and one which causes increase and decrease. Let us begin by discussing movement, because it is the primary kind of change.

a11   Everything in motion is moved either by itself or by something else. Now, where self-movers are concerned it is obvious that the moved object and the agent of movement are contiguous; after all, the immediate agent is within the thing moved, so there is nothing in between. As for things that are moved by something else, there are four ways in which this can happen, because there are four kinds of movement which are imparted by an external agent—pulling, pushing, carrying, and rotating. All changes of place are reducible to these four. ᵃ17 For instance, pushing along is a kind of pushing (namely where the agent moves something away from itself, by pushing and following up the push), and pushing away is a kind of pushing too (when the agent does not follow up the push), and so is throwing (when the movement away from itself that the agent causes is stronger than the natural motion of the thrown object and continues as long as the imparted motion is controlling the object). Then again, pushing apart is a kind of pushing 243ᵇ3 away, and pushing together is a kind of pulling; pushing apart is a kind of pushing away, because pushing away is moving something away either from oneself or from something else; and pushing together is a kind of pulling, because pulling may involve movement towards either oneself or something else. The same analysis also applies to all the species of pushing together and pushing apart, such as packing the weft with a beater and separating the warp with a heddle (the first being a kind of pushing together, the second a kind of pushing apart). And the same goes also for every other kind of com- ᵇ7 bination and separation (they will all prove to be species of pushing together and pushing apart), except for those that are involved in coming to be and ceasing to be.* (At the same time, however, it is clear that combination and separation do not constitute a further genus of movement, since *all* kinds of movement may be distributed among one or another of the genera already mentioned.) Also, breathing in is a kind of pulling and breathing out is a kind of pushing. And the same goes for spitting and all other movements which expel or take in things by means of the body—some are kinds of pulling and others are kinds of pushing away.

All other kinds of change of place should be similarly ᵇ15 reduced as well, since all of them fall under the four headings I have mentioned. But carrying and rotating are in their turn reducible to pulling and pushing. Carrying is subsumed under one of the other three genera first because the carried object moves coincidentally (since it is in or on something that is moving), and second because the object that is doing the carrying is either being pulled or pushed or rotated;* so carrying belongs to all three genera at once. And rotating is a combination of pulling and pushing; after all, if *A* rotates *B* it must be both pulling and pushing *B*, since it is sending one part of it away from itself and drawing the other part towards itself. And the upshot of all this is that if something pushing is contiguous with what is being pushed and something pulling is contiguous with what is being pulled, it is clear that, as far as change of place is concerned, there is nothing between the agent of change and the changed object.

244ᵃ7   In fact, the relevant definitions show this as well, since pushing is movement away from either oneself or something else towards something else, and pulling is movement away from something else towards either oneself or something else, where the movement of the object doing the pulling† is faster* than the movement separating the continuous objects from

ᵃ11 each other, so that the other object is pulled along. (It might be thought that there is also another way in which pulling happens —that this is not the way in which wood draws fire* to itself, for instance. But it makes no difference whether the object doing the pulling is moving or at rest: it pulls things either to where it is or to where it was.) And it is impossible to move something either from oneself to something else or from something else to oneself without being in contact with it. It is clear, then, that there is nothing between that which undergoes change of place and that which causes change of place.

244ᵇ2   Nor, in fact, is there anything between the agent of alteration and the altered object, as a survey of examples would show: in every case, it turns out that the final agent of alteration and the first object altered are contiguous. The point is that when we talk about alteration, we are presumably talking about things which are altered by being affected in respect of their affective qualities* (as we call them). After all, every body differs from every other in its perceptible qualities:* it differs in more such qualities or in fewer, or it has the same qualities but to a greater or less degree. But anything which is altered is also altered *by* these perceptible qualities, since

there is an underlying quality of which they are affections.*
For instance, we say that something is altered by being heated   b7
or sweetened or compressed or dried or made pale, and it
makes no difference whether we are talking about living things
or inanimate objects, or again, where living things are concerned, whether we are talking about their insensitive parts
or the sense-organs themselves. (Even sense-organs are altered in a way, because an actual perception is a bodily change
which affects the organ of perception in some respect.) So
living things are liable to all the alterations which affect inanimate things, but inanimate things are not liable to all the
alterations which affect living things, because they are incapable of alteration in sense-organs. And living things are aware
of being changed, whereas inanimate things are not. However, it is perfectly possible for a living thing to be unaware
as well, when any alteration that takes place is not in the
organs of perception.

Anyway, since any case of alteration involves the agency of   245a2
perceptible things, the final agent of alteration and the first
object of alteration are obviously contiguous in all these cases.*
For instance, there is continuity between the air and the agent,
and also between the body which is the object of alteration
and the air. Again, there is continuity between the colour and
the light, and between the light and the organ of sight. The
same goes for hearing and smell as well, since in these cases
too the air is the immediate agent of change as far as the
changed object is concerned. Matters are similar in the case
of taste too, because the flavour is contiguous with the organ
of taste. And inanimate, insensible objects are no different in
this respect from living things. The upshot is that there is
nothing between the object and the agent of alteration.

There can also be nothing between the agent of increase   a11
and the increasing object either. For the immediate agent of
increase causes increase by being added on in such a way that
the whole becomes a single unit. And again, the agent of
decrease* causes decrease by some part of it being subtracted.
So both these agents (of increase and of decrease) must be
continuous with the changing objects, and things that are
continuous have nothing between them.

ᵃ16 Obviously, then, there is nothing intermediate between the object of change and the immediate agent of change (which is, in relation to the changing object, the final agent of change).

3. *Only perceptible qualities can be altered or can alter other things**

245ᵇ3 The following considerations will show the truth of the idea that everything which is altered is altered by things which are perceptible to the senses, and that only those things which are said to be affected in their own right by perceptible things are subject to alteration. The point is that the most plausible alternative candidates for liability to alteration are, first, figures and shapes and, second, states (and in either case the acquisition and loss of them). In fact, however, neither of these things are subject to alteration.

ᵇ9 First, when something has acquired its final shape or structure, we do not describe it as what it is made out of; we do not, for instance, describe a statue as bronze* or a candle as wax or a bed as wood. Instead, we modify the word and say

ᵇ12 that they are bronzen, waxen, and wooden. However, when something has been affected and altered in some way, we do describe it in those terms: we say that the bronze or the wax is liquid or hot or hard, and furthermore we also describe that which is liquid or hot as bronze—in other words, we describe both the matter and the affection in the same terms.* So where figures and shapes are concerned, that which has come into being (i.e. that in which the shape is) is not described as that from which it came; but it is so described in the case of affections and alterations. Obviously, then, the coming into being of these shapes or figures is not alteration.*

246ᵃ4 Besides, that way of speaking—saying that a person or a house (or anything else that had come to be) had been altered —would be regarded as ridiculous. Even if it is true that a necessary prerequisite for anything to come into existence is the alteration of something (for example, the condensation, rarefaction, heating, or cooling of matter), it still remains the case that for something to come to exist is not the same as for it to alter, and its generation is not alteration.

Second, states (both of the body and of the mind) are not ^{a}10 alterations* either, because states are always either good or bad, and neither goodness nor badness is an alteration. Goodness is a kind of completion: it is when something becomes as good as it may be that we say that it is complete, because that is when it pre-eminently conforms with its nature. A circle, for instance, is complete when it is pre-eminently a circle and when it is as good a circle as there could be. Badness, on the other hand, is the dissolution of and departure from this completion. So just as we do not describe the completion of a house as an alteration (since it would be ridiculous for the coping-stone and the tiling to constitute an alteration, or for the house to be altered rather than completed by being coped and tiled), the same goes also for good and bad states, and for those who have them or gain them. A good state is a completion, a bad state is a lack of completion, and so neither of them is an alteration.

Moreover, all good states, in our opinion, are in one way 246^b3 or another relative to something.* Good bodily states like health and fitness, for instance, we take to depend on the blending of hot and cold internal elements in such a way that they are in harmony either with one another or with their environment. The same goes also for all other good and bad states of the body, such as beauty and strength. They all exist by being in one way or another related to something, and they all dispose a possessor of them either well or badly with respect to the relevant affections—that is, those affections which naturally bring the state into being or destroy it.* Since related things are not themselves alterations, then, and since they are not subject to alteration or generation or any kind of change at all, obviously neither bodily states nor the acquisition and loss of them are alterations. However, it may be ^{b}14 that a necessary prerequisite for their coming into existence and ceasing to exist (just as it is a necessary prerequisite in the case of shape and form) is the alteration of something, such as the hot and cold or dry and moist bodily elements, or whatever it may be which bodily states primarily depend on. After all, any good or bad state is said to have as its sphere what will naturally alter one who has the state; for a good state makes its possessor either unaffected by certain things or affected in a particular way, and a bad state makes its possessor liable to be affected or unaffected in the opposite way.

b20 The same goes for mental states as well. They too all exist because they are in some way or another relative to something; and again good states are conditions of completion while bad states are departures from completion. Moreover, a good mental state disposes its possessor well, and a bad mental state disposes its possessor badly, with respect to the affections relevant to it. And the consequence is that they too cannot be alterations, and nor can the acquisition and loss of 247ᵃ6 them. However, a necessary prerequisite for them to come into existence is the alteration of the organ of perception. But this must be altered by perceptible things, because all moral virtue is concerned with bodily pleasure and pain, and this is found either in actions or in memory or in anticipation. So those pleasures and pains which are found in actions rely on perception in the sense that they are aroused by something perceptible, and those which are found in memory and anticipation are based ultimately on perception, because people feel pleasure either when they remember past experiences or ᵃ13 when they anticipate future ones. All such pleasure, then, must be produced by perceptible things. Now, since it is the presence of pleasure and pain that determines the presence of both good and bad states in a person (since such states concern pleasure and pain), and since pleasures and pains are alterations of the organ of perception, then obviously the alteration of something is a necessary prerequisite for the loss and acquisition of these states too. The upshot is that there is alteration involved in the generation of mental states, but they themselves are not alterations.

247ᵇ1 States of the intellectual part of the mind are not alterations either, nor are they generated. The point is that, in the case of the knowing part of the mind, we have a particularly strong claim that it is in some way relative to something. It is also clear that states of this part of the mind are not generated,* since a potential knower becomes a knower without having changed himself; what makes the difference is the presence of something else. For when a particular appears, the knower somehow knows the universal by means of the particular. And as with the previous cases we have considered, actual use of the faculty of knowledge is not generated, any more than a sight of something, or a contact, is generated. The actual use of the faculty of knowledge is similar to these cases.

The original acquisition of knowledge is not a case of gener- [b]9 ation or alteration, because it is when the thinking part of the mind has come to a rest and is not active that we are said to know and understand, and generation never has rest as its end-point, since (as I explained earlier*) change in general is not subject to coming into being. Moreover, when someone [b]13 has passed from being drunk or asleep or ill to the opposite condition, we do not say that he has become knowledgeable again, despite the fact that he was incapable of making use of his knowledge before; by the same token, we do not say that he has become knowledgeable when he first acquires the state, because understanding and knowledge come about as a result of the mind quietening down from its natural disturbance. This also explains why children are not as good as [b]18 older people at learning or at forming judgements on the basis of their sensory experience; it is because their minds are filled with disturbance and movement. Now, the mind naturally quietens itself down and brings itself to a state of rest in some cases, but external forces are needed in other cases. In both kinds of case, however, a prerequisite is the alteration of some of the body's parts, just as it is when a person has become sober or has woken up and then starts to make actual use of the knowing part of his mind.

These arguments show that being altered and alteration [248]a6 occur in perceptible things and in the part of the mind which is capable of perception, and only coincidentally in anything else.

4. *When is one change faster than another?*

It might be wondered whether or not every kind of change is [a]10 comparable with every other kind. If it is, and if what it is for two things to have the same speed is for them to change the same amount in the same time, then the circumference of a circle will be equal to, or longer or shorter than, a straight ᵃ13 line.* Also, an alteration will be equal to a movement, if one thing completes an alteration and another completes a movement in the same amount of time; that would make an affection equal to a distance, which is impossible. But surely it is true that when something changes the same amount in the same time as something else, they have the same speed? An affection cannot be equal to a distance, however, which means that an alteration cannot be equal to a movement, or less than it, which means that changes are not always comparable.

ᵃ18   How will this conclusion fare in the context of the circle and the straight line? It would be absurd if the movement of something on the circumference of a circle could not be similar to the movement of another thing on a straight line, and if it was just inevitably faster or slower than the one on the straight line, as if one were a downhill movement and the ᵃ22 other were an uphill movement. In fact, it does not make any difference to the argument to claim that it just is inevitably faster or slower, because the circumference can be both greater and smaller than a straight line, which means that it can also be equal. For if it takes time A for one to traverse B and the other to traverse C, then B might still be greater than C; after all, that is what we mean when we call something 'faster'. So something is also faster if it takes less time to traverse the same distance. Consequently, there will be some part of A in which B will traverse a part of the circle equal to the distance C takes the whole of A to traverse. But if the paths are comparable, the conclusion I have already mentioned follows, 248ᵇ6 and a straight line is equal to a circle. But the paths are not comparable, and therefore the movements are not comparable either. No, only things with the same unambiguous description are comparable. Why, for example, can we not compare whether a pen, some wine, or the note sounded by the highest string on a lyre is sharper? They are not comparable, because they are 'sharp' in different senses of the word. However, the note of the high string is comparable to the note of the next string down, because 'sharp' has the same meaning in both cases. So does 'fast' not have the same meaning when it refers to movement on a circle as when it refers to movement on a straight line, and is it even less univocal in the case of alteration and movement?

Perhaps, in the first place, it is not true that things which ᵇ12 are described in non-ambiguous terms are always comparable. For instance, 'plenty' has the same meaning when it refers to water and to air, but plenty of water and plenty of air are not comparable. And even if 'plenty' is ambiguous, 'double' is not, since it is the ratio 2 : 1; but water and air are still not comparable in these terms. Or perhaps the same argument applies to these cases too. For indeed 'plenty' is ambiguous, and sometimes even definitions are ambiguous. Suppose, for example, one were to say that 'plenty' is 'so much and then some extra'; 'so much' differs in different cases. Likewise, 'equal' is ambiguous, and then straight away 'one' probably becomes ambiguous;* and if 'one' is ambiguous, 'two' is as well. For why would some things be comparable and others not, if in each case there was just one nature?

Or is incomparability due to a difference in the things which ᵇ21 immediately bear the properties? So a horse and a dog are comparable in terms of pallor because the immediate bearer of the property is the same in both cases—namely, their skin. Similarly, they are comparable in terms of size. However, water and speech are not comparable in terms of clarity or quantity,* because the immediate bearers are different. But it is clear that this line of argument would enable us to identify every single usage of each term, and to claim that they are simply the properties of different bearers in each case, and so 'equal', 'sweet', and 'clear' will be identical, but differentiated by being in different bearers. Besides, it is not just anything which can be the bearer of just any property; a single property has only one immediate bearer.

But is it the case that for things to be comparable they 249ᵃ3 must not only be describable unambiguously, but there must also be no specific difference either in the attribute in question or in what bears it? What I mean, for example, is that colour is divisible into species, and that is why there is no comparison in respect of being colour (we cannot ask, for instance, which of two colours is more of a colour; this makes sense as a question about a specific colour, but not as a question about colour *qua* colour), but one can compare things for whiteness. The same goes for change as well: what it is for two things to have the same speed is for them to change the

ᵃ9 same amount in the same time. So if one thing has altered in such-and-such a part of its length and another thing has moved the same distance, is the alteration then equal to and of the same speed as the movement? Surely not: the idea is absurd. And the reason it is absurd is because change has different species. But then if things which move an equal distance in an equal time have equal speeds, our straight line and our circum-

ᵃ13 ference are equal. What is causing the difficulty here? Is it because movement is a genus or because lines are a genus? The time is the same in each case, but if the lines are specifically different kinds of lines, then the movements are specifically different too, because movements will be of different species if their paths are of different species (and sometimes if the means of motion are different: for instance, if feet are used, the species of movement is walking, while if wings are used, it is flying. Or perhaps this is wrong,* and movement is only differentiated by the shape of the path?) And so things which move the same distance in the same time have the same speed, but 'same' means that no different species of line and no different species of movement can be involved.

ᵃ21 So we had better look into the question of what differentiates one kind of change from another. And our argument suggests that the genus is not a unity, but is additionally an unnoticed plurality. Some ambiguous terms have meanings which are far apart from one another, some have meanings which are somewhat similar, and some have meanings which are almost identical (either because they belong to the same genus or because they are analogous), and that is why the terms do not seem to be ambiguous, although they really are.

ᵃ25 So when is there a different species? Is it when different bearers have the same property, or when different bearers have different properties? And how do we define difference? (Or again, how do we judge that clarity and sweetness are or are not the same?) Is it because something seems different when it is the property of a different bearer, or is it because it is completely different?

Where alteration is concerned, then, is it possible for one ^{a}29 alteration to have the same speed as another? Assuming that recovering health is an alteration, it is certainly possible for people to finish being cured either quickly or slowly, and sometimes at the same time, which would mean that alterations can have the same speed as one another, since they take the same amount of time to finish altering. But what kind of alteration has been finished? We cannot talk of equality in this context: similarity is the equivalent here of equality in the category of quantity. But let us assume that something has 249^b4 the same speed if it performs the same change in the same amount of time. So should we compare the bearer of the affection or the affection itself? In the present case, it is the fact that health is a single thing that enables us to grasp that it is present to a similar degree, rather than to a greater or lesser degree. However, if different affections are involved— for example, if two things are altering by becoming white and healthy respectively—there is no sameness or equality or similarity at all here, because these different affections immediately generate different species of alteration, and no single alteration is involved (which is the conclusion we reached in the case of movement too).

So we had better try to see how many different kinds of ^{b}11 alteration there are, as well as how many different kinds of movement there are. Now, there will be a different species of change if the changing objects are of different species (that is, the objects which are changing in their own right, rather than coincidentally), and the same goes for generic and numerical difference too. But in order to see whether alterations have the same speed, should one consider the affection, to see whether it is the same or similar, or should one consider the altering objects, to see, for example, to what extent each of them has become white? Or should one consider both the affection and the objects, in the sense that the alteration is the same or different depending on whether or not the same affection is involved, and equal or unequal depending on whether or not the objects involved are equal or unequal?

ᵇ19   And we had better ask the same question about coming
into existence and ceasing to exist as well. Is it possible for
one instance of generation to have the same speed as another?
Yes, if they take the same amount of time and if the same
objects are involved—indivisibly the same objects, that is,
such as men and not just animals. And one instance of generation is faster than another if a different result is produced
in the same amount of time; I say merely a 'different' result
because in this case we do not have a pair of terms for the
two different things as we do have the pair 'more' and 'less'
ᵇ23   in the case of dissimilarity. (Alternatively, if substance is
number,* one instance of coming into existence is faster than
another if it produces a greater and smaller number of things
of the same kind in the same amount of time.) But there is
no word for what is shared by the two things generated, or
for each of them individually, corresponding to 'more' for a
larger or excessive affection and 'greater' for a larger quantity.

5. *Concerning the proportion: power acting is to weight moved as distance moved is to time taken*

ᵇ27   An agent of movement always moves something, does so in
something, and does so to some extent. By 'in something' I
mean 'in some time', and by 'to some extent' I mean 'over a
certain amount of distance', because at one and the same
time it is causing movement and has caused movement,* so
that there will always be a certain distance which has been
moved and a certain period of time that has been taken. So
let *A* be the agent of movement, *B* the object moved, *C* the
distance it has moved, and *D* the amount of time it has taken.
The ratios will be preserved, then, if in an equal amount of
time an equal power *A* moves half *B* double the distance *C*,
and moves half *B* over *C* in half *D*. And if the same power
moves the same object just such a distance in just such a time,
and half the distance in half the time, then half the power will
take an equal amount of time to move an object half the
weight over an equal distance. For example, let *E* be a power
which is half *A* and *F* be an object which is half the weight

of *B*. The two examples are similar—the power and the weight are in the same ratio in both cases—and so the two powers will move their respective objects over an equal distance in an equal time.

Also, if *E* moves *F* over *C* in *D*, it does not necessarily 250ª9 follow that *E* will take an equal time to move double the weight of *F* over a distance which is half *C*. So if *A* moves *B* the distance *C* in *D*, *E* (i.e. half *A*) will not move *B* in *D* or in any part of *D* over a part of the distance *C* which bears the same ratio to the whole of *C* as that which obtains between *A* and *E*. It may well be that *E* will not move *B* at all. After all, the fact that a given power as a whole has moved an object such-and-such a distance does not mean that half the power will move it any distance in any amount of time. If it did, one man could move a ship, since the power of the haulers and the distance which they all moved the ship together are divisible by the number of haulers. That is why Zeno is wrong ª19 in arguing that the tiniest fragment of millet makes a sound; there is no reason why the fragment should be able to move in any amount of time the air which the whole bushel moved as it fell.* (In fact, the fragment in the bushel does not move even that much of the whole of the air as it would move if it were by itself, because within the whole bushel no fragment even exists, except potentially.*) However, if there are two agents of movement, and each of them moves such-and-such a weight in such-and-such an amount of time, the two powers added together will move the combined weights over an equal distance and in an equal time, because in this case the ratios are preserved.

So does the same go for alteration and increase? Yes, because ª28 there is a certain agent of increase and a certain object increased, and the one causes increase and the other is increased in a certain amount of time and to a certain extent. Likewise for the agent of alteration and the object of alteration too: there is something which is altered to a certain extent (defined in terms of degree) and in a certain amount of time. Twice as much alteration takes twice as much time, and an object twice the size takes twice as much time to alter; an object half the size takes half as much time to alter, and in half the time

there will be half as much alteration and in the whole time
250^{b}4 an object half the size will alter twice as much. However, the
fact that the agent of alteration or of increase causes such-
and-such an amount of alteration or increase in such-and-
such an amount of time does not make it inevitable that it
will alter or increase an object half the size in half the amount
of time, and will cause half as much difference in half the
amount of time; no, it may well be that it will cause no
alteration or increase at all, which was what we found in the
case of weight.*

VIII

THE ETERNAL AND UNCHANGING CAUSE OF ALL CHANGE

1. *Change always has existed and always will*

Was there a time when change came into existence, while it 250^b11
did not exist before, and does it cease to exist again, so that
everything stops changing? Or does change neither come into
existence nor cease to exist, in which case it always was and
always will be, and is an imperishable and unfailing property
of things? This would make it the life, as it were, of all
naturally constituted things.

Now, all those who treat of nature claim that there is such $^b15$
a thing as change, because they are concerned with how the
world was created and they focus exclusively on coming to be
and ceasing to be, for which the existence of change is a
necessary prerequisite. Some, however, claim that change is
eternal; this is the view of all those who say* that there are
infinitely many worlds and that some of these worlds are
coming into existence, while others are being destroyed (for
change must be involved in the processes of coming into
existence and being destroyed which the worlds undergo). On
the other hand, those who claim that there is only one world
make it either eternal or not eternal, and then they make
corresponding assumptions about change as well.

Suppose, then, that it is possible for there to be no change $^b23$
at all at some time. There are only two ways in which this
could possibly be the case, which have been proposed respectively by Anaxagoras and Empedocles. Anaxagoras claims
that everything was mixed together and at rest for an infinite
amount of time, and then intelligence instilled change and
separated one thing from another. Empedocles claims that
things are alternately changing and at rest—that they are
changing whenever love is creating a unity out of a plurality

or hatred is creating a plurality out of a unity, and that they
b29 are at rest in the times in between. He writes:

> In that the one has learnt to grow from many,
> And the many flourish in their turn as one dissolves,
> This way they come to be, unstable in their life.
> But as they never cease from this exchange,
> That way they are for ever unchanging in a cycle.*

For one must suppose that by 'this exchange' he means the
change from one state to the other.

251a5   So we had better try to discover the truth about whether or
not change is eternal, because doing so is relevant not only to
our study of nature, but also to our enquiry about whether
there is a first principle. Some points we established earlier,
during our lectures on nature, can start us off. Our view is*
that change is the actuality of the changeable, in so far as it is
changeable. So, for each kind of change, there must be things
a11 with the capacity for that change. In fact, leaving aside our
definition of change, everyone would agree that for each kind
of change it must be something which is capable of change
that changes; for example, it must be something capable of
alteration that alters and something capable of change of
place that moves. So before anything burns there must first be
something capable of being burnt, and before anything causes
burning there must first be something capable of causing
burning. So these things too must either have come into being
at some time, when they did not exist before, or be eternal.

a17   Now, if every one of those things which are capable of
change came into existence, then the given change must be
preceded by some other change—namely, the coming into
existence of that which is capable of change or of that which
is capable of causing change. The alternative is that things
with these capacities were eternally pre-existent without there
being any change; however unreasonable this idea may seem
at first sight, it is bound to appear even more so on further
a23 consideration. Suppose that there exist things capable of being
changed and things capable of causing change, and that there
is a time when one of the things that are capable of causing
change is acting as a first agent of change (and when there is

a first changing object too), and a time when none of them is, but when it is at rest. If it is at rest, it must previously have changed, because rest is the privation of change and so there must have been some cause of its rest.* We will find, then, that there is a change preceding the first change. The point is _a28 that some things cause only a single kind of change, but others cause opposite kinds of change; fire, for instance, causes heat rather than cold, whereas knowledge, despite being single, seems to be of both opposites at once.* Nevertheless, there seems to be something similar in the first kind of case as well, because something cold can cause heat by turning aside, as it were, and departing, just as someone with knowledge can deliberately make a mistake, when he uses his knowledge perversely. Anyway, it is only under special conditions, not 251^b1 all conditions, that the capacity to act and be affected, or to cause change and be changed, is realized: the objects in question have to be close to each other. So it is when they are close to each other that one thing causes change and another thing is changed, and when they are disposed in such a way that the one is capable of causing change and the other is capable of being changed. So if it is not the case that change ^b5 was eternally happening, it obviously follows that they were not so disposed as to be capable of being changed and causing change respectively. No, one or the other of them must have undergone a change, because where relatives are concerned that is inevitable. For instance, if *A* is now double *B*, when it was not before, then one or the other of *A* and *B*, if not both of them, must have undergone a change. It follows, then, that there must be an earlier change preceding the first change.

Besides, how could there be such a thing as 'earlier' and ^b10 'later' if there is no such thing as time? And how could there be such a thing as time if there is no such thing as change? So if time is a number of change, or is* a certain kind of change, then if time is eternal, change must be eternal too. But with a single exception everyone is clearly in agreement about time: they all say that time does not come into being. In fact, Democritus even uses this to disprove the notion that everything comes into being; after all, he says, time does not come into being. Plato* is the only one who claims that time has an origin; he says that it came into being at the same time as the heavens (which in his view came into being). Anyway,* if time cannot either exist or be conceived apart from the now, and if the now is a sort of mid-point, in the sense that it simultaneously includes a beginning (of future time) and an end (of past time), then time is bound to be eternal. For the final point of any stretch of time that we take must be a now (because apart from the now there is nothing in time to take hold of); but the now is both a beginning and an end, and therefore there must always be time on both sides of it. But if there must always be time on both sides of it, there must always be change on both sides of it as well, given that time is a kind of property of change.

$^{b}28$ The same argument* shows that change does not cease to exist either. Just as in the case of change coming into existence we found that there is a change preceding the supposed first change, so in the present case we will find that there is a change subsequent to the supposed last change. The point is that something does not stop changing and being capable of change at the same time.* Take burning and being burnable, for instance: it is possible for something burnable not to be burning. Nor does a thing stop causing change and being capable of causing change at the same time. If A, then, is capable of causing B to cease to exist, A will still need to cease to exist* when it has caused B to cease to exist, and then that which is capable of causing A to cease to exist will in its turn need to cease to exist later, and so on. For ceasing to exist is a kind of change. Obviously, then, given the impossibility of the alternative, change is eternal and does not exist at one time but not at another time. In fact, to say that it does looks very like a fantasy.

$^{a}5$ The same goes also for the view that this is how things naturally are and that we should regard change as a principle. This is apparently Empedocles' view, when he says that the fact that love and hatred take it in turns to be the dominant factor and to cause change is one that belongs to things of necessity, and that there is rest for the period of time in between. Anaxagoras and others who, like him, claim that there is only

a single principle would probably share this view.* However, ᵃ11
nothing natural—nothing due to nature—is disorderly, because in all things nature is responsible for order. Also, infinity is not comparable with infinity in terms of rational
proportion, but all order is a proportion. As for there being
rest for an infinite amount of time and then change at some
time, and there being no differentiation involved in this (to
explain why it should happen now rather than earlier) and,
again, no order either, this cannot be the work of nature. I
mean, anything natural is either in a simple state—not sometimes in one state and sometimes in another (as fire, for instance, naturally moves upwards and does not sometimes do
so and sometimes not do so)—or if it is not in a simple state,
then it is subject to proportion. That is why it would be ᵃ19
better to follow Empedocles and anyone else who has proposed the same theory as him and say that the universe is
alternately at rest and then changing again; for there is a
certain order in that kind of universe. But even proponents of
this view cannot just assert it, but must supply a cause for it;
they should not set up a mere hypothesis or make any unjus-
tified assumptions, but should argue either from particular
instances or demonstratively. For in themselves the factors ᵃ25
proposed by Empedocles are not causes of the cycle, and that
is not their essential nature; no, the essential nature of love
is to combine and of hatred to separate. If he adds to the
definition that they do so in turn, then he should say in what
cases this happens,* as he does say, for example, that there
is something which brings people together (i.e. love) and that
enemies avoid one another. After all, his assumption that this
also happens in the universe as a whole is based on its mani-
festation in particular cases. And then he needs an argument
that they occupy equal periods of time. And, in general, it is ᵃ32
an unsound inference to think that if something either always
is so or always happens in such-and-such a way, then this is
sufficient evidence of a principle. Democritus' explanation*
of natural phenomena comes down to the fact that things
used to happen like that earlier as well as now, but he does
not bother to look for a principle to explain this 'always',
and so he is correct in some cases, but wrong to say that it

explains all cases. For instance, the angles of a triangle always add up to two right angles, but all the same there is a further reason why this is eternally so, whereas there is no further reason for eternal principles.

252^b5 So much for our arguments to show that there neither was nor will be a time when there was not or will not be change.

2. Refutation of objections to the preceding arguments

^b7 The arguments which suggest the opposite are not difficult to deal with. The idea that it is possible for there to be change at some time, when there was no change at all before, is based above all on the following considerations.

^b9 First, it seems as though no one change is eternal, because it is the nature of every change to have a starting-point and an end-point; consequently every instance of change is inevitably limited by the opposites between which it happens, and nothing goes on changing *ad infinitum*.

^b12 Second, it is evidently possible for an object to change although it is not already changing, either as a whole or in its parts. An inanimate object, for example, none of whose parts is moving and which is not moving as a whole—which is, in other words, at rest—can start to move at some time. But if change does not come into being when it did not exist before, we would expect the object either to be always moving, or not to move ever.

^b17 But this phenomenon is particularly evident in the case of living things. For instance, we may on some occasions be quiet, with no movement going on inside us, but still start to move at some time; a movement arises within us which is started by us ourselves, even if nothing external has caused us to move. We do not see the same thing happening in the case of inanimate objects: their movement is always initiated by something external to them and different from them. We describe animals, however, as self-movers. So imagine an animal which is completely at rest at some time: in this case, movement can arise within a motionless object out of the object itself, rather than being due to some external agent.

But if this can happen in an animal, why should it not also be true of the universe? If it happens in a microcosm, why not in the macrocosm as well? And if it is true of the world, it might be true of the infinite too,* if the infinite is such that it can be in motion and at rest as a whole.

Now, as for the first of this list of objections, it is true that ᵇ28 a change which involves opposites does not continue as one and the same change for ever. In fact, this may well be necessarily so, if a change of one and the same object cannot always be one and the same change. I am referring, for example, to the question whether a single string always emits one and the same note, or if it is a different note on different occasions,* even though the condition and the movement of the string is the same. But whatever the answer to this question, it is perfectly plausible for a certain kind of change to be single in the sense that it is both continuous and eternal. I will explain this matter more later,* however.

There is nothing strange in something moving which is not 253ᵃ2 already moving, if the external mover is there at one time rather than another. What we have to do, however, is try to find out how this can happen—I mean, how a single object can sometimes be moved, and sometimes not be moved, by another single object with the capacity of moving it. For the sole point of this objection is to raise the question* why some things are not always at rest and other things always moving.

The most problematic objection might seem to be the third ᵃ7 one, which points to what happens with living things—that is, to how change occurs where it was not present before. The point is that an animal which was at rest earlier is subsequently walking, when apparently nothing external has set it in motion. But this is false, because we can see that there is always change in at least one of the animal's natural parts, and the cause of this change is not the animal itself, but possibly the animal's environment. Also, when we say that an animal ᵃ14 initiates its own change, we do not mean every kind of change, but only change of place. So there is nothing against the view, which may even be inevitable, that the environment causes a lot of changes within the body, and that some of these stir the animal's thoughts or desires, which then cause the whole animal to move. This is what happens, for instance, in sleep: although there is at the time no change occurring in the animal as a result of perception, nevertheless there is some change in it, and so it wakes up again. But this is another matter that will be clarified later.*

3. *There are things which are sometimes changing and sometimes at rest*

ᵃ22 The problem I have already mentioned—why some things are sometimes changing and at other times are at rest again—can serve as the starting-point of our enquiry. Now, it is necessarily true that either everything is always at rest, or everything is always changing, or some things are changing and other things are at rest. Again, in the last case, it is necessarily true either that the objects which are changing never do anything but change and the objects which are at rest never do anything but rest, or that all of them are naturally capable of both change and rest, or (a remaining third possibility) that some things never change, and some things never do anything but change, and some things are capable of both change and rest. This third possibility is the one we should argue for, because it provides us with a solution to all the difficulties and brings this treatise to a conclusion.

ᵃ32 First, the idea that everything is at rest.* For people to ignore the evidence of their senses and look for an explanation for everything being at rest is feeble-minded: it engages the issue
ᵃ35 at a general level rather than disputing particulars. Also, it is hardly an exaggeration to say that the claim affects not just natural science, but every branch of knowledge there is, and all received opinions too, since none of them would exist
253ᵇ2 without change. Besides, just as in mathematical discussions objections about principles are not the province of a mathematician (and the same goes for every other science too), so also the objection we are currently considering is not the province of a natural scientist, for whom it is a basic assumption that nature is a principle of change.

ᵇ6 We may also say that the idea that everything is changing is false, but it does not contradict our investigation to the same extent as the previous idea. For although we established in our lectures on nature that nature is a principle of rest as much as of change, nevertheless change is a natural phenomenon. And according to some people* it is not the case that some things are changing, while others are not, but rather that everything is changing all the time; they claim, however, that this fact goes unnoticed by our senses. But it is not ᵇ11 difficult to withstand this theory, even though its proponents do not define what kind of change they are talking about or whether they mean every kind. For example, neither increase nor decrease can go on continually: there is also something which intervenes between them.* Our argument here resembles the one about stones being worn away by dripping water or being split by plants; the fact that the dripping water has displaced or removed a certain amount of the stone does not mean that it removed half of that amount of stone in half of that time. No, it is no different from the case of men haul- ᵇ18 ing a ship:* although so many drops displace so much stone, a proportion of them may not be able to displace that much stone in *any* amount of time. It is true that the amount of stone which has been moved is divisible into a plurality of parts, but none of them was moved on its own; the point is that they were all moved together. Clearly, then, the fact that the stone removed is infinitely divisible does not necessarily mean that at any given time some part of it is being removed; all we have is that at some time all of it was removed.

The same goes for any kind of alteration as well. The fact ᵇ23 that the object undergoing alteration is infinitely divisible does not mean that the alteration is infinitely divisible too; the alteration may well happen all over, as freezing does. Also, when something is ill, there has to be a time when it will get better; the change from being ill to being well does not take place instantaneously, but the only possible end-point of the change is health. So to say that alteration goes on and on continually is a drastic way of disputing obvious facts. The point is that alteration has an opposite as an end-point. Besides, the stone in our illustration does not become harder or softer.*

As for movement, it would be strange if we failed to notice ᵇ31 the downward motion of a stone; nor do we fail to notice that it is at rest on the earth. Besides, the earth and everything else is bound to be at rest when it is in its proper place, and to move away from its proper place only when forced to do so. So if some things are in their proper places, it also cannot be the case that everything is always changing place.

254ª1 These and other similar arguments should convince one of the impossibility of everything always changing or always being at rest. But it is also impossible for some things to be always at rest and the others always changing, so that nothing is sometimes at rest and sometimes changing. Similar arguments to the ones we have already deployed give us grounds for asserting the impossibility of this idea. After all, we can see these objects—the same ones I used as examples before—undergoing the kinds of changes we have already discussed.

ª8 And we can add the point that anyone who argues for this idea has obvious facts to contend with, because the idea leaves no room for increase or for forced change (the latter because it makes it impossible for something to be at rest and then to undergo an unnatural change). So it leaves no room for coming to be and ceasing to be, and it is a view almost universally held that change is a process of coming to be and ceasing to be, on the grounds that the end-point of a change comes to be, or the thing comes to be in it, and the starting-point of the change ceases to be, or the thing ceases to be in it. It is clear, therefore, that some things are sometimes changing and sometimes at rest.

ª15 We must now get to grips with the view that everything is sometimes at rest and sometimes changing and apply the arguments used above* to it. We had better make our starting-point once again the same as it was before—in other words, the distinctions we drew not long ago. The point is that either everything is at rest, or everything is changing, or some things are at rest and some things are changing. And if some things are at rest while others are changing, it necessarily follows that either everything is sometimes at rest and at other times changing, or some things are always at rest while the others are always changing, or some things are always at rest, while others are always changing, and still others are sometimes

ª23 at rest and sometimes changing. Now, although we have

already shown* the impossibility of everything being at rest, let us repeat the point now. For if it is really true, as some people* claim, that being is infinite and unchanging, it remains the case that this is not what our senses tell us and that many things do seem to change. So if there is such a thing as false belief, or even any kind of belief, then there is also such a thing as change. There must also be such a thing as change if there is such a thing as imagination, or if things seem to be different at different times. After all, imagination and belief are held to be changes of a certain kind. But to look into this issue, and to try to find an argument to explain things in respect of which we are already too well placed to need an explanation, is a sign of being a poor judge of what is better and what is worse, of what is and what is not trustworthy, and of what is and what is not a starting-point for argument.

By the same token, it is also impossible that everything is $^a33$ changing, or that some things are always changing and others always at rest. An adequate and convincing response to all these theories is to point out that we can *see* that some things are sometimes changing and sometimes at rest. Obviously, then, it is just as impossible for everything to be continually at rest and for everything to be continually changing as it is for some things to be always changing and the others to be always at rest.

We are left, then, with the job of seeing whether everything 254^b4 is capable of both change and rest, or whether some things are like that, while others are always at rest and still others always changing, which is the view whose truth we have to demonstrate.

4. *Everything that changes is changed by something*

Now, agents of change may cause change, and objects of $^b7$ change may be changed, either coincidentally or in their own right.* They are coincidental agents or objects of change if they are properties of the things which are causing change or being changed, or if they cause change or are changed because one of their parts is causing change or being changed. They are agents or objects of change in their own right if it is not because they are mere properties of the things which are causing change or being changed that they cause change or are changed, and if it is not in virtue of some part of themselves that they cause change or are changed.

b12　Some things that are changed in their own right are changed by themselves, while others are changed by external agents; sometimes their change is natural, while at other times it is forced and unnatural. The change of anything that is changed by itself is natural; this is the case with all animals, for example. For animals are self-movers, and we say that everything which has its own inner source of change is changed naturally. That is why the self-movement of an animal as a whole is natural, but its body may undergo either natural or unnatural movement, depending on the kind of movement it happens to be undergoing and what kind of element the animal

b20　consists of. And the motion of anything that is moved by an external agent may be either natural or unnatural. Examples of unnatural movement are something earthy moving upwards and fire moving downwards. It is also common for the movements of animals' parts to be unnatural: they derive their unnaturalness from the position the animal has adopted or from the kind of movement involved.

b24　The fact that an object in motion is being moved by something is particularly apparent in the case of things which are moved unnaturally, because it is obvious that they are being

b27　moved by something other than themselves. After unnatural movements, among things whose movements are natural, the most obvious case is things such as animals which are moved by themselves; for what is unclear in such cases is not that the thing is being moved by something, but how we should tell which part of the thing is causing the movement and which part is being moved. After all, it seems likely that the distinction between mover and moved should obtain for animals just as much as it does for things such as ships which are not created by nature, and that this division explains how the creature as a whole moves itself.*

b33　However, the remaining half of the last of our divisions is particularly difficult to explain. We said that in some cases movement by an external agent is unnatural movement, but there are other cases which we still need to consider, and to contrast with these because their motion is natural. These are the cases where it might be difficult to see what it is that moves the thing. I am thinking of things such as light objects and heavy objects, because although it takes force to move them in the opposite direction, they move naturally to their proper places; heavy things naturally move downwards and light things naturally move upwards. The point is that when they are moved unnaturally, it is obvious what they are moved by, but this is not obvious in the case of their natural movements.

We cannot say that they are moved by themselves, because 255ᵃ5 this is a special property of animals and living things, and because there is no way in which they could make themselves stop moving (I mean, for instance, that if something is responsible for its own walking, it is also responsible for stopping itself walking). So if the upward movement of fire is within its own power, then downward motion should obviously be so as well. At any rate, if these things are self-movers, it does not make any sense for them to be moved by themselves only in one way.

Also, how is it possible for something which is continuous ᵃ12 and homogeneous to move itself? For in so far as it is a single continuum (as opposed to being held together merely by contact), it cannot be affected by itself; it is only in so far as there is separation within it that one part of it is naturally capable of acting and another part of being affected. It follows, then, that none of the things in question is a self-mover, because they are homogeneous substances, and that nothing else that is continuous is a self-mover either. No, there has to be a distinction within the object between mover and moved— the kind of distinction we see when some living thing causes something inanimate to move.

It turns out, however, that these things too are always ᵃ18 changed by something; this would become obvious if we distinguished between the various agents of change. It is possible to apply the distinctions we mentioned to the agents of change as well as to the changed objects. For some things, to be capable of causing change is contrary to their nature; a lever, for example, is not naturally capable of moving the weight it moves. Other things, however, are by their natures capable of causing change; something which is actually hot, for instance, is capable of changing something which is poten-

^a24 tially hot. And the same goes for all similar cases. By the same token, then, something which is potentially of a certain quality or quantity or in a certain place may by its nature be capable of being changed, when it has within itself, and not coincidentally, the source of the relevant kind of change. (I say 'not coincidentally' because the same thing may be of a certain quality and of a certain quantity as well, where the one is a coincidental concomitant of the other and not a property that belongs to it in its own right.) So fire and earth are forcibly moved by something when they are moved unnaturally, but are moved naturally by something when the result is the actuality of what they already possessed in potential.

^a30   The difficulty in knowing what causes changes such as the upward movement of fire and the downward movement of earth is due to the ambiguity of 'potential'. Someone who is learning something knows it potentially in a different sense from someone who already has that information but is not actually putting it to use.* It is always the case that when something capable of acting and something capable of being affected come together, then what is potential becomes actual. Consider, for instance, someone who has passed from the state of potential he was in as a learner to a different potential state (the point being that someone who possesses knowledge but does not have it consciously in mind knows it potentially, but not in the same sense that he knew it potentially before he had learned it); when he is in this state, as long as nothing stops him, he will actualize his knowledge* and have it consciously in mind, and if he does not he will prove to be in the

255^b5 contradictory state, that of ignorance. The same goes for natural objects as well: something cold is potentially hot, and then it changes and becomes fire and burns things, as long as nothing stops it and prevents it from doing so. The same goes for heavy and light as well: something becomes light instead of heavy—becomes air, say, instead of water (for water is the first thing to be potentially light)—and it is at once light. So it will immediately actualize its potential, unless something stops it from doing so. But the actuality of anything light is to be in a certain place—that is, high up—and it is being stopped if it is in the opposite place. And the same goes also for quantity and quality.

But this is exactly what we are trying to discover—why ᵇ13 light things and heavy things move to their own places. And the reason is that it is their nature to tend in certain directions, that this is what it is to be light and heavy; what it is to be light is defined by an upward tendency and what it is to be heavy is defined by a downward tendency. However, as ᵇ17 I have already said, there are a number of different ways in which something may be potentially light and potentially heavy. When something is water, it is potentially light in one sense, and there is also a sense in which it is still potentially light when it is air, because it may be prevented from being high up. If whatever is preventing it from being high up is removed, however, it becomes actually light and continually rises higher and higher. The way in which something of a certain quality changes to being actual is similar: a knower immediately has his knowledge consciously in mind, unless prevented from doing so. And something of a certain quantity spreads out unless prevented from doing so.

Now, if someone moves the obstacle or hindrance, there is ᵇ24 a sense in which he is causing the object to change and a sense in which he is not. For example, someone who pulls away a supporting pillar or who takes the stone out of a wineskin under water is only coincidentally causing the object to move; likewise, when a ball bounces off a wall, it is not the wall but the thrower of the ball who causes it to move. Although it is clear, then, that none of these things is a self-mover, each of them does contain with itself a source of movement; it is a source which enables them to be affected,* however, rather than to cause movement or to act.

Everything that changes, then, does so either thanks to its ᵇ31 own nature or because it is forced to do so, contrary to its nature; everything which is forced to change contrary to its nature is changed by something—that is, by something other than itself; and as for things which are changed naturally, those which are changed by themselves are changed by something and so are those (like light things and heavy things) which are not changed by themselves. For you could say that they were changed either by whatever it was that produced them and made them light or heavy, or by whatever it was that got rid of the obstacles and hindrances. It follows, then, that everything that changes is changed by something.

5. There must always be a first agent of change, which is not itself changed by anything else

256ᵃ4 However, there are two ways in which this can happen. In the first case, the agent of change is not itself responsible for the change, but there is something else which causes the agent to change and so is responsible for the change. In the second case, the agent is itself responsible for the change, and there are again two ways in which this can happen: the agent is either immediately next to the final object or there are a number of intermediate agents in between. An example of the latter is when a stone is moved by a stick, which is moved by a hand, which is moved by a person, who no longer causes movement by being moved by something else.

ᵃ8 We describe both the final and the first agents as agents of change, but a first agent of change is, properly speaking, more of an agent of change, because it is responsible for changing the final agent, and not vice versa, and because the first agent is needed for the final agent to be an agent of change, but the converse is not true. The stick, for example, will not cause movement unless the person causes it to move.

ᵃ13 Now, everything that changes has to be changed by something, and this something must either be changed by something else or not; if it is changed by something else, there must eventually be a first agent of change which is not changed by something else, whereas if the immediate agent of change is the first agent of change, there need not be any other agent. The point is that it is impossible for there to be an infinite series* of agents of change which are themselves changed by something else, because an infinite series has no first term. Therefore, if everything that changes is changed by something, and if the first agent is changed, but not by something other than itself, it necessarily follows that it is changed by itself.

There is also another way to put the same argument. Every [a]21 agent of change not only changes something, but does so by means of some instrument. That is, any agent of change uses either itself or something else to cause change; a person, for instance, may use himself or a stick to cause movement, and the wind knocks something down either by itself or by driving a stone along. But no change can be caused unless there is something that causes change by using itself to change the instrument of the change. Now, if it causes change by itself, [a]26 there is no need for there to be anything else by means of which it causes change; but if there is something else by means of which it causes change, there must eventually be something which will use itself rather than anything else to cause change, because otherwise the series will be an infinite one. So if something causes change by being changed, there must be an end to the series; it cannot be an infinite series. I mean, if the stick causes movement by being moved by the hand, the hand causes the stick to move; and if something else uses the hand to cause movement, then there is something distinct from the hand which causes the hand to move too. In other words, if at any given stage change is imparted [a]32 by something making use of an instrument which is distinct from itself, this stage must be preceded by a stage in which something uses *itself* to cause change. So if this latter agent of change is undergoing change, but there is nothing else which is changing it, it necessarily follows that it is changing itself. And so it follows from this argument too that either an object of change is being changed immediately by something which changes itself, or at some stage of the series we find something which changes itself.*

In addition to these arguments, the following considera- 256[b]3 tions also yield the same conclusion.* If everything that changes is changed by something which is itself changing, either this is a coincidental property of things (so that although it is a changing thing that causes change, it is not *because* it is changing that it is an agent of change), or this is not the case and it is a property that belongs to things of its own right.

b7 First, then, if it is a coincidental property, an agent of change is not bound to be changing. But if this is so, it obviously follows that it is possible for there to be a time when nothing in the world is changing,* because what is coincidental is not inevitable, but may not be the case. Now, anything that follows from a possible premiss cannot be impossible (even though it may be false). But since we have already proved that there must always be change, the non-existence of change is an impossibility.

b13 In fact, this is a reasonable conclusion to reach. For there must be three things—the object which is being changed, the agent of change, and the instrument by means of which it causes change. Now, the changing object must be changing, but it need not cause change. The instrument of change must not only cause change but also be changing, because it changes together with the changing object and in the same place as it; this is clear in the case of change of place, where it is necessary that the objects touch each other up to a certain point.* And that which causes change in a way that makes it different from the instrument of change must be unchanging.

b20 Since we can see the final object in the series—that is, that which is capable of being changed, but does not contain within itself a cause of change—and since we can also see that which is changed, but by itself rather than by something else, it is reasonable, not to say necessary, that there is also a third thing, which causes change while being itself incap-
b24 able of change. That is also why Anaxagoras is right in saying that intellect (which is a cause of change, according to him) is not affected by or mixed in with anything else; for this is the only way in which it can cause change while being itself changeless, and in which it can control other things while not being mixed with them.

b27 Second, if the change undergone by the agent of change is not a coincidental property, but an essential property (in the sense that, if it were not changed, it would not cause change), it necessarily follows, in so far as it is changed, that the change it undergoes is either the same species of change as the change it causes, or belongs to a different species. In other words, it is either the case that a heater is also being heated and a healer is being healed and a mover is being moved, or that a healer is being moved (say) and a mover is being increased. But either alternative is clearly impossible. I mean, on the first alternative, the distinction of different species of change has to be continued until we reach indivisible species, and so if someone is teaching geometry, he is also being taught the very same aspect of geometry that he is teaching, or if he is throwing something, he is also being thrown in exactly the same way. On the other alternative, one kind of 257ª3 change comes from another, and so if a mover is being increased, for example, the agent of this increase is being altered by something else, and the agent of this alteration is undergoing some other kind of change. But there must be an end to this series, because there is a finite number of kinds of change.* And to bend the series round and say that an agent ª7 of alteration is being moved is no different from saying straight away that a mover is being moved and a teacher is being taught. (The point is that, obviously, everything that is changed is changed not only by the immediate agent of change but also by any agent of change that is further back in the series, and the further back the agent of change is, the more of an agent of change it is.) But this is impossible, because it means that a teacher is learning* things which he as a teacher is bound to know, since that is what distinguishes him from the learner.

An even more absurd consequence is that if everything that ª14 is changing is being changed by something else that is changing, then everything that is capable of causing change is capable of being changed. To say that it is capable of being changed amounts to saying that everything with the capacity for healing has the capacity for being healed, and that everything with the capacity for building has the capacity for being built, whether an agent acts upon it immediately or through a number of intermediaries. By 'through a number of intermediaries' I mean, for example, if everything that is capable of causing change is capable of being changed by something else, but the change which it is capable of undergoing is not the same change as that which it imparts to anything that comes into contact with it, but a different one (so that, for example, something with the capacity for healing has the capacity for being taught); but, as I pointed out earlier, the same species of change will recur at some stage further back

a23 in the series. So the one suggestion is impossible and the other* is a fantasy. I mean, it would be absurd if there was some necessity that what is able to alter other things should itself be capable of increase. The series will come to an end, then, and therefore there is no necessity that a changing object is changed by something else which is itself changing. And the upshot is that the first thing to be changed must either be changed by something which is at rest,* or it must change

a27 itself. Even if it were necessary, however, to look into the question of whether a cause and source of change was something that is itself responsible for changing itself, or something that is changed by something else, everyone would choose the former option, because that which is a cause in its own right is always prior to what is also itself a cause, but only derivatively.

A first agent of change must itself be unchanging

a31 We must make a fresh start, then, and try to find out how and in what sense a thing does change itself. Now, everything that changes must be divisible into infinitely divisible parts, given that (as we proved earlier,* during our lectures on nature in general) everything that changes in its own right is a continuum. So anything which is the agent of its own change cannot cause itself to change everywhere,* because if it could, then—although it is one thing, belonging to one indivisible species—the whole of it would be both undergoing and causing the same movement or the same alteration, and so it would be simultaneously teacher and learner of the same information, prescriber and follower of the same medical treatment.

257b6 Moreover, we have established* that it is something with the capacity for change that changes. Now, this is something which is potentially rather than actually changing. Anything potential is on the way to actuality, and the process of change is the actuality of the capacity for change, but it is incomplete. An agent of change, however, is already actual; it is something hot, for instance, that produces heat and, in general, it is something which is in possession of a form that imparts it. This means that a single thing would be hot and not hot at the same time and in the same respect; and the same goes for every other case where the agent of change has to have the quality it imparts.* It follows that in anything which is the agent of its own change the part which is the agent of change is different from the part which undergoes the change.

However, the following arguments show that it is not ᵇ13 possible for a thing to change itself in such a way that *each* of the two parts is changed by the other part. In the first place, this would mean that each part would change *itself*, and then there would be no first agent of change.* (After all, the closer a cause is to being first, the more it is a cause and agent of the change, compared to the next agent in the sequence; for we found that there are two kinds of agent of change, one which is changed by something else and one which is changed by itself, and any agent which is further away from the object of the change is closer to the source than any intermediary agent.) In the second place, there is no ᵇ20 necessity that what causes change should itself be changed, unless it is changed by itself, so it can be only coincidentally that one part changes the other in return. It is possible, then, that one part may not cause the other to change. If we assume that this possibility is the case, we end up with a part that is changed and a part that causes change while remaining unchanged itself. In the third place,* there is no necessity that ᵇ23 the agent of change be changed in return; what is necessary, however, for change to be a constant feature of the world, is that there has to be either an agent of change which remains unchanged itself or one which is changed by itself. In the ᵇ25 fourth place, if each of the two parts were changed by the other part, each part would undergo the same change as it causes, so that a heater will be heated.

In fact, however, no primary self-changer can have either ᵇ26 a single part which changes itself or a number of parts each

of which changes itself. The point is that there are only two
ways in which a whole can be changed by itself: it can either
be changed by some one of its parts or by itself as a whole.
Now, if the whole undergoes change because one of its parts
is changed by itself, then this part must be the primary self-
changer, because if it were isolated from the whole, it would
continue to change itself, whereas the whole would stop

$^{b}32$ changing. On the other hand, if the whole is changed by itself
as a whole, the self-changing of the parts must be only coincidental. So since it is not necessary that they change themselves, we can assume that they do not do so. But if that is
the case, part of the whole will cause change while remaining
unchanged, and part will be changed; for this is the only way
in which anything can be a self-changer. Moreover, if the whole
changes itself, one part of it must be the agent of change,
while a different part must be undergoing the change. In
other words, AB will be changed by itself and by A as well.

$258^{a}5$ The agent of change may be changed by something else or
may be unchanging, and the object of change may or may not
cause change in anything. That which changes itself, then,
must consist of something which is unchanging but which
causes change, and also of something which is changed but
does not necessarily cause change—it may or may not do so.
So let A be something which causes change but is unchanging; let B be something which is changed by A and which also
causes change in C; and let C be something which is changed
by B but does not cause change in anything. (Since C will
eventually be reached* even if more than one intermediary is
involved, we can take the case where only a single intermediary is involved.)

$^{a}12$ Now, ABC as a whole is a self-changer. But if I isolate C,
AB will change itself, with A being the agent and B the object
of the change, but C will not change itself and will not be
changed at all either. Also, BC will not change itself either
without A, because B causes change by being changed by
something else, not by being changed by one of its parts. So

$^{a}18$ only AB changes itself. For something to change itself, then,
it must have a part which causes change but is unchanging,
and a part which is changed but which need not cause change

in anything; and there must be contact between both parts or from one to the other.* If the agent of change is continuous (as the changed object is bound to be), they will both be in contact with each other. It is clear, then, that it is not because one of its parts is capable of changing itself that the whole changes itself; no, it changes itself as a whole—it is both the agent of change and the object of change—by virtue of the fact that one of its parts is the agent of change and the other is that which is changed. It does not cause change as a whole, and it is not changed as a whole either, but *A* causes change and *B* is changed.

But if *A*, the unchanging agent of change, is a continuum,* $^a27$ what would happen if something was taken away from it? Or what would happen if something was taken away from *B*, the object of change? Will what is left of *A* continue to cause change, or will what is left of *B* continue to be changed? If so, *AB* would not be a primary self-changer, because what is left of *AB* is continuing to be a self-changer even after something has been taken away from *AB*. But there is nothing to prevent each of them—or anyway the part that is changed— from being potentially divisible, though actually undivided, so long as, if it were divided, it would cease to be what it now is.* In other words, there is no reason why self-change should not be a property primarily of divisible things.

These arguments show that the primary agent of change is 258^b4 unchanging, because there are two possibilities. The sequence of things being changed, and changed by something, either ends in immediate contact with a first member of the series which is unchanged, or it ends with something which is also changed, but which initiates and ends its own change. On either alternative, it turns out that the primary agent of change, in all cases of change, is unchanged.

6. There is a first agent of change which is eternal, and is not changed even coincidentally

Since change must always exist without failing, there must be $^b10$ a first agent of change (or perhaps more than one) which is eternal and unchanging. Now, it is irrelevant in the present

context whether or not each of those things which are unchanging but which cause change is eternal, but the following
considerations make it clear that there has to be *something*
that is itself unvarying and utterly free from both unqualified
and coincidental change,* but which is capable of causing
^b16 change in something else. We can freely grant that it may be
possible for some things to exist at one time and not at
another without undergoing any process of coming into existence* and ceasing to exist; indeed, if it is a thing which has
no parts that exists at one time and not at another, it may
even be necessary for it, and for everything like it, to exist at
one time and not at another without changing. We can also
accept that some unchanging agents of change may exist at
one time and not at another. But this cannot be true of all
such agents, because it is obvious that there must be something which is responsible for self-changers sometimes existing and sometimes not. The point is that since nothing that
lacks parts is liable to change, a self-changer must have
magnitude, but nothing we have said makes it necessary for
an agent of change to have magnitude.

^b26 Now, it cannot be anything unchanging but non-eternal
that is responsible for things coming into existence and ceasing to exist, and doing so continuously; nor can some cases
of coming into existence be caused by one unchanging agent
of change and some by another. For the eternity and continuity of the process could not be explained by any one such
cause or by all of them together. After all, the situation to be
explained is eternal and necessary, whereas the series of causes
would be infinite, and its members would not all exist at the
^b32 same time. So it is clear that even if it happens thousands of
times that some unchanged agents of change cease to exist
and are succeeded by others, and even if the same happens
to many self-changers (so that there are different unchanged
agents for different changes), still there is something which
includes them all and is separate from every one of them, and
it is this that is responsible for some things existing and other
things not existing, and for the continuity of change. And
while this is responsible for their change, they are responsible
for other things changing.

Since change is eternal, then, the first agent (if there is just 259ª6 one) or agents (if there are more than one) of change must also be eternal. (We should assume that there is one first agent rather than a plurality, and a finite number rather than infinitely many, because where it makes no difference to the outcome it is always preferable to choose a finite number of causes. After all, what is finite—and therefore better—will be present in nature wherever possible.) In fact, we do not need to assume that there are more than one; and since it is eternal and the first unchanged agent of change, it will be the source of change for everything else.

Here is another proof that the first agent of change must ª13 be single and eternal. It follows from the eternity of change, which we have already demonstrated, that it must be continuous, because what is eternal is continuous (as opposed to what is successive, which is not continuous). But if it is continuous, it is single. A change is single, however, if both the agent that causes it and the changed object are single, since if now one agent was involved and now another, the change as a whole would be successive rather than continuous.

We can be certain, as a result of these arguments, that ª20 there is something which is primary and unchanged; we will also come to the same conclusion from looking once more at the starting-points of our discussion. It is clear that there are things that are sometimes changing and at other times are at rest. Indeed, the fact that there are things which vacillate in this way and have the ability to change and to be at rest proves this, and has already enabled us to eliminate the alternatives—that everything is changing, that everything is at rest, and that some things are always at rest while other things are always changing. Everyone is well aware, however, that there ª27 are things that are sometimes changing and at other times are at rest; but we also wanted to prove that there are some things which are eternally unchanging and other things that are eternally changing, and to demonstrate each of their natures. In pursuit of this goal, we established that everything which changes is changed by something, and that this something is either unchanging or changing, and that if it is changing it is at each stage changed either by itself or by something

else. And so we reached the point of grasping that for changing things the cause among changing things is that which
changes itself, but that the cause of the change of *everything*
259ᵇ1 must be something unchanging. Now, it is a plain fact of
experience that there are things which change themselves—
living things, for instance, and animals—and these things
suggested that it might be possible for change to arise where
it had been completely absent before, since this is what we
see happening in their case: at one time they are immobile
and then they are in motion again, apparently. So what we
have to understand is that their self-changing involves only
one kind of change,* and that even this kind of change is not
strictly a self-change, because it is not due to the animals
themselves, but to other changes which occur naturally in
them and for which they are not responsible; I am thinking
of changes like growth, wasting away, and breathing, which
are going on in every animal even when it is at rest and not
ᵇ11 in the process of any self-initiated change. Responsibility for
these other changes lies with the environment and often with
the things which are ingested by living things. Food, for instance, is responsible for some changes; animals sleep while
food is being digested, and then wake up and become self-
changers while food is being distributed around their bodies,
and so the original source of the change is external to them.
That is why they are not always being changed continuously
by themselves; the agent of change is something other than
them, which is itself moving and changing in relation to each
of these self-changers. And in all these cases the first agent of
change, which is responsible for them being self-changers, is
itself changed—but only coincidentally so, in the sense that
the body changes place, and so that which is in the body
changes place too, as it moves itself by means of leverage.*
ᵇ20 We can be certain, therefore, that anything which is unchanging but also changes itself coincidentally cannot be an
agent of continuous change. And so, since it is necessary that
there should be continuous change, the first agent of change
has to be something which does not change even coincidentally,
if there is to be in the world an unfailing and imperishable
change (as we put it), and if the universe is to remain contained

within itself and in the same state; for if the cause is stable, the universe must be stable too, since it is continuous with its cause. (However, coincidental self-change is different from being coincidentally changed by something else; some sources of change among the heavenly bodies—i.e. those whose movement is complex*—are coincidentally changed by something else, whereas coincidental self-change is restricted to mortal things.)

How all changes depend upon the eternal first agent of change

Also, if there always is something which causes change but is $^b32$ itself unchanging and eternal, then whatever it is that is immediately changed by it* must also be eternal. This is also clear from the fact that nothing would come to be and cease to be or change at all without there being something which causes change by being changed itself. For the fact that this unchanging thing stands in a relation to something which changes does not mean that it itself changes at all, and so it will always cause change in the same way and will always cause the same change. On the other hand, anything which 260^a5 is changed by something which, though changing, is changed directly by that which is unchanging,* causes different kinds of change, because it stands in various different relations to different things. Because it occupies contrasting places or takes on contrasting forms, it will produce contrasting changes in every one of the other things it affects, and will cause them to be at rest sometimes, and at other times to change.

As a result of these arguments, we can now see our way $^a11$ through the difficulty we raised at the beginning; that is, we can see why it is not the case that everything is either changing or at rest, or the case that some things are always changing while others are always at rest, but instead some things are sometimes changing and sometimes are not. The reason for this is now clear. It is because some things are changed by an unchanging and eternal agent (which is why they are always changing), whereas other things are changed by something which is itself changing, with the result that they too are bound to change. As has been pointed out, however, because the unchanging changer remains simple and identical and in the same state, the change it causes is single and simple.

7. *The primary kind of change is movement, that is, change of place*

a20 Still, a fresh start* will clarify these issues for us even more. We should try to find out whether or not any change can be continuous, and, if so, what kind of change this can be, and also what kind of change is primary. The point is that if change is necessarily eternal and if a particular kind of change is primary and continuous, it clearly follows that this is the kind of change which the first agent of change causes—that is, the kind which is necessarily single, identical, continuous, and primary.

a26 Of the three kinds of change—change of size,* change of quality, and change of place (or in other words, movement)—it is movement that is bound to be primary. After all, alteration is a necessary prerequisite for increase, since anything that is increasing is in a sense being increased by something similar to itself, but in a sense by something dissimilar to itself—by something dissimilar because opposites are said to be food for each other, and by something similar because all increase involves something becoming like what it is added a33 to. There is a change from one opposite to another, then, and this is necessarily alteration. But if there is alteration, there must be an agent of alteration, which makes a thing actually hot instead of being potentially hot. Now, obviously the agent of change is not always in the same state, but is sometimes closer to and sometimes further away from that which is being altered—but this is impossible without movement. So the necessary eternity of change means that the primary kind of change, movement, is necessarily eternal too, and also that the primary kind of movement is necessarily eternal, if there are primary and secondary kinds of movement.

260b7 Besides, condensation and rarefaction constitute the origin of all qualities: heavy and light, soft and hard, hot and cold—they are all supposed to be kinds of condensation and rarefaction. But condensation and rarefaction are combination and separation* (i.e. the processes which are said to explain the coming to be and ceasing to be of substances), and combination and separation inevitably involve change of place. Moreover, the size of things which increase and decrease changes place.

Another way of coming to see the primacy of movement is ^b15 as follows. The point is that 'primary' is as ambiguous a word when it refers to changes as it is in other contexts. A thing is said to be prior when its existence is a prerequisite for the other things to exist, but not vice versa, and there is also priority in time and priority in form.

The following reasoning establishes the primacy of move- ^b19 ment in the first sense.* Change must exist continuously; either continuous change or successive change may exist continuously, but there is more continuity to continuous change and it is better for there to be continuous change rather than successive change, and, wherever we can, we assume that what is better obtains in natural things; it is possible for there to be continuous change (this will be proved later,* but for the time being we can take it for granted); and the only kind of change which can be continuous is change of place. So change of place must be primary, because neither increase nor decrease nor alteration, nor again coming to be or ceasing to be, are necessary prerequisites for movement, but the continuous movement which the first agent of change imparts is a necessary prerequisite for the existence of these other kinds of change.

Movement is also primary in time, because things which ^b29 are eternal cannot experience any other kind of change. It may be that movement is the final kind of change experienced by any single one of the things that are subject to coming into existence (because once they have come into existence, alteration and increase immediately follow, and movement is a change proper to things only after they have attained completion); all the same, priority has to be given to something else—something which moves and which will prove to be responsible for the generation of things that come into existence, but not by coming into existence itself, as a father is responsible for the coming to be of his child. It might seem ^{261a}3 as though coming into being is the primary kind of change because the first thing any object does is to come into being; but although this is so in the case of any one of the things that come into being, still before anything comes into being something else must be in motion—something which is not coming into being, but which already exists—and this in turn must be preceded by something else. And since coming into being cannot be primary (because then ceasing to exist would be a property of everything that is liable to change), it is clear that none of the changes which are consequences of coming to be can be prior to movement; by the changes which are consequences of coming to be, I am referring to increase and then alteration, decrease and ceasing to be. The point is that they all come after coming into being, and so, given that coming into being is not prior to movement itself, it follows that none of the other changes are either.

ᵃ13   Thirdly, there are no exceptions to the evident rule that anything which is coming into being is incomplete, and is in progress towards its cause; and so it is that which comes *later* in the process of coming into being that is prior in nature. Now, the last thing gained by everything which is in the process of coming into being is movement. That is why although some living things (such as plants and a number of kinds of animals) lack what it takes to move and are utterly immobile, others—the ones which have attained completion—do have the ability to move. And so, if movement belongs more to things which have more completely attained their nature, then this kind of change must be prior to all the other
ᵃ20 kinds in terms of form. There is another reason for this as well: for a changing object, there is less loss of form involved in motion than in any of the other kinds of change; in fact, it is the only one which involves no change of form, in the sense that quality changes during the process of alteration and quantity changes during the processes of increase and decrease.

ᵃ23   The most important point, however, is that this kind of change—change of place—is clearly the kind which, strictly speaking, a self-changer causes. But in our view it is a self-changer which is, out of all the things which are changed and

which cause change, the original source and the primary cause
of the change of changing things.

No other kind of change can be continuous and eternal

These arguments demonstrate that movement is the primary ᵃ27
kind of change; now we have to show which *kind* of movement is primary. Our investigation into this question will at
the same time justify the assumption we have been making,
both just now and earlier,* that there is a kind of change
which can be continuous and eternal. The following considerations prove that continuity is impossible for any other
kind of change. The point is that all variations and changes
are between opposites: for example, being and not being are
the starting-points and end-points of coming to be and ceasing to be, as opposite affections are the limits in the case of
alteration, and either largeness and smallness or a thing's
complete size and incomplete size are the limits in the case of
increase and decrease; and opposite changes are those with
opposite end-points. Now, if something has not always been 261ᵇ1
undergoing a particular kind of change, but did exist earlier,
then it must earlier have been at rest. Evidently, then, the
changing object will be at rest while it is in either opposite
state. The same goes for changes as well as variations, since
coming to be and ceasing to be are opposed in both their
unqualified and specific forms. And so, since it is impossible
for anything to be undergoing opposite changes at the same
time, the change in question cannot be continuous,* but there
will be a stretch of time between one change and its opposite.

It does not make any difference whether or not contradic- ᵇ7
tory changes are opposites, nor is the issue relevant to the
argument; all that matters is that it is impossible for the same
thing to be undergoing them at the same time. Nor does it
make any difference even if it is not necessary for the object
to be at rest in each contradictory state, or if change is not
opposite to a state of rest; for the end-point of ceasing to be
is non-existence, and presumably a non-existent thing is not
at rest. All that matters is that a stretch of time occurs between the changes, because it is this that guarantees that the

change is not continuous. In the previous cases too, the relevant point was not that the two changes were opposite to
b15 each other, but that they could not coexist. Nor is there any need to be disturbed by the fact that a single thing will prove to have more than one opposite, as change, for instance, is opposed both by rest and by the change with the opposite end-point. All we have to do is understand that there is a sense in which the change is opposed both by the opposite change and by the state of rest (just as an equal or moderate quantity is opposed to both a larger quantity and a smaller quantity), and that neither opposite variations nor opposite
b22 changes can coexist. Moreover, in the case of coming to be and ceasing to be, it would also seem completely ridiculous for something to have to cease to be, without persisting for any time, as soon as it has come to be. So we can confidently conclude, therefore, that the same goes for the other kinds of change as well, since it accords with nature that all of them should be alike.

8. Only circular movement can be continuous and eternal*

b27 We can now go on to claim that it is possible for there to be an infinite change, which is single and continuous, and that this is circular movement. Every moving object moves either in a circle or in a straight line or in a motion compounded of these. It follows* that if one or the other of straight or circular motion is not continuous, the compound motion cannot be continuous either.

b31 Now, continuous motion is obviously impossible for something that is moving over a finite straight line. It has to reverse its direction, and anything which does so on a straight line is involved in opposite movements, since in terms of place upward is the opposite of downward, forward is the opposite of backward, and to the left is the opposite of to the right; these are the oppositions which place admits. But we have already defined* single and continuous change as change which affects a single subject in a single period of time and in a specifically indivisible respect; for three things are involved

—the changing object (for instance, a man or a god), when (i.e. the time), and thirdly in what respect (i.e. a place or an affection or a form or a magnitude). Opposites, however, are different in species and do not constitute a unity; and the distinctions mentioned are differentiae of place.

Evidence of the fact that movement from *A* to *B* is the 262ᵃ6 opposite of that from *B* to *A* can be found in the fact that they make each other stop and come to a halt if they occur at the same time. The same goes for circular motion too; the movement from *A* to *B* is the opposite of that from *A* to *C*, because they make each other stop even if they are in themselves continuous movements with no change of direction, because opposites cancel each other out and obstruct each other. However, sideways movement is not the opposite of upward movement.

But the impossibility of continuous movement on a straight ᵃ12 line is shown particularly clearly by the fact that anything which reverses its direction has to come to a standstill, not only if it is moving on a straight line, but also if it is moving on a circle. Moving *in* a circle is different from moving *on* a circle, because in the latter case the object may either do the same movement again or return to its starting-point and reverse direction. Rational argument, as well as the evidence of ᵃ17 our senses, will convince us that it is necessary to stop when reversing direction. The first point to note is that there are three things—beginning, middle, and end—and the middle is both a beginning and an end, depending on which of the other two it is taken in relation to; so although it is numerically one, it can be seen as two. Then there is the distinction ᵃ21 between potential and actual. This means that any of the points between the extremes of the straight line is potentially a mid-point, but it is not actually a mid-point unless the moving object divides the line there—that is, comes to a stop and then begins to move again. That is how the middle becomes both a beginning and an end; it is the starting-point for the subsequent movement and the end-point for the preceding movement. This is what happens if, say, a moving object *A* stops at *B* and then moves again towards *C*. However, if the motion is continuous, *A* cannot have reached the ᵃ28 point *B* or have left it; it can only *be* there in a now. It is not there in any period of time (except that it is there within the whole time of which the now in question is a division). The claim that it has reached *B* and has left it amounts to having *A* at a constant standstill as it moves, because it is impossible for *A* simultaneously to have reached *B* and to have left *B*. So the two events must have happened at different points of time,* in which case there will be a stretch of time between these two points; and so *A* will be at rest at *B*, and equally at all the other points on the line, since the same argument 262ᵇ5 applies to them all. So when a moving object *A* treats the mid-point *B* as both an end and a beginning, the fact that it has created this duality (as one might also create it in thought) means that it is bound to have come to a standstill. But in fact, it has left the point *A*, the starting-point, and has reached the point *C*, when it has completed its journey and come to a standstill.

ᵇ8 We can also meet the difficulty that arises at this point, which is as follows. Take a line *E*, equal in length to another line *F*, and suppose an object *A* is in continuous motion over *E* from its very beginning to a point *C*; suppose also that when *A* is at the point *B*,* an object *D* is simultaneously in uniform motion from the beginning of *F* towards *G* and is moving at the same speed as *A*; *D* will then reach *G* before *A* reaches *C*, because the thing which sets out and leaves first is bound to arrive first. The reason *A* falls behind is because it did not simultaneously reach *B* and leave it; if it did, it would not fall behind; but it is bound (so the argument goes) ᵇ17 to come to a standstill there. The consequence is that we should avoid saying that at the time when *A* reached *B*, *D* was moving from the beginning of *F*, because if *A* did reach *B*, then it also left *B*, and that cannot happen simultaneously. We should say instead that *A* was at *B* at a division of time, and not for any period of time.

ᵇ21 Therefore, in cases like this—that is, when the motion is continuous—we cannot speak of reaching and leaving a point. However, we do have to speak that way when there is a reversal of direction. For if an object *G* was in motion towards *D* and then reversed its direction and travelled back down again, it did treat the extremity *D* as both an end and a beginning and make this single point two. That is why it is bound to have been at a standstill; it did not simultaneously reach *D* and leave it, because in that case it would have been there and not have been there in the same now. But we $^b28$ cannot use the same argument that we used to solve the earlier difficulty, because we cannot say that *G* was at *D* at a division of time and did not reach *D* or leave it. After all, in this case *G* must come to an actual final point, not merely a potential one. What the mid-points are potentially, *D* is actually; it is an end-point on the journey from the bottom and a starting-point on the journey from the top, and by the same token it is a starting-point for the downward movement and an end-point for the upward movement. Anything that reverses its direction on a straight line must, then, come to a standstill. So it is impossible for there to be continuous, eternal movement on a straight line.

We should make the same response to anyone who uses 263^a4 Zeno's argument to ask whether it is always necessary to traverse half the distance first, and points out that there are infinitely many half-distances and that it is impossible to traverse infinitely many distances; or then there are others who put the same argument another way and maintain that, as one moves over a half-distance, one has to count it before completing it, and has to do so for each half as it happens, and so traversing the whole distance turns out to involve having counted an infinite number, which is admittedly impossible.

Now, originally,* during our discussion of movement and $^a11$ change, we solved these difficulties by taking into account the fact that time contains within itself an infinite number of parts; after all, there is nothing strange in someone traversing infinitely many distances in an infinite time, and infinity is a property of time in the same way that it is a property of length. But although this solution is adequate as a response to the question (since the question was whether it is possible to traverse or count infinitely many things in a finite time), it will not do as a response to the actual facts of the matter. For $^a18$ if our questioner were to ignore distance and whether infinitely many distances can be traversed in a finite time, and were to ask the same question about just the time itself (given that time is infinitely divisible), the same solution would no longer constitute an adequate response. No, we would have to draw on the true account we have just given, and say that anyone who divides a continuous line into two halves is treating the single point at which the division occurs as two points, because he is making it both a starting-point and an end-point; and counting out halves is no different from dividing

a26 into halves. But to make these divisions is to destroy the continuity of the movement as well as the line, because continuous movement is movement over a continuum, and although there are infinitely many halves in any continuum, these are potential, not actual. Any actual division puts an end to continuous movement and creates a standstill. It is obvious that this is what happens when someone counts successive halves, because he inevitably counts a single point as two, since the consequence of counting two halves rather than a continuous line is that a single point forms the end of

263b3 one half and the beginning of the other half. So the reply we have to make to the question whether it is possible to traverse infinitely many parts (whether these are parts of time or of distance) is that there is a sense in which it is possible and a sense in which it is not. If they exist actually, it is impossible, but if they exist potentially, it is possible. I mean, anyone in continuous motion has coincidentally traversed infinitely many distances, but he has not done so in an unqualified sense; it is a coincidental property* of a line that it contains infinitely many halves, but it is not essential to what it is to be a line.

b9 It is also clear, however, that if one does not always attach the point of time which divides earlier from later to the later state of the object involved, the same thing will at the same time both be and not be so-and-so, and will not be so-and-so at the time when it has become so-and-so. So although the point is common to both earlier and later time, and although it is the same numerically single point, it is conceptually different from itself (because it is the end of one period of time and the beginning of the other) and always belongs to the

b15 later affection of the object involved. Take a stretch of time *ACB** and an object *D*, which is white in time *A*, but not

white in time *B*. In *C*, then, it is both white and not white.
For if the object was white for the whole of *A*, then it is true
to say that it is white in any part of *A*, and if it is not white
for the whole of *B*, it is true to say that it is not white in any
part of *B*, and *C* is a part of both *A* and *B*. The solution is $^{b}20$
not to grant that it is white for the whole stretch of time, but
to say that it is white for the whole stretch of time except the
final now, namely *C*, which is already part of the later stretch
of time. Whether during the whole of *A* it was coming to be
white or ceasing to be white,[†] these processes are complete at
C. It follows that *C* is the first point of time at which it is true
to say that the object is white or not white; otherwise it will
not be what it has become, and it will be what it has ceased
to be, or it will be simultaneously both white and not white
and, to put it generally, it will necessarily both be so-and-so
and not be so-and-so.

Also, that which is something which it was not before must $^{b}26$
have been becoming that, and it is not what it is becoming.
It follows that time cannot be divided into indivisible stretches
of time. For suppose an object *D* was becoming white in time
A, but in time *B* (another indivisible stretch of time, consecutive to time *A*) it has become white and so is white; since in
time *A* it was becoming white and so was not white, but in
time *B* it is white, there must be some process of coming to
be between *A* and *B*,* and therefore there must also be a
period of time in which the process took place.

The thing to notice is that the same argument cannot be $264^{a}1$
brought to bear against those who deny that there are indivisible periods of time; instead, on this view, it has become
white, and is white, in the final point of the very stretch of
time during which it was becoming white; and there is no
other point to which this final point is either consecutive or
successive. Indivisible stretches of time, however, are successive. But it is clear that if time *A* as a whole is the time during
which it was becoming white, the time in which it was becoming white and became white is no greater* than all the
time in which it was becoming white.

These arguments* (and others like them) are particularly ap- $^{a}7$
propriate to the subject in question, so we can have confidence

in them; but the same conclusion would also seem to emerge from the following considerations, which are more abstract. Take any object that is moving continuously and is not deflected from its path; anything it comes to in the course of its movement must be something it was moving towards before. For example, if it comes to *B*, it was also moving towards *B*, and—since there is no difference in this respect between one phase of its journey and an earlier phase—it was moving towards *B* right from the start of its journey, not just when it was close. The same goes for the other kinds of change too.

a14 Now, on the hypothesis we are considering, when something which is moving from *A* comes to *C*, it will go back to *A* without breaking the continuity of its motion. So when it is moving from *A* to *C*, it is also moving to *A* from *C* at the same time, so that it is moving with opposite motions simultaneously (since movements from opposite ends of a straight line are opposite movements). It also follows that it will be changing from a state which it is not in.* But this is impossible, so it follows that it must stop at *C*. Its movement is not single, then, since any movement which is interrupted by stopping is not single.

a21 There is another more general argument, applicable to every kind of change, which also makes the point. We found that our list of kinds of change was exhaustive,* so that every instance of change is an instance of one of the kinds of change we have mentioned, and every instance of rest is an instance of a state of rest that opposes one of these kinds of change; also, an object which is undergoing a particular change, but does not always do so—I mean one of the specifically different kinds of change, not a change which is merely part of a whole*—must have previously been in the state of rest which is the opposite of this particular kind of change, because rest

a28 is the privation of change. Therefore, if movements from opposite ends of a straight line are opposite movements, and if it is impossible to undergo opposite movements at the same time, something which is moving from *A* to *C* cannot at the same time be moving from *C* to *A*. Since it is not moving from *C* to *A* at the same time, but *will* do so in the future, it must be at rest at *C* before doing so; for this is the state of rest that is opposed to movement from C. So this argument shows that the movement in question is not continuous.

Moreover—and here is another argument which is more 264^{b}1 particularly appropriate to the issue—something that has ceased to be not white has simultaneously become white. So if it is continuously altering to white and from white, without resting for the slightest stretch of time, then there are no different times for it to have ceased to be white, become not white, and become white, and so it did all three at the same time.*

Besides, the fact that time is continuous does not imply ^{b}6 that change is continuous rather than merely successive. How could opposites like whiteness and blackness share a limit?*

However, movement on the circumference of a circle will ^{b}9 be single and continuous. I mean, no impossible consequence follows from the idea, because something which is moving *from A* will simultaneously be moving *to A* on the same forward path (for it is moving towards any point that it will reach), without simultaneously undergoing opposite or contrary movements. For movement towards a point is not always the opposite or the contrary of movement away from the same point; the two movements are opposites if they are on a straight line (the points at either end of a diameter, for instance, are as far apart as they can be and so are opposites in terms of place) and are contraries if they are on the same length. So there is no reason why circular motion should not be continuous and uninterrupted, because its starting-point and its end-point are identical, whereas the end-point of movement on a straight line is different from the starting-point.

Moreover, circular movement never covers the same points ^{b}19 twice,* whereas movement on a straight line often does. Now, a movement that is always happening at different points at different times can be continuous, but one which recurs at the same points cannot be continuous, because the moving object would necessarily be undergoing opposite movements at the same time.* It also follows that there cannot be continuous movement on a semicircle or on any other curve either, since the moving object is bound to move recurrently over the same points and to move in opposite ways, because the end and the beginning of the movement do not coincide. However, in circular movement they do coincide, and so circular movement is the only kind of movement that is complete.

b28   This distinction shows that no other kind of change can be continuous either, since they all turn out to involve recurrent change over the same points. Alteration, for instance, involves recurrent change over whatever lies between the extremes, change of quantity involves recurrent change over the intermediate sizes, and a similar analysis applies to coming to be and ceasing to be. It does not make any difference whether we find that there is a large or small number of intermediate stages of the process of change, or whether we add or subtract any such intermediate stage; in either case the upshot is that there is recurrent change over the same points.

265a2   We can also see now that those natural scientists* who say that everything perceptible is constantly changing are mistaken. The change involved must be one of the kinds of change we have distinguished, and in fact, with their talk of everything flowing and decaying, they have alteration in mind above all, and they even describe coming to be and ceasing to be as alteration. However, our argument has now reached the point of making the general claim about every kind of change that none of them except circular movement can be continuous; and so these thinkers are wrong to attribute continuity to either alteration or increase.

a10   So much for arguing that no kind of change except circular movement is either infinite or continuous.

9. Circular movement is the primary kind of movement

a13 As for the fact that circular movement is the primary kind of movement, this is obvious. Every movement (to repeat*) is either circular or rectilinear or a combination of the two, and because the first two make up the latter kind, they must have priority over it. Also, the higher degree of simplicity and completeness possessed by circular movement means that it

a17 has priority over rectilinear movement. In the first place, it is impossible to move over an infinite straight line, because there can be no such thing as a straight line which is infinite in this sense;* also, even if there were such a thing, it would not be traversed by anything, because the impossible does not happen and it is impossible to traverse something which is infinite in extent. In the second place, movement on a finite straight line ᵃ20 can either reverse direction or not; if it does, it is a composite of two movements, and if it does not, it is incomplete and must cease to exist. But where priority in nature, in definition, and in time* are concerned, the complete is prior to the incomplete and that which does not cease to exist is prior to that which does. Besides, a movement that can be eternal is prior to one which cannot. Now, circular movement can be eternal, but no other kind of movement, and no other kind of change either, can be eternal, because they are bound to involve rest, and the presence of rest means that the movement or change has ceased to exist.

It is reasonable for us to have concluded* that circular ᵃ27 rather than rectilinear motion is single and continuous. After all, movement on a straight line has a determinate beginning, end, and middle, which are all parts of the straight line, so that there is a point from which the movement will begin and a point where it will end; for at the limits—the starting-point and the end-point—nothing is ever in motion. However, there are no determinate points in circular movement. For why should any given point on the line be a limit? Any and every point is equally a beginning, a middle, and an end, and so anything which is moving around a circle is both always and never at a starting-point and at an end-point. (That is why a sphere is both moving and at rest, in a sense; it is because it occupies the same place.) The reason for this is that the centre 265ᵇ2 of the circle has all these attributes: it is the beginning, middle, and end of the circle's magnitude. The consequence is that, since the centre is not on the circumference, there is nowhere for a moving object to be at rest in the sense of having completed a journey, because there is no point to act as the limit of its movement,* but instead it is constantly in motion around the centre. Also, because the centre does not move, the whole is in a sense always at rest as well as being in continuous motion.

ᵇ8 We can also draw another conclusion, which is convertible. Because circular movement is a measure of change, it must be primary (since everything is measured by what is primary), and because it is primary, it is a measure of the other movements.

ᵇ11 Moreover, uniformity is possible only for circular movement. For there is no uniformity between the way things move on a straight line* when they are leaving the starting-point and the way they move when they are approaching the end-point, since they always accelerate the further they get away from a state of rest. Only circular movement is such, by its nature, that its beginning and end are not included within itself, but come from elsewhere.

ᵇ17 All those who have treated of change* bear witness to the fact that change of place is the primary kind of change, in the sense that they all assign the origins of change to things which cause this particular kind of change. Separation and combination, for instance, are movements, and this is the kind of change imparted by love and hatred, because hatred

ᵇ22 causes separation and love causes combination. Then there is Anaxagoras, who says that intelligence causes separation and

ᵇ23 is the original agent of change. Even those who deny the existence of any such cause, but claim that change is due to the presence of void, belong to the same category, because they too think that change of place is the kind of change experienced by natural things (I mean, the kind of change which is caused by the void is movement, and is as it were change of place*), and they do not attribute any of the other kinds of change to their elements, but only to things compounded of these elements. For they claim that things increase, decrease, and alter as a result of the separation and combina-

ᵇ30 tion of atoms. The same also goes for those who explain coming to be and ceasing to be by means of condensation and rarefaction, because they use combination and separation

ᵇ32 to regulate these processes. We can also add those who attribute change to the mind, since they say that the source of change is that which changes itself, and the kind of self-change initiated by animals and all living things is change of place. Also, we use the word 'change'* properly only when we refer to change of place; if a thing is at rest in the same place, but happens to be increasing or decreasing or altering, we say that it is changing *in a certain way*, but we do not say that it is changing *tout court*.

We have argued* that there always was and always will be   266ᵃ6 change; we have also shown what the source of eternal change is, and in addition which the primary kind of change is, and which is the only kind of change that can be eternal; and we have proved that the first agent of change is unchanging.

10. *The eternal first agent of change has no magnitude, and is located at the outer edge of the universe*

Now we have to argue that the first agent of change has no   ᵃ10 parts and no size, and we should start by establishing the relevant premises.* One of these is that nothing finite can cause change for an infinite time. The point is that there are three things: the agent of change, the changed object, and thirdly that in which the change takes place, that is, the time. Now, either they are all infinite or they are all finite or some are finite (two of them, perhaps, or just one of them). So let   ᵃ15 *A* be the agent of change, *B* be the changed object, and *C* be an infinite stretch of time.* Let *D* be that which causes change in *E* (a part of *B*); it will not take a time equal to *C* to do so, because it takes more time to change a greater object. Therefore, the time it takes (*F*) will not be infinitely long. Now, by adding to *D* I will exhaust *A* and by adding to *E* I will exhaust *B*, but I will not exhaust *C* by constantly subtracting equal amounts from it, because *ex hypothesi* it is infinite. It follows that the whole of *A* will cause change in the whole of *B* in a finite amount of time—that is, some part of *C*. It is impossible, therefore, for anything to have an infinite change imparted to it by something finite.

Obviously, then, something finite cannot cause change for   ᵃ23 an infinite time; and the following argument will demonstrate the validity of the general principle that it is impossible for infinite power to be possessed by a finite magnitude.* We can assume that a greater power always achieves an equal result in less time, whatever kind of change is involved—it may be heating, for example, or sweetening or throwing. It also necessarily follows that anything which is acted upon by an agent which may be finite, but which possesses infinite power, is affected in some way, and is affected more than it would be by something else (since infinite power is greater than any other power). But this cannot take any time at all. Suppose A is the time taken by the infinite power to heat or propel an object, and suppose it takes a time AB for a finite power to achieve the same result;* then if I add to this finite power and constantly make it greater and greater, I will eventually arrive at the position of having caused the change in time A. For if I go on and on adding to a finite quantity, I will eventually exceed any determinate quantity, and if I go on and on subtracting from a finite quantity, I will eventually fall below any determinate quantity. It follows that the finite power will take the same amount of time as the infinite power to cause the change; but this is impossible, and so nothing finite can possess infinite power.

It is also impossible for an infinite magnitude to possess finite power. (Despite the fact that a smaller magnitude may possess a greater force, it is more probable that a greater force will be possessed by a greater magnitude.) Let AB be an infinite magnitude. Then BC (a part of AB) has a certain amount of power, which can cause a change D in a certain amount of time which we can call EF. Now, if I take a power which is double that of BC, it will cause the change in FH— that is, an amount of time which is half EF (for we may suppose that this is the proportion). If I carry on like this, I will never exhaust AB,* and I will constantly be taking smaller and smaller parts of the given time. The power will be infinite, then, because it exceeds every finite power (after all, where any finite power is concerned, the time it takes to achieve its result must be finite too, because if a certain amount of power causes change in a certain amount of time, a greater power will take a shorter, but still determinate, amount of time to cause the change, in inverse proportion). Any power which exceeds every determinate power is infinite, as is also the case with number and magnitude.

The point can also be proved by taking a certain power

(the same in kind as that possessed by the infinite magnitude, but possessed in this case by a finite magnitude) of which the finite power possessed by the infinite magnitude is a multiple.

So these arguments show the impossibility of a finite mag- $^b25$ nitude possessing infinite power and of an infinite magnitude possessing finite power. But before going any further it would be a good idea to resolve a certain difficulty concerning movement.* Given that, with the exception of self-movers, every moving object is moved by something, how is it that some things—things that are thrown, for instance—have continuity of movement when that which initiated the movement is no longer in contact with them? If the mover also causes some- $^b30$ thing else to move—the air, for instance, which causes movement by being in motion itself—it remains equally impossible for the air to be in motion when the first cause of movement is no longer in contact with it or causing it to move. No, all the things that are moving must move at the same time as the first mover and must have stopped moving when the first mover stopped imparting motion, and this is so even if, like a loadstone, the first mover makes what it has moved capable of causing movement itself. So what we have to say is that 267^a2 although the first cause of movement imparts the ability to cause movement to the air or the water (or whatever else it may be that is, by its nature, capable of causing movement and of being moved), nevertheless the air or water or whatever does not stop causing movement and being moved at the same time as the first mover stops; it may stop being moved as soon as the cause of movement stops imparting movement, but it retains its ability to cause movement. That is why it imparts movement to something else which is consecutive to it, and the same goes for this in turn. The process of stopping $^a8$ begins when each consecutive member of the series has less power to cause movement, and the motion finally comes to an end when the previous member of the series no longer makes the next one a cause of movement, but only makes it move. The movement of these last two members of the series, the mover and the moved, necessarily ends simultaneously, and so the whole movement comes to an end.

So this kind of movement occurs in things which are capable $^a12$

of sometimes being in motion and sometimes being at rest. Despite appearances, it is not continuous motion; for the objects are either successive or in contact, since no single mover is involved, but a number of movers, one after another. That is why this kind of movement, which some people call mutual replacement, occurs in air and water. But the way I have described is the only way to resolve the difficulty; mutual replacement fails to do this since it makes everything cause movement and be moved simultaneously, and so also stop simultaneously. As things are, though, the appearance is of a single thing which is moving continuously. So the question arises: by what is it being moved? And we find that it is not being moved by a single mover.

$^a21$ Continuous motion must be a feature of the real world, and continuous motion is single—which is to say that it is the property of a magnitude (since what has no magnitude cannot move) and of a magnitude which is single and which is moved by a single agent (otherwise the motion will not be continuous, but a series of consecutive, distinct motions). Now, if the mover is single, it is either a moving or an unmoving $^a25$ mover. If it is moving, it will necessarily accompany the object it moves and itself move, and it will also have to be moved by something; this leaves us with a series that will come to an end when a point is reached where movement is caused by something which does not itself move. For a mover which does not itself move is bound not to be involved in the motion it causes; it will never fail to be capable of imparting movement, because imparting movement without being moved is effortless; and the fact that the mover is not in motion means that, if this is not the only kind of movement to be uniform, 267^b5 it is at any rate more uniform than any other kind. But if the movement is to be regular, the moved object must also not change its relation to the mover, and the mover must be either in the centre or on the circumference of the circle, because these places are the sources of the circle. But the nearer things are to the mover, the faster they move; the fastest motion in this instance is the movement of the circumference;* so that is where the mover is situated.

$^b9$ It might be wondered whether it is possible for something which is itself moving to cause continuous movement,* rather than the kind of movement,† like that which is caused by pushing something again and again, which is continuous only by being successive. In this instance, the pushing (or pulling, or combination of the two) must be done either by the original mover itself, or by something other than itself which inherits the movement and passes it on to something else in turn (which is the analysis we developed earlier when discussing throwing, given that the air, being divisible, causes movement by its parts being moved one after another). In either case, however, the movement cannot be single; it must be a series of consecutive movements. It is only something which does not itself move, then, that causes continuous movement, because since it is always the same itself, its relation to whatever it moves will remain the same and continuous.

With these points in place, it is easy to see that the first ᵇ17 mover, which is itself unmoving, cannot have any magnitude. If it has magnitude, it must be either finite or infinite in extent. Now, we proved earlier,* during our lectures on nature, that there cannot be an infinite magnitude, and we proved just now that it is impossible for something finite to have infinite power. The first agent of movement, however, causes eternal movement and does so for an infinite time. It is clear, then, that it is indivisible, and has no parts or magnitude.

EXPLANATORY NOTES

BOOK I

184ᵃ25 *anything general is a kind of whole*: Aristotle's overall message is that natural bodies, considered as wholes, are familiar to us, but our task is to discern their fundamental constituents. Here he apparently compares this with the way that a general property may be analysed by a definition into its 'logical constituents'.

184ᵇ17 *the natural scientists*: Aristotle is thinking here of the group called 'the Ionians'. For these, and for the others mentioned in this paragraph, see Introduction, §2.

184ᵇ21 *Democritus' principles*: Democritus supposed that there were atoms of *all* shapes, and hence of infinitely many shapes.

184ᵇ25 *whether being is single and unchanging*: none of Aristotle's predecessors had held that there is just one unchanging *principle*, so what he actually discusses is the Eleatic thesis that there is just one *thing*, and it never changes. (But, as he points out at 185ᵃ3–5, this thesis implies not that there is one principle but that there are none.)

185ᵃ7 *the Heraclitean thesis*: Aristotle thinks that Heraclitus held that contradictory statements could both be true at the same time. Cf. 185ᵇ20–5. (The idea that being is a single person is Aristotle's own suggestion of an absurd thesis; cf. 185ᵃ24.)

185ᵃ16 *squaring the circle*: this is the problem of how to construct a square that is equal in area to a given circle. The problem was not shown to be insoluble until 1882 (F. Lindemann). The usual view is that what Aristotle calls 'the attempt by means of segments' was an attempt by the mathematician Hippocrates of Chios (*fl. c.*450–430 BC), based on some perfectly good results about the squaring of lunes. (Lunes are areas shaped like a new moon, i.e. formed by intersecting two circles of different sizes.) Antiphon the sophist was a contemporary of Socrates, and the usual view is that his attempt involved an infinite series of ever closer approximations. On the topic generally, see T. L. Heath, *History of Greek Mathematics* (Oxford: Clarendon Press, 1921), i. 183–200, 220–35.

185ᵃ22 *all things are one*: Aristotle apparently takes this as short for 'all things are one *being*', and so goes on to ask what kind of being is intended.

185^b16 *no such things as quantity or quality*: Aristotle's argument is that if a quantity is to exist, then there must also exist a substance which has that quantity, and hence being must be divisible into a substance and a quantity. If being is not so divisible, then no quantity exists, and so being cannot be of any quantity. It cannot, then, be either limited or unlimited.

185^b28 *Lycophron*: a sophist about whom we know little.

186^a3 *either potentially or actually*: Aristotle means that what is actually one whole may also be many parts potentially, so it is both one and many, but 'not in conflicting ways'. Similarly he seems to mean that what is pale and what is educated may be the same thing coincidentally but will be different things in definition. So again these things are both one and many, but not in conflicting ways.

186^a10 *Melissus' argument*: it appears that Aristotle took Melissus' argument to be this. He assumed that every created thing has a 'beginning', in the sense of a first part to be created. He assumed also that every such thing has an 'end', in the sense of a last part to be created. By an illegitimate conversion he inferred that what is not created has no first and last parts, and hence that it is spatially infinite. So, since the universe was not created, it is spatially infinite. (Note that Aristotle himself holds that the universe was not created and is spatially finite.) Aristotle objects first that the inference is fallacious and second that anyway the assumption is false. For example, when water freezes, the freezing may happen all over at the same time; there does not have to be a first part that freezes (cf. 253^b23–6).

186^a33 *it means just being*: Aristotle has pointed out, just above, that even though the word 'pale' is unambiguous, still it is used *both* to describe pale objects *and* to name the attribute pallor. (This is true in Greek, since the Greek for 'pallor' is literally translated as 'the pale'.) He infers that the same will be true of the word 'being', even if we confine it to a single meaning (i.e. to the meaning in which it applies to substances, presumably). It too will be used *both* to describe the various things that are (i.e. exist) *and* to name the attribute being (i.e. existence). So he concludes that if Parmenides is to reach the result that only one thing exists, he must suppose that the word 'being' has *only* the second use. The suggestion that 'being' means just being is therefore to be interpreted as the suggestion that it applies only to the attribute existence, and to nothing else. (I should perhaps add that, if this suggestion is intended as a reconstruction of Parmenides' own thinking, it is hopelessly anachronistic.)

186ᵇ8 *Therefore, what is pale has no being*: the assumption is that something, say pallor, is an attribute of existence. (It is not clear why Aristotle feels himself entitled to this assumption.) Supposing further that pallor is not identical with existence, the premiss that only existence exists then implies that pallor does not exist. But Aristotle seems to infer from this that no pale thing exists, and it is not clear how this inference is warranted.

186ᵇ14 *what just is a being*: this phrase translates the same Greek words as does the phrase 'what just is being', used above. But Aristotle here changes the use of the words, perhaps feeling that the argument above requires this. In fact from now on any substance is counted as something which just is a being.

186ᵇ18 *A thing is said to be coincidental if . . . or if*: the second condition given here is more usually given as a condition of what is *not* coincidental but is in its own right (see e.g. *Posterior Analytics* 73ᵃ37–ᵇ3, *Metaphysics* Δ, 1022ᵃ29–32).

186ᵇ35 *Does it then follow . . . indivisible entities?*: the argument began by claiming that the definition of 'man' as 'two-footed animal' divides one substance into two others. The alternative was said to be that 'animal' and 'two-footed' should each be coincidental attributes, but that alternative has now been refuted, so the original claim must stand. But how does that claim imply that the universe is made of indivisible entities? Well, if the claim is that *every* substance divides in this way into two (or more) others, the result is an infinite regress of definitions, which Aristotle would certainly regard as a vicious regress. (See e.g. *Posterior Analytics* 82ᵇ37–83ᵃ1.) He must conclude, then, that some substances cannot be thus divided, and it might seem natural to say that the universe is 'made of' these.

It should be observed that this argument concerns what one might call the 'logical' division of a universal (in the category of substance) into two others. Yet it seems to have taken the place of an exposition of Zeno's argument for an infinite regress generated by the *spatial* division of one magnitude into two others. (See Introduction, §14.) For this is what Aristotle refers to in the next two sentences as 'the dichotomy'. (It is also what had been heralded at 186ᵇ12–14.) Presumably Aristotle would think that the spatial regress is not vicious, for he does accept the infinite divisibility of bodies and of space (even if he wishes to add, in Book III, that this infinity is only potential). Certainly, he does *not* agree that the universe is made of indivisible *bodies*, i.e. atoms.

I see no explanation of why Aristotle should substitute an argument about logical division, with a conclusion he accepts, for an analogous argument about spatial division, with a conclusion he rejects.

187ᵃ1 *Some people*: the atomists. Aristotle thinks that (i) in order to resist Parmenides' argument they were forced to conclude that non-being exists (i.e. empty space, which they did call 'non-being'), and (ii) in order to resist Zeno's argument they were forced to posit indivisible magnitudes (i.e. atoms).

187ᵃ13 *one of the three*: water (Thales), or air (Anaximenes), or fire (Heraclitus).

187ᵃ17 *Plato describes his 'great and small'*: this forms part of Plato's so-called 'unwritten doctrines' (cf. 209ᵇ15), and does not occur in his published dialogues. Our knowledge of these doctrines is somewhat precarious, and it depends mainly on what Aristotle tells us in allusions such as these.

187ᵃ25 *the homoeomerous substances*: that is, the various different kinds of stuff. (The word means 'having parts like the whole'. For example, gold is homoeomerous, since every part of a piece of gold is itself gold; but a face is not homoeomerous, since the parts of a face are not themselves faces.)

187ᵃ29 *they make statements*: though Aristotle says 'they', it is Anaxagoras in particular that he is thinking of in this paragraph. (For example, 'everything was originally mixed together' is a quotation from Anaxagoras.)

187ᵃ37 *anything which comes into being*: Aristotle is thinking of the generation of this or that kind of stuff (e.g. flesh or bone or hair) and not of the whole creature of which these are parts.

187ᵇ30 *some definite size*: that is, the size which is the smallest possible size for any piece of flesh. (Aristotle has already argued, at 187ᵇ13–18, that there is such a size.)

188ᵃ16 *water and air are made of each other*: in English, Aristotle's point is better put in this way: a house is made *from* bricks, and is also made *of* bricks, whereas air comes *from* water but is not made *of* it.

188ᵃ20 *even Parmenides*: Parmenides' work contained both a 'Way of Truth', which claimed that reality was single and unchanging, and a 'Way of Seeming', which admitted change and plurality. Here Aristotle must be meaning to cite the Way of Seeming, but (*a*) the citation is inaccurate, and anyway (*b*) Parmenides clearly says that his Way of Seeming is not true.

188ᵃ23 *position, shape, and arrangement*: these are the three ways in which Democritus' atoms can differ from one another.

189ᵃ1 *the same list*: the list is given in full at *Metaphysics A*, 986ᵃ23–6, as follows:

limit	unlimited
odd	even
one	many
right	left
male	female
at rest	in motion
straight	curved
light	darkness
good	bad
square	oblong

It is Pythagorean in origin.

189ᵃ13 *every genus contains just one opposition*: Aristotle is thinking in particular of the *ultimate* genera, i.e. the categories. For example, one might say that the one fundamental opposition in the category of quantity is that between more and less. But the case relevant to the argument is the category of substance, because our ultimate concern is with the principles of natural objects, and these are all substances. It appears from the following chapter that Aristotle's doctrine must be that in this category the fundamental opposition is that between form and privation. Cf. 201ᵃ4–5. (But one may observe that in Ch. 7 the notion of a 'form' is applied to properties in *all* the categories in which change is possible, and is not confined to what are called 'substantial forms'.)

The argument here is repeated at 189ᵇ22–7.

189ᵃ21 *there is some reason*: note that the three arguments at 189ᵃ20–ᵇ1 are tentative, and it is not clear that Aristotle himself subscribes to them. This may be connected with the fact that he is speaking here in terms of the traditional opposites, whereas his own position will adopt just 'form and privation' as the fundamental opposites. Since these forms *do* include substantial forms, the third argument will then become irrelevant, and the second will at least need a qualification.

189ᵇ1 *both arguments*: the argument of Ch. 5 that opposites must be included, and the argument just given that opposites by themselves are not enough.

189ᵇ11 *as I have already said*: 187ᵃ16–17.

189^{b}14 *some more recent thinkers*: Plato, in his 'unwritten doctrines'. Cf. 187^{a}16–20.

189^{b}19 *if there are . . . two oppositions*: Aristotle's first argument against this supposition is obscure. Apparently he dismisses the first alternative, that 'each pair will need a separate extra thing as an intermediate', on the ground that, as he has just said, 'we do not need more than just the one to be acted on'. But then he dismisses this second alternative too, on the ground that it implies that 'each pair is capable of generating out of the other', and that this will make one or other pair 'redundant'. It is not clear what this mutual generation is supposed to be, nor why it would be implied by a common matter for both oppositions. It is pertinent here to recall that Aristotle's *own* position will be that the four sublunary elements share a common matter and are characterized by two oppositions, i.e. hot vs. cold and wet vs. dry. (But, to be fair, one should add that Aristotle does *also* say that only the first opposition provides an active force which generates changes, namely at *Meteorologica* iv. 1–2.)

190^{a}11 *what is not educated . . . does not persist*: one can clearly see here a feature of Aristotle's way of thinking that in fact affects all of his discussion. He supposes that, when we have an uneducated person, the two expressions 'the person' and 'the uneducated thing' do in one way refer to the same object, though there is also another way in which they do *not*. It is because there are in a sense two *different* things being talked of that he can say that one persists whereas the other does not. (Useful general discussions of this aspect of Aristotle's thought are G. B. Matthews, 'Accidental Unities', in M. Schofield and M. Nussbaum (eds.), *Language and Logos* (Cambridge: Cambridge University Press, 1982), 223–40; and F. A. Lewis, 'Accidental Sameness in Aristotle', *Philosophical Studies*, 42 (1982), 1–36.)

190^{a}25 *and not of bronze becoming a statue*: I presume that Aristotle means not 'not' but 'not only'. (See A. Code, 'The Persistence of Aristotelian Matter', *Philosophical Studies*, 29 (1976), 357–67.)

191^{a}20 *not yet clear*: this appears to promise a future discussion. If so, it seems best to suppose that the reference is to the *Metaphysics*, where Aristotle does discuss the claim of matter to be substance (notably in ch. 3 of book *Z* and in ch. 1 of book *H*).

191^{b}12 *nothing else is*: that is, nothing besides being itself.

191^{b}22 *but it already has the property of being an animal*: this is a mistake. A dog is not an animal before it becomes a dog, i.e. before it is conceived. The truth is that Aristotle's 'analogy' (191^{b}19) does not at all illustrate the point that he should be making, which is this: a dog comes from something which is not a dog but is something which exists, namely an ovum. (In the present context it is irrelevant that what turns the ovum into a dog is itself (the sperm of) another dog.)

191^{b}29 *elsewhere*: the reference is not clear, but is perhaps best taken to be to the very full discussion of potentiality and actuality in book *H* of the *Metaphysics*. (On the other hand, the solution briefly canvassed here is rejected as unhelpful at 317^{b}13–33 of the *De Generatione et Corruptione*.)

191^{b}35 *others*: Aristotle evidently means those who followed Plato's 'unwritten doctrines' (cf. 187^{a}17–20, 189^{b}14–16). His accusation is that they assigned to the 'great-and-small' the two distinct roles of a persisting matter and a non-persisting privation. This, he says, led them to call it non-being (^{a}7), because this description fits the privation, and also to say that it is what persists and combines with form (^{a}13), because this description fits the matter. (Considered as matter, it also fulfils the role that Plato had assigned to space in the *Timaeus*, and Aristotle is no doubt thinking of this when he describes it as a 'mother' (^{a}14; cf. *Timaeus* 50d, 51a).)

192^{a}14 *the other aspect of the opposition*: that is, the privation. It is not clear whether Aristotle is reporting a Platonist view when he describes this as 'pernicious', and as 'desiring' the form. At any rate it appears to be his own view that matter (but not the privation) does 'desire' the form. If this seems overfanciful, we should recall the important role that he assigns to the form of a living thing (Introduction, §7(iii)).

192^{a}23 *a woman longing for a man*: it is pertinent to recall that Aristotle's theory of sexual reproduction is that the female provides the matter and the male provides the form. From our contemporary point of view, Aristotle is irredeemably sexist.

192^{a}35 *first philosophy*: this is Aristotle's name for metaphysics, and much of the *Metaphysics* is indeed devoted to the principles of form (principally books *Z*, *H*, *Λ*).

192^{b}2 *the following expositions*: this phrase probably includes *all* of Aristotle's many writings on nature, and not just the succeeding books of the *Physics*. (In fact only Book II of the *Physics* fits the description given here.)

BOOK II

193ª9 *nature and substance*: the substance of a thing is what it is to be that thing (194ᵇ21), i.e. what is given in its definition. (Note that Aristotle gives no reason for implicitly identifying this with its nature.)

193ª12 *Antiphon*: see note on 185ª16.

193ª17 *if on the other hand*: the thought is that if the nature of a bed is its wood, since that remains when the bed ceases to be a bed, then equally we should be able to say that the nature of wood is what remains when it ceases to be wood (some earth and water, say). Aristotle then adds the further suggestion that the nature of wood will *also* be the nature of things made of wood. Pressing this line of thought, the nature of any natural object will be the *ultimate* matter of which it is made, which may well be the same for all things. But we started with the thought that a thing's nature would be its *proximate* matter, e.g. wood for wooden objects and bronze for bronze objects.

193ª29 *the first matter*: this phrase is ambiguous between ultimate matter and proximate matter. See previous note.

193ᵇ11 *form too is nature*: the argument is that father and son have the same form, since both are men.

193ᵇ12 *'nature' in the sense of a process*: this is a play on words which cannot be reproduced in English. The same Greek word (*physis*) means both 'nature' and 'growth'. (At least, that is what Aristotle supposes, though scholars think he is wrong about the meaning 'growth'.) The argument is that a growing thing is growing *towards* its proper form, but Aristotle does not explain here why he thinks that this goal is also the inner cause of the growth, and hence the nature of the object. See Introduction, §7(iii).

193ᵇ21 *consider later*: it is usually supposed that the reference is to the *De Generatione et Corruptione*, which has a longish discussion of coming into being in ch. 3 of book I. But the awkwardness is that that discussion refers back to an earlier and allegedly fuller discussion (317ᵇ13). The truth seems to be that the most relevant passage is actually Chs. 7–8 of the *first* book of the *Physics*.

193ᵇ36 *those who say that there are forms*: Aristotle is speaking not of his own forms but of Plato's forms. He commonly complains that the Platonist is mistaken in supposing that a form *exists* independently of any material objects of that form. Here his complaint seems to be somewhat different: it makes no sense even to *consider* such a form in abstraction from matter (whereas it does make sense to *consider* geometric forms abstractly, even though they too cannot *exist* except as forms of material objects).

194^{a}6 *like a snub nose*: this is a favourite example. Aristotle's point is that you cannot define snubness without mentioning the special kind of matter that is capable of being snub, namely a nose. (Cf. e.g. *Metaphysics E*, 1025^{b}30–1026^{a}1.)

194^{a}31 *the end for which he was born*: that is, death. (It is not known which poet is being quoted.)

194^{a}36 *my dialogue On Philosophy*: this was one of Aristotle's published works, and only a few fragments of it have survived. The two meanings of 'what a thing is for' are 'that which is aimed at' and 'that which is benefited'.

194^{b}8 *is already present*: the argument is that in the case of art there is a reason for there to be two separate studies, one concerned only with matter, but this does not apply in the case of nature.

194^{b}12 *separable in form*: it is not clear what Aristotle means by this phrase. A natural suggestion might be that the forms to be studied are always embodied in matter, though they may be *considered* in abstraction from matter, or may be *defined* without bringing in matter. But actually the opening of this chapter has denied that possibility.

194^{b}14 *first philosophy*: see note on 192^{a}35.

195^{a}18 *premises (from which a conclusion comes)*: see Introduction, n. 13.

195^{a}20 *others are causes*: Aristotle means not 'others of the things just listed as causes', but 'others of the things just mentioned', and in fact those just listed as *effects*. He equates 'x comes from y' and 'y underlies x', and in each case counts y as material cause and x as formal cause.

195^{a}34 *Polyclitus*: a famous sculptor of the second half of the fifth century BC. His most celebrated work was the chryselephantine statue of Hera at Argos.

195^{b}12 *there are six kinds of cause*: Aristotle means that, with regard to *each* one of the four types of cause distinguished earlier, there are six different ways in which it can be described, but his explanation hardly justifies the figure six. For example, the efficient cause of a particular statue may be described (i) by the description which properly reveals its role as cause (e.g. 'this sculptor'), or (ii) by a description coincidental to this role (e.g. 'Polyclitus'). Moreover, these descriptions may either single out the particular cause (as illustrated), or may give its species or its genus (e.g. 'an artist', 'an animal', respectively). This is apparently counted as producing four different ways of describing the cause. Finally, the descriptions may be simple (as illustrated), or they may be combined. But it is not clear why the possibility of combination should be thought to produce just two further ways of describing the cause.

195^{b}18 *at the same time*: Aristotle supposes that the time when the cause is acting, and the time when the effect is occurring, are the same time (see Introduction, pp. lxv–lxvi). But he also supposes that an *actual* cause is a cause-when-acting (e.g. a builder building) and this cannot be the same thing as what he calls a *potential* cause (e.g. a builder), since the latter exists at times when the former does not. This is a way of admitting that the cause of a happening is not really an object (the agent) but another happening (the agent acting). But while we can appreciate this thought when it is applied to efficient causes, it is difficult to see how it could apply to the other three kinds of cause.

195^{b}36 *Some people*: see next note.

196^{a}24 *there are others*: in both this and the above case the reference appears to be to Democritus. His view is that everything whatever happens 'of necessity', and hence that what we speak of as a chance event is not in fact due to chance. The argument at 196^{a}1–7 may well be due to him (or to Leucippus). He equally ascribed the origin of our world to 'necessity', meaning by this that it was the inevitable result of many atoms colliding with one another. But Aristotle thinks of him as holding that it is due to chance, since he does maintain that it did not occur for any purpose. (We have no firm evidence that Democritus himself spoke of 'chance' at this point, but that may well be an accident; only scattered fragments of Democritus' writings survive.)

196^{a}25 *this world of ours and all the worlds*: the atomist theory is that space is infinite, that atoms are to be found throughout space, and hence that this world that we inhabit is only one of infinitely many worlds.

196^{b}5 *there are those*: no particular thinker is here referred to, but merely popular superstition. (There was even a cult of Chance.)

196^{b}18 *and some are not*: namely, events which serve the purposes not of men but of nature. (On these, see Introduction, §7(iii).)

196^{b}32 *later*: Ch. 6.

196^{b}33 *Here is an example*: the example is apparently the same as that at 196^{a}1–7.

197ᵃ25 *good luck*: the Greek phrase is more literally 'good chance'. Similarly the Greek word translated as 'good fortune' is derived from the same word 'chance'.

197ᵃ27 *That is also why*: the explanation is that if I 'just miss' a harmful outcome, I think of myself as having actually been in a harmful situation and then escaping from it. Such an escape would genuinely be 'good fortune'.

197ᵇ8 *it cannot exercise choice*: on Aristotle's account, choice requires deliberation (*Nicomachean Ethics* iii. 2–3).

197ᵇ10 *Protarchus*: a pupil of the orator Gorgias, who is a speaker in Plato's dialogue *Philebus*.

197ᵇ20 *when the cause . . . is external*: the rationale for this condition is obscure. The traditional elucidation is that if the cause is internal to the object affected, then the event will be something that happens always or usually (e.g. if a cubical box falls in such a way that it can be sat on). But see note on 197ᵇ33.

197ᵇ22 *the term 'pointless'*: the Greek word translated as 'spontaneous' (*automaton*) resembles the phrase translated as 'in itself pointless' (*auto matēn*, ᵇ30). Aristotle suggests that the former is derived from the latter, but there is no real connection, and his explanation is wholly unconvincing. (His idea is that what fails of its usual effect is 'in itself pointless', but it may have another and unintended effect, which will be 'spontaneous'. However, his examples of spontaneous effects are *not* examples where a usual effect fails.)

197ᵇ33 *in the case of naturally occurring events*: Aristotle's general doctrine is that a chance or spontaneous event must be one that does serve a purpose, though that purpose does not explain why it occurred. Where the purpose is one that a rational being might have intended, we have what is properly called a *chance* outcome, and Aristotle (rashly) takes it to follow from this that only a rational being can do something by chance. But where the purpose in question is one of 'nature's purposes' we can only say that the outcome is *spontaneous*. (This is clearly the view presented at 196ᵇ18–22, 197ᵇ1–9, 20–2, 198ᵃ2–7.) However, his examples of spontaneous outcomes have all been examples where the result might have been intended by a person, though it is not easy to see how they might have been 'intended by nature' (197ᵇ15–18, 30–2). In this passage he does seem to envisage something that might have been 'intended by nature', but was not so 'intended', since in fact it was 'unnatural'. But unfortunately he gives no examples, and moreover he says that such examples are not after all spontaneous, since the cause is not external.

The Greek commentators conjecture that he is thinking here of the unnatural birth of 'monstrosities' (199ᵇ1–4), but in that case we do not have an example of something that does serve nature's purposes. Ross conjectures instead that he is thinking of the spontaneous generation of living things, without seed, which he thinks happens with shellfish and some other creatures. (The account of this in *De Generatione Animalium* iii. 11 does assign an external cause, namely the warmth of the sun. The much briefer account in *Metaphysics* Z9—which *may* have other examples in mind—denies an external cause.) But in any case the result is that Aristotle gives no examples of a natural purpose being accomplished, but not as nature intended, and in this paragraph he *seems* to imply that any such examples would *not* after all be spontaneous, since the cause would not be external.

198ᵃ11 *more primary causes*: note that Democritus could readily accept this result, since on his account the formation of the world will be due to the *nature* of the atoms involved.

198ᵃ24 *the last three . . . come to the same thing*: Aristotle is claiming that this equation holds for living things. In fact he holds that the form of a thing is *both* what is given in its definition, *and* what its purpose is, in the case of *any* object that has a purpose. This includes not only whole living things but also their various parts (e.g. the eyes, the feet, the kidneys), and of course artefacts. In ch. 7 of *Metaphysics Z* he also argues that the efficient cause of an artefact again has the same form as it, since he there takes the efficient cause to be the form already existing in the *mind* of the maker. But it is not clear quite what he would say about the efficient causes of particular parts of the body. Indeed, it is not clear whether each has its own efficient cause.

198ᵃ27 *everything which is changed itself when initiating change*: this is a strong claim, namely that in *all* cases of efficient causation (save perhaps one; see note on 198ᵃ36), the object which is the cause already has the form which it produces in the effect. (The claim is repeated at 202ᵃ9–12; Aristotle inherited it from Plato's *Phaedo*: see 101b and 102b–106e.) We have seen in the previous note how the principle is supposed to apply to the generation of animals and artefacts, but Aristotle wishes to apply it yet more widely. For example, he holds that the efficient cause of movement must itself be moving with the motion that it communicates (see Introduction, §16), and he is fond of remarking that only an object that is hot itself can make other things hot (e.g. in Book VIII at 251ᵇ29–30, 255ᵃ21–3, 257ᵇ9–10).

198ᵃ29 *three areas of study*: namely (i) God (or Gods), (ii) the heavenly bodies, (iii) everything else.

198ᵃ36 *not itself a natural object*: matter contains within itself a cause of its own change (e.g. its natural upwards or downwards motion), and so does an ordinary efficient cause (construed as a material object possessing a form, not as the form itself). Aristotle implies that if we merely pursue the usual question 'What comes after what?' we shall be led only to these causes. But there are also causes which are not natural objects, namely forms (as both formal and final causes) and God. In book *Λ* of the *Metaphysics* God is himself conceived as causing change in the way that a final cause does, but the distinction drawn here suggests that God's causation is being viewed as a special kind of efficient causation (as would seem to be the view of our Book VIII).

198ᵇ5 *from this there necessarily comes that*: this *appears* to refer to what are called 'the laws of matter' in §7(iv) of the Introduction. (But Ross interprets it as a reference to the efficient cause, which would otherwise be omitted.)

198ᵇ7 *what must be present if the thing is to exist*: this presumably refers to the conditional necessity illustrated in Ch. 9: a given form will require matter of a specific kind.

198ᵇ8 *this is what it is to be the thing*: this refers to the formal cause, as given by the definition.

198ᵇ8 *because it is better that way*: this refers to the final cause, but in a way that *ought* to distinguish it from the formal cause. See Introduction, §7(iii).

198ᵇ32 *as Empedocles says*: Fragment 61.

199ᵃ2 *dog-days*: the hottest period of the year, counted from the rising of the dog-star (i.e. Sirius, in the constellation Canis Maior).

199ᵃ12 *a naturally occurring house*: Aristotle imagines that such a house would be constructed in the same order as we employ, i.e. first the foundations, then the walls, then the roof. Presumably his ground is that a different order would not be possible, and he thinks that this substantiates his claim that there is a correspondence between nature's way of doing things and human ways of doing things. But here he has failed to take into account that in natural generation things *grow* (from seed), whereas human production never takes this form.

199ᵇ9 *'And first whole-natured . . .'*: this is a quotation from Empedocles (Fragment 62), who says that the first things to be produced are 'whole-natured' (i.e. homogeneous) mixtures of earth and water.

199^b23 *as I have already explained*: Ch. 5.

199^b28 *skill does not make plans*: this is a surprising statement. Aristotle is apparently thinking that, once you have learnt how to make something, you no longer have to plan how to do it. But one will of course object that in nature there is neither planning *nor* a learning process. (For 'naturally occurring ships', cf. on 199^a12.)

200^a4 *earth*: that is, mud-bricks.

200^a14 *the necessity is in the matter*: this apparently means that the matter *is necessitated* (sc. by the end), not that the matter itself *necessitates*. Similarly the later statement that what is necessary is the matter (200^a13). By contrast Aristotle implies that the end, i.e. the form, is not necessitated by anything.

200^a23 *a chain of reasoning*: in human production the end to be realized is the starting-point of the reasoning which tells us how to achieve it. There is no such reasoning in nature (199^b26–8), but Aristotle apparently implies that there is something analogous. (No doubt what lies behind this is his thought that *our explanations* must start from the end to be realized.)

200^b4 *what is necessary*: i.e. the matter. The suggestion is that one might include as part of the definition of a saw not only its purpose but also the material (i.e. iron) without which the purpose could not be achieved.

BOOK III

200^b12–25 *Introduction to Books III–IV*: Book II has identified nature with a principle of change (Ch. 1), and Book III opens by promising to devote attention to change. It then introduces the programme to be followed in Books III and IV. As a matter of fact all of Books V–VIII are also concerned with change—and more directly concerned with it than most of Books III and IV—but the present passage says nothing of them. (In particular the present passage says that infinity will be discussed, because that notion is important for the analysis of continuity. But it does not say that continuity will also be discussed, though in fact Book VI is devoted to this topic.)

200^b26 *only actually*: only God contains no potentiality.

200^b33 *substance or quantity or quality or place*: in Ch. 2 of Book V Aristotle will argue that these are the *only* categories in which change takes place. See next note.

201ᵃ8 *as many kinds of change as . . . of being*: it appears (from the previous note) that Aristotle cannot quite have meant this. But in the context it would be adequate for him to claim that the kinds of change are *no more* than the categories of being.

201ᵃ27 *without being changed itself*: that is, God.

201ᵃ34 *clear in the case of opposites*: Aristotle's point appears to be that where the same thing has two opposite potentialities it is easy to see that the thing (e.g. the bronze) is not the same as the potentiality (e.g. of being a statue). The actuality he is talking of is not the actuality of the thing, but the actuality of its potentiality (which is what he is trying to indicate by his locution 'the actuality of that which exists potentially, in so far as it is potentially this actuality').

201ᵇ20 *alternative classifications*: it is not clear that anyone before Aristotle tried to *define* change. His reference to 'inequality' has been taken to point to Plato's *Timaeus* (57e–58c), and otherwise it has been held that he has some unknown Pythagoreans in mind. (For the ground behind this conjecture, see next note.)

201ᵇ25 *the second list*: the reference is to the list of pairs of opposites cited in the note on 189ᵃ1. This is of Pythagorean origin.

201ᵇ32 *but an incomplete one*: Aristotle's thought is that while a change is going on—i.e. while it actually exists—it is not yet complete. It becomes complete only when it is finished, i.e. when it has ceased to exist.

202ᵃ3 *as I stated earlier*: 201ᵃ23–7.

202ᵃ6 *it takes contact to do this*: this will be argued in Ch. 2 of Book VII.

202ᵃ11 *that which is potentially a man*: on our theory, what Aristotle is referring to is the unfertilized ovum. (On his theory it is the menstrual fluid in general, without differentiation.)

202ᵃ15 *because*: it is difficult to discern any argument in the considerations that follow.

202ᵃ19 *uphill and downhill*: this slightly cryptic example is taken from Heraclitus, Fragment 60: 'The road up and down is one and the same'.

202ᵃ21 *At an abstract level*: Aristotle probably means: at a level which takes into account views that are generally held, but which ignores the definition of change just given, and the 'argument' based upon it.

202ᵃ30 *The upshot*: Aristotle treats both alternatives of this 'upshot' as impossible, for neither would be at all plausible as a view about God, who causes change in other things without changing himself.

202ᵃ36 *there must be a single actuality*: Aristotle treats the previous two
arguments as conclusive (ᵃ25–36), and thus as establishing that the
doing and the being done to must indeed be 'a single actuality'. In
the present paragraph he puts an objection to this conclusion, and
then in the succeeding paragraphs he argues that the objection is not
cogent.

202ᵇ8 *a single, identical actuality*: the thought is that two different
potentialities, e.g. the potentiality for teaching long division and the
potentiality for learning it, may nevertheless have the same actuality, i.e. the process which is at the same time both an actual teaching of long division and an actual learning of it. But it does not
seem to add to the explanation to say that these two potentialities
are the same 'only in the way that what is potential is related to
what is actual'.

202ᵇ13 *'mantle' and 'cloak'*: a common example in Aristotle of two
synonymous terms.

202ᵇ15 *do not have all the same properties*: elsewhere Aristotle claims
that things may be coincidentally the same but differ in their properties, giving the example that the man approaching may be Coriscus,
and yet it is true that you know who Coriscus is, but not true that
you know who the man approaching is (*Sophistical Refutations*
179ᵃ32–ᵇ4). But he does not need to rely on such a controversial
claim here. The argument that, if any case of teaching is at the same
time a case of learning, then to teach is the same as to learn, does
not rely on the principle that if *A* and *B* are the same then they must
share all their properties.

202ᵇ20 *absolutely identical*: that is, identical in definition.

202ᵇ21 *of one thing on another, and of one thing by another*: that is,
the capacity to teach, which is exercised *on* the learner, and the
capacity to be taught, which is realized *by* the teacher.

203ᵃ3 *they all*: this is an exaggeration, as 203ᵃ18–19 shows.

203ᵃ4 *the Pythagoreans and Plato*: in Greek 'infinite' and 'unlimited' are
the same word. Both the Pythagoreans and Plato applied this word
to an undifferentiated stuff, waiting for a limit (or shape, or form)
to be imposed on it. (The same *may* be true of Anaximander's use
of this word.) However, Aristotle himself uses the word in a different sense, which he explains as meaning 'not traversable' (204ᵃ3–6).

203ᵃ14 *successive gnomons*: a gnomon is a right-angled shape, as in a
carpenter's set-square. What is now the usual interpretation of this
passage is that given by Ross, which contrasts the two figures shown
in the diagrams.

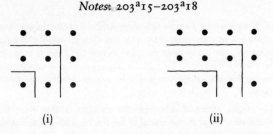

(i)   (ii)

In (i), successive right-angled shapes are put around an initial 'one', but however many you put, the result is always a figure of the same shape, namely a square. In (ii) the successive right-angled shapes are put around an initial 'two', and the result is a succession of rectangles of infinitely many different shapes (i.e. with sides related as n to $n + 1$).

Diagram (i) evidently does give the right interpretation of 'successive gnomons placed around unity', but there are two obvious objections to diagram (ii). First, it results from putting gnomons 'around two', whereas Aristotle himself says not 'around two' but 'apart from unity'. Second, it contains gnomons which each have an *even* number of dots, so it is difficult to see how it could be evidence for the claim that infinity results 'when an even number is enclosed and limited by an *odd* number'. (For those who wish to pursue this question, relevant evidence is given in Ross's note and in T. L. Heath, *History of Greek Mathematics* (Oxford: Clarendon Press, 1921), i. 77–84.)

203ᵃ15 *Plato . . . has two infinites*: as Aristotle himself says elsewhere (e.g. 192ᵃ11–12), Plato used *two words*, i.e. 'great' and 'small', to form a combined phrase which he conceived as referring to *one thing*. (But see also 206ᵇ28–33, where Aristotle tries to support his point.)

203ᵃ16 *the natural scientists without exception*: Aristotle is here not counting either the Pythagoreans or Plato (or, indeed, the Eleatics) as natural scientists. Even so, there is an exception. See next note.

203ᵃ18 *those who posit a finite number*: we should understand 'a finite number' as *excluding* the number one. So Aristotle claims that each of the Ionians made their single element infinite, though in fact this claim seems not to be borne out by the surviving evidence. (Anaximander did indeed *call* his one element 'infinite', but he may not have meant by this what Aristotle means. See note on 203ᵃ4.) He also goes on to claim that Anaxagoras and Democritus, each in their different way, posited an infinite variety of elements. The exceptions whom he notices here evidently include Empedocles (and, in a way, Aristotle himself).

203ᵃ22 *the atoms with all their different shapes*: see note on 184ᵇ21.

203ᵃ26 *are like that*: that is, both come from the same original mixture, i.e. food (or, in particular, bread; cf. Fragment 10).

203ᵇ12 *those . . . who do not recognize any other causes*: the exceptions noted are Empedocles (who recognized love as a cause) and Anaxagoras (who recognized intelligence). No doubt we should also include the atomists as exceptions (for they invoked necessity as a cause). So what Aristotle is thinking of is the Ionians. The phrase 'steers everything' apparently comes from Heraclitus (Fragment 41); the phrase 'contains everything' from Anaximenes (Fragment 2); we have no direct evidence that the others propounded anything similar.

203ᵇ15 *people*: of the five considerations that follow, all but the third appear to be attributed to 'people in general', with no special reference to the natural scientists. But Aristotle may well think that the third is due to Anaximander in particular.

203ᵇ30 *no difference between being possible and being actual*: even if this is restricted to 'eternal things', still the principle cannot be defended, for it may be possible both that p and that not-p, but both cannot be actual. (On this topic in general, see J. Hintikka, *Time and Necessity* (Oxford: Clarendon Press, 1973), esp. ch. 5.)

203ᵇ34 *some thing or things . . . infinite in number?*: Aristotle *may* be thinking that being infinite *in number* cannot be an attribute of any *one* thing, or he *may* be thinking that being infinite could be a 'coincidental' attribute of something, rather than an attribute 'in its own right'.

204ᵃ6 *infinite by addition or by division or both*: as emerges from what follows, a thing is 'infinite by addition' if by continuing to add the same amount, again and again, one does not exhaust it. That is to say that it is infinitely extended. (But a new sense enters at 206ᵇ3 ff., where the amounts added do not have to be equal.) A thing is 'infinite by division' if it is—potentially—infinitely divisible. Aristotle will argue in Ch. 5 that nothing can be infinitely extended, but will concede in Ch. 6 that many things are infinitely divisible.

204ᵃ8–34 *The infinite is not itself a substance*: Aristotle rejects the view he has credited to the Pythagoreans and Plato (but mistakenly; see note on 203ᵃ4). He introduces it in Platonic terms, as the view that the infinite is 'just itself', i.e. that it has no essential nature other than to be infinite.

204ᵃ25 *a plurality of infinites*: from a modern perspective there is nothing wrong with this. For example, the infinite set of whole numbers consists of two infinite parts, the set of even numbers and the set of odd numbers. Now of course Aristotle is not thinking of sets, but of extended magnitudes, but even so his claim is rash. For example, a straight line that is infinite in both directions may be divided into two parts, one infinite in one direction and the other infinite in the other direction. (Contrast 204ᵇ19–22.)

204ᵃ31 *the air . . . or the even numbers*: the identification of 'the infinite' with air is Pythagorean, as is the very different identification with even numbers (for which see note on 203ᵃ14).

204ᵇ1 *things which are intelligible*: Aristotle here presupposes a Platonic division of reality into (i) perceptible things, (ii) mathematical entities, (iii) other intelligible things which have no magnitude, i.e. the Platonic forms. (Note that mathematical entities are themselves regarded as intelligible rather than perceptible; for example the 'intelligible body' of 204ᵇ6 is a mathematical solid, e.g. a cube, and the 'number that exists apart from perceptible things' of 204ᵇ7 is a 'mathematical number', e.g. the number ten itself rather than—say—ten horses.) The Platonic theory is that these 'intelligible' objects exist independently of their perceptible instances; Aristotle denies this (cf. 193ᵇ31–194ᵃ12), but he continues to describe them as intelligible rather than perceptible.

204ᵇ12 *a finite . . . number of elements*: Aristotle fails here to consider the alternative of an infinite number of elements. But he has argued against this alternative in Ch. 4 of Book I (and in *De Caelo* iii. 4).

204ᵇ23 *as some say*: Anaximander.

205ᵃ1 *apart from any of them being infinite*: Aristotle is assuming the point argued earlier (204ᵇ10–19, 25–9) that if one element is infinite it will destroy all the others, so that the whole universe is composed just of this one element. He here adds that that cannot happen, whether or not the one element is infinite. (But the reason he offers, i.e. that all change is between opposites, hardly establishes his point.)

205ᵃ25 *as I have already explained*: see 204ᵇ10–19.

205ᵃ30 *infinitely many elements*: see note on 204ᵇ12.

205ᵇ34 *objective divisions of the universe*: this is a strange claim. In Aristotle's universe the distinction between up and down is indeed objective, but not that between left and right, or that between forward and back. In the *De Caelo* (285ᵇ15–286ᵃ2, 287ᵇ22–288ᵃ12) Aristotle tries to derive objective meanings for them from another

genuinely objective feature of his universe, namely the direction of
the rotation of the fixed stars. But the attempt is hardly successful.

205^b35 *place cannot be infinite*: the argument that follows seems to
take this premiss in the sense of 'there cannot be an infinite variety
of different places, each the natural place for some variety of body'.

206^a17 *indivisible lines*: for reasons for believing in such lines see Introduction, §14. (At *Metaphysics A*, 992^a20–2, Aristotle ascribes this
belief to Plato.)

206^a20–1 *will . . . will*: we may soften this 'will' to 'may' without affect-
ing the argument.

206^b15 *in the same way that matter does*: see note on 207^a21.

206^b27 *Plato . . . two infinites*: see note on 203^a15.

206^b32 *he has number end at ten*: This must again be a statement about
Plato's 'unwritten teachings', for there is nothing of the kind in his
dialogues. I find it difficult to take the information seriously. (But
the Pythagoreans did have a special reverence for the number ten,
and they apparently thought of all other numbers as repetitions of
the first ten.)

207^a1 *'that which always has something beyond itself'*: there is an ambiguity in the Greek at this point, for the phrase could also be
translated 'that which always has some further part'. The first
meaning is strongly suggested by the contrast that Aristotle draws
here, but the second better fits his account of infinity. The same
ambiguity affects the phrase translated as 'some further part' at
207^a8.

207^a9 *that which has no part missing*: the definition is taken from Plato
(*Theaetetus* 205a, *Parmenides* 137c).

207^a17 *'to join flax to flax'*: a proverbial expression.

207^a21 *the infinite is the matter*: the comparison between the infinite
and matter was introduced at 206^b15, apparently on the ground
that each of them can be regarded as a potentiality which is never
realized. In the case of infinity this is straightforward: it means that
there is no one time at which all the members of an infinite series
exist. In the case of matter Aristotle's thought is, first, that matter
is potentially all kinds of substance, so it never is any one substance
'in its own right'—for the same matter could persist though the
substance was destroyed—and then, second, he ignores the qualification 'in its own right'. So he reaches the absurd position of holding that matter is potentially what it cannot be actually.

In the present passage Aristotle is thinking of the infinite as what is infinitely divisible, and he claims quite reasonably that this is the matter of an object rather than its shape or form. But he still retains the mistaken idea that the matter of an object *is* that object potentially, but *cannot be* that object actually (ᵃ22). He further adds that matter is unknowable, and this idea arises through the same fallacious reasoning: it is true that matter has no form 'in its own right'—for the same matter changes from one form to another—but the qualification 'in its own right' is essential. It is a mistake to say blankly that 'matter has no form' (ᵃ26), and to infer from this that it cannot be known. In any case, what is 'unknowable' about an infinite division specified by an infinite series such as:

$$\left\{ \frac{1}{2}, \frac{1}{4}, \frac{1}{8}, \frac{1}{16}, \ldots \right\}?$$

207ᵃ30 *the great and the small ought to contain intelligible things*: Aristotle is (wrongly) supposing that Plato's great-and-small was supposed to be a version of what *he* calls 'the infinite' (see note on 203ᵃ4), and is noting that on Plato's theory this 'infinite' is defined, limited, and contained by other things, not vice versa.

207ᵇ6 *the number one is indivisible*: standard Greek practice does not admit fractions as numbers. (But see note on 220ᵃ30.)

207ᵇ8 *'three' and 'two' are derivative*: Aristotle's point is that numerals used as nouns derive from numerals used as adjectives, since 'three' as a noun is short for 'three ones'.

207ᵇ15 *the number of time*: this presumably refers to the number of days, or years, since some given date. But if it means the number of days (or years) that there have been altogether, then it points to a problem with Aristotle's account. For already (in his view) there have been infinitely many days, and this appears to be *a kind of* 'completed' infinity.

207ᵇ22 *primary . . . and dependent senses*: in general, Aristotle wishes to derive the properties of time from those of change, and the properties of change from those of magnitude. (Cf. 219ᵃ14–19, 235ᵃ13–ᵇ5.) So here he says that infinity—i.e. infinite divisibility—applies in the first place to magnitudes, and is transferred from there first to change and then to time. But in fact the main thrust of his analysis has been in the opposite direction, for the foundation of his account is that only a *process* can be infinite in the basic sense, and this straightforwardly includes the infinity of time, but not the infinity of any spatial stretch.

207ᵇ25 *later*: the account of what time is will come in Book IV (Chs. 10–14), but the account of what change is has apparently been given already in Chs. 1–3 of this book. Either, then, Aristotle means to refer to his further account of change in Book V, or he is thinking of explaining, not time and change themselves, but what infinity comes to in their case. This is presumably the same as his explanation of why every magnitude is divisible into magnitudes, which is given in Book VI (primarily in Chs. 1–2).

208ᵃ5–23 *Response to the arguments for an actual infinite*: the five responses answer to the five reasons given at 203ᵇ15–30 for supposing that the infinite must exist. In detail: (i) answers 203ᵇ18–20, (ii) answers ᵇ20–2, (iii) answers ᵇ22–30, (iv) answers ᵇ16–17, (v) answers ᵇ17–18.

BOOK IV

208ᵇ14 *not just relative to us*: see note on 205ᵇ34.

208ᵇ22 *Geometrical figures show this*: the shape drawn on paper has a right and a left relative to an observer, and also has a place. It represents a 'perfect' geometrical shape which has no place, but can still be said to have a right and a left, relative to us, which is inherited from the right and left of the drawing. Thus 'right' and 'left' can be applied, relative to us, to objects which certainly do not have a right and a left in nature. Aristotle apparently thinks that 'this shows' that objects which do have a place do have a right and a left in nature.

208ᵇ29 *Hesiod*: the earliest Greek didactic poet (eighth century BC). This quotation is from his *Theogony* (lines 116–17).

209ᵃ5 *by which every body is defined*: it is a standard idea that body may be defined as 'what is extended in three dimensions'. (Cf. 204ᵇ20, *De Caelo* 268ᵃ7.)

209ᵃ11 *we cannot differentiate*: why not? After all, it seems fair to say that the point that is the tip of an arrow is changing its place all the time that the arrow is in flight. Aristotle *may* be relying on an argument in Plato's *Parmenides* (138a3–7) which aimed to show that what has no parts cannot be in contact with anything, or at least not in a way which allows it to be *surrounded* by something. (Cf. also 231ᵃ26–ᵇ4 in Book VI.)

209ᵃ18 *things which are merely intelligible cannot constitute an object with magnitude*: since Aristotle is prepared to count mathematical solids as intelligible (see note on 204ᵇ1), this claim requires support. Aristotle would have done better to speak not of magnitudes but of locations. See also the next note.

209ᵃ19 *not one of the four causes*: a thing's natural place may surely be counted as the end or goal of its natural motion. With that in mind, Aristotle has earlier suggested that a place has a 'power' (208ᵇ10), and elsewhere it leads him to say that the natural place of an element is its 'form' (e.g. *De Caelo* iv. 310ᵃ33).

Neither this argument nor the previous one receives any attention in what follows. The remaining four arguments recur at 212ᵇ22–9, where Aristotle claims that his positive account of place resolves them.

209ᵃ27 *It follows*: from what is this supposed to follow? The thought *seems* to be that if every body is in a place, and if every place contains a body, then the body and its place must fit one another exactly. This is indeed something Aristotle believes (211ᵃ2), though it hardly follows from the premises stated. Then from this it does indeed *seem* to follow that a body's place must expand and move as it does.

209ᵃ31 *sometimes derivatively*: Aristotle's thought is that what he calls the 'immediate' place of a body is described 'directly' as its place, and larger places containing this are also said to be its place, but derivatively.

209ᵇ12 *in the Timaeus*: see *Timaeus* 48e–52d. Plato speaks of space as a 'receptacle' but he does not call it matter. Aristotle, however, thinks that it is fair to call it matter, since it is assigned the same role as *he* assigns to matter, namely to be receptive of form. (There is, however, this obvious difference: Aristotle's matter can move from one place to another; the parts of Plato's space cannot.)

209ᵇ15 *his unwritten doctrines*: on Aristotle's account, in these doctrines Plato assigned to the 'great and small' the same role as he had assigned to space in the *Timaeus* (cf. 209ᵇ33–210ᵃ2). Plato does indeed identify place and space in the *Timaeus* (52a–b); Aristotle apparently implies that he did so also in the unwritten doctrines. It is because of this identification that Aristotle can say that Plato tried to say what *place* is, for he did indeed try to say what *space* is. (For more on this aspect of Plato's unwritten doctrines, see i. 9.)

209ᵇ25 *as we have already said*: 208ᵇ2–8.

210ᵃ10 *air's place is destroyed*: it is not clear whether this argument is supposed to be a further argument against the suggestion that place is either matter or form. If so, then presumably the thought is that when air turns to water its form ceases to exist and its matter shrinks, and Aristotle will be committing *himself* to the view that places can do neither of these. But it appears that on his *own* theory a place can cease to exist. So it is perhaps better to suppose that Aristotle is here reverting to the approach of Ch. 1, and adding one more to the list of six prima-facie problems there given.

210ᵃ19 *any part*: Aristotle has not quite said what he means, which is that any part *of the definition* of the species—either the genus or any of the differentiae—may be said to be 'in' the species.

210ᵃ21 *and, in general, that form inheres in matter*: this fifth sense of 'in' is apparently supposed to cover any case of a property being *in* an object that has it, e.g. as pallor may be said to be *in* a body (or, more strictly, *in* the surface of the body). Cf. 210ᵃ34–ᵇ8, ᵇ25–7.

210ᵃ32 *the jar of wine*: Aristotle's thought is that one can say 'the jar of wine', and mean by this just the wine; one can also say 'the jar of wine', and mean by this just the jar; hence one can truly say 'the jar of wine is in the jar of wine', because one can mean by this that the wine is in the jar.

210ᵇ19 *two things . . . in the same place*: the argument is this. Suppose x is in x. Suppose also that there exists something y other than x, which is in x, *and* is such that x is its *immediate* place—i.e. y fills all the space in x. Then x and y are two different objects in the same place, i.e. in x, which—Aristotle says—is impossible. Notice that one might agree with Aristotle that the conclusion is impossible, but might nevertheless hold that 'x is in x' is sometimes true (e.g. where x is the whole universe), because one rejects (in this case) the auxiliary assumption about y.

210ᵇ25 *not . . . implying location*: suppose that the wine is in the jar. Then Aristotle's thought is that you can also say that the *place* of the wine is in the jar, but that this 'in' does not give the place of the place; rather, it says that the place 'belongs to' the jar in much the same way as a property belongs to an object. (On the theory of place that Aristotle will give in the next chapter, the place of the wine is the inner surface of the jar, and this does 'belong to' the jar, but does not have the jar as its place. For on Aristotle's account it is only bodies, extended in all three spatial dimensions, that have places.)

211ᵃ12 *The first point*: while this 'first point' is no doubt perfectly sound in itself, it is not clear why Aristotle should give it such prominence here, since it will play no part in the positive account that he is working up to. Indeed, it can be made the basis of an *objection* to his account (see Introduction, §10). The thought is similar to that in 208ᵇ2–8.

In fact neither this paragraph at ᵃ12–23 nor the following paragraph at ᵃ23–9 adds any useful new material. The point made at ᵃ29–ᵇ1 is a helpful preliminary, since it puts a restriction on the things that have places. But the main argument does not really begin until ᵇ5.

211ᵃ30 *continuous*: in Book V Aristotle will explain that *x* is continuous with *y* when their limits not only touch but also 'are identical' (227ᵃ10–17). He means by this that the boundary between them is not marked by any physical distinction between what is on one side of it and what is on the other; as we might say, it is a 'merely notional' boundary. So what he is suggesting here is this. Suppose we have a glass of water, and we consider some inner part of that water, say the part which contains all the water half an inch or more from the boundary of the water. Then this part is continuous with what contains it, i.e. the rest of the water, so it has no real boundary of its own. Consequently we do not say that it has a place, for it is merely a part of all the water in the glass. But the whole of that water will have a place.

211ᵇ21 *every bit behaves in the same way*: the passage is obscure, but the meaning *seems* to be that as the water is removed from the jug, the smaller and smaller volumes of water remaining in the jug will each have a place. On Aristotle's own theory each place in this infinite series of ever smaller places will exist only at the instant at which it is occupied. But on the theory he is criticizing, a place is 'capable of independent and permanent existence', so all the places exist simultaneously. We thus get infinitely many places all in the same jug.

211ᵇ23 *place will not in fact be a stable entity*: as the next sentence shows, Aristotle is thinking of what happens when the jug moves as a whole, and he claims that the place of the water in the jug will on this theory be a place that changes place. So in this way too we shall get several places coinciding, for the place of the water, and the place which that place occupies, will coincide. (Aristotle goes on to imply that the truth is that when the jug moves as a whole the water in it stays in the same place. But later he realizes that this will not do. See 212ᵃ14–21.)

211ᵇ25 *any given part is moving*: the situation Aristotle is imagining is one in which (*a*) the jug as a whole is moving from one place to another, *and at the same time* (*b*) the water in it is being poured out, to be replaced by air, or the water in it is rotating, so that one part of it is being replaced by another. His point is that we have to supply a place for this *latter* change to be happening in, and it evidently will not do to take this to be the place that the parts *will* occupy when the jug as a whole has stopped moving.

211ᵇ29 *the place of the whole world*: Aristotle will claim that there is no such thing as the place of the whole world, since nothing surrounds it (212ª31–ᵇ22). *Perhaps* this final sentence, which seems otherwise pointless, is intended to draw to our attention that the theory Aristotle is criticizing *does* assign a place to the whole world. (But, if so, then Aristotle's description of the theory at 211ᵇ7–9 was misleading. See Introduction, pp. xxxviii–xxxix.)

211ᵇ31 *not separate but continuous with its surroundings*: Aristotle has claimed at 211ª29–ᵇ1 that such a thing does not have a place, since it has no real limits. His suggestion here seems to be that someone might insist that the thing ought to have a place even in this case, and he would then find that the thing's matter was the only possible candidate.

212ª26 *the containing limit which lies in the direction of the centre*: by this Aristotle presumably means the limit which is the outer boundary of the earth, and the inner boundary of water and air. If this is right, then *it* is not stable—for I can alter it by picking up a handful of earth—though the centre itself is stable.

By the phrase 'the containing limit which lies in the direction of the periphery' Aristotle may mean either the outer boundary or the inner boundary of the natural place of fire, but it is difficult to see how the phrase could cover both (as, apparently, it should do).

212ª32 *if water were such a body*: that is, if water were a body with no body outside it. (It is not clear what Aristotle thinks is gained by imagining water to be as the whole universe is actually. His attention returns to the whole universe in the next sentence.)

212ᵇ3 *as I explained earlier*: 211ª17–ᵇ1. (Undetached parts of a continuous whole are in place potentially, since they will be in place if they become detached.)

212ᵇ8 *as I have said*: the reference is apparently to 212ª31–2.

212ᵇ22–213ª11 *This account resolves the problems*: the first paragraph responds very briefly to four of the six problems about place raised in Ch. 1 (209ª2–30). The next tries to show that it is reasonable for there to be natural places, which was introduced as a requirement at 211ª4–6. The final paragraph aims to develop this thought further, but in fact it is no longer relevant to the notion of a natural place.

212ᵇ28 *the limit is in the thing which is limited*: see note on 210ᵇ25.

212ᵇ31 *are akin to one another*: Aristotle means that earth and water share the property of being cold, water and air share the property of being wet, and air and fire share the property of being hot. (We may add that earth and fire share the property of being dry, but these are not successive elements, naturally in contact with one another.) He alleges that these 'kinships' make it 'reasonable' for the natural places to be arranged as they are, and for each element to move to, and rest in, its natural place.

212ᵇ31 *If two things have fused together*: this means that there is no real boundary that separates them. Aristotle is perhaps thinking of homogeneous mixtures of the elements.

212ᵇ35 *someone moving a portion of water or air*: Aristotle's point seems to be that if you thrust aside some water (e.g. in swimming) or some air (e.g. in walking), the water or air will as soon as possible refill the space it vacated. This is taken as showing that it had a natural tendency to remain where it was.

213ᵃ5 *later*: the most relevant passages seem to be *De Caelo* iv. 3–5 and *De Generatione et Corruptione* i. 3.

213ᵃ7 *one potentially and the other actually*: Aristotle's meaning is that water is actually matter (the matter from which air will be made), and is potentially 'actuality', i.e. form (the form which air has). His view is that the series 'earth, water, air, fire' progresses from what is more in the nature of matter to what is more in the nature of form.

213ᵃ21 *and thirdly any shared opinions*: the present chapter evidently gives the views of those who claim that there is a void, and of those who claim that there is not. Aristotle is apparently thinking of the beginning of the next chapter (213ᵇ30–214ᵃ16) as giving the 'shared opinions' here referred to.

213ᵃ26 *wineskins*: that is, wineskins inflated with air, which will not collapse however much you jump on them, stretch them on the rack, and so on.

213ᵃ27 *a water-thief*: the instrument is for lifting water from a bowl. It has a number of small holes at the bottom, which water will flow through, and a hole at the top which can be closed by the thumb, thus preventing the flow. (The pipette is our nearest equivalent.) In our phrase, the instrument demonstrates that 'nature abhors a vacuum', and so it is more relevant to the dispute than Aristotle allows.

213ᵇ1 *plenty of other natural scientists*: we do not know of whom (if anyone) Aristotle is thinking.

213^{b}12 *many unequal objects . . . too*: the thought is that a whole brick
can be made to coincide with half a brick if one *first* divides the
whole brick into two halves, and *then* puts all three halves into the
same place. (But note that the first step destroys the whole brick, so
the second step does not get *it* to coincide with the half-brick.)

213^{b}17 *also the wineskins*: the supposed fact is that if you start with a
cask of wine, and draw off all the wine into wineskins, then you can
put back into the cask all the same wine *and* the skins it is now in.
The moral is that the wine must be compressed, so that it takes up
less space, when it is put into skins. We do not know how it came
to be believed that this is a fact.

213^{b}19 *the phenomenon of growth*: the thought is that if food is to be
absorbed by a growing body, then the body must contain empty
spaces which can receive that food. (But the thought is surely illogi-
cal, for if the food absorbed merely fills a space that was hitherto
empty, then the expected result should be an increase in density
with *no* growth.)

213^{b}21 *as much water as the empty vessel can*: again, we do not know
how it came to be believed that this is a fact.

213^{b}22 *The Pythagoreans*: Aristotle is our main source for these rather
quaint theories of the Pythagoreans. (The last sentence of the paragraph presumably connects with the obscure Pythagorean doctrine
that in some way 'all things are numbers'.)

214^{a}13 *some people*: no doubt Aristotle is thinking of Plato. See notes
on 209^{b}12 and ^{b}15.

214^{a}19 *or inseparable*: inseparable void is discussed at 216^{b}33–217^{a}10.
The Introduction offers a conjecture on what Aristotle means (p.
xliii).

214^{a}27 *something Melissus overlooked*: cf. 186^{a}16–18.

214^{b}5 *get in their own way*: Aristotle means that in each case the suggested explanation does not actually work, and he goes on to argue
this for the case of growth. He claims (i) that *all* parts of a body
grow, since the body maintains the same shape, and he means to
include here even the smallest parts. He also claims (ii) that in this
case the growth does take place by absorbing extra matter (in contrast to 214^{b}1–2 just above). And he points out (iii) that the explanation was supposed to avoid having to say that two bodies can
occupy the same place. Thus the first three alternatives that he
offers may be ruled out, and so this explanation is committed to the
fourth alternative, that *all* the body is void! The reason is that *every*
part grows, so *every* part absorbs extra matter, which (on this

theory) means that *every* part, however small, contains a void. But that cannot happen unless every part *is* a void. (I take it that this last step depends upon the point that what we are considering is a 'separate' and not an 'inseparable' void. Even so, the conclusion does not follow if voids may be infinitely many and arbitrarily small.)

To apply the same objection to the case of the ashes absorbing water one would need to be given as a premiss that *every* part of the ash, however small, absorbs some water. It is not clear why Aristotle should think that he is entitled to this premiss.

214^{b}20 *those who take place to be a separate something*: this refers to the theory of place as 'mere extension', discussed at 211^{b}14–29. It is not at once clear why this account of what place is could not accommodate Aristotle's own theory of 'natural' places and 'natural' motions. Cf. note on 214^{b}33.

214^{b}27 *no part will be in a place, but in the whole*: at 211^{a}29–31 Aristotle has claimed that an undetached part, with no real boundary of its own, is not said to be in a place, but rather to be in the whole of which it is a part. Here he apparently implies that those who think of places as 'separate and persisting' would *wish* to say that an undetached part does have its own place, but are prevented (by common usage?) from doing so.

214^{b}31 *because of the equilibrium of things*: this was Anaximander's view, and Plato's too (*Phaedo* 109a, *Timaeus* 62d).

214^{b}33 *the void . . . contains no differentiation*: the implication is that Aristotle's own natural places *do* contain differentiation, and apparently this must be due to the matter in them, i.e. to the fact that they are already occupied by earth or water or air or fire. This makes it seem that his own gravitational theory is really a theory of like attracting like. But in fact that is not his own theory (as is clear from several places, e.g. 205^{b}1–18). In his theory the natural places are differentiated by their situation, and it is clear that void places *could* be differentiated in just the same way.

215^{a}9 *there is no distinction*: I take it that Aristotle must mean: since the atomists' space is infinite, up and down cannot be distinguished by their situation, and since it is void, they cannot be distinguished in any other way either. (Notice that Aristotle has here slipped into arguing against the atomists in particular, since they are the main proponents of the void. But clearly one could believe that space may be empty without also believing it to be infinite.)

215^{a}14 *when things are thrown, they continue to move*: Aristotle thinks that this is possible in air, but would not be possible in a void. The theory of replacement is found in Plato's *Timaeus* (79a–80c); it proposes that the projectile pushes aside the air in front of it, which therefore sweeps round to fill the space behind it, and so pushes the projectile on further. Aristotle gives his own theory in Book VIII, Ch. 10, at 266ᵇ27–267ᵃ20. (Notice how the next two paragraphs show clearly that Aristotle regards inertial motion as an impossibility.)

215ᵃ25 *one moving body*: except for 215ᵃ14–19 Aristotle has been confining his attention to *natural* motion. He is still confining his attention in this way here and in what follows.

215ᵇ15 *cannot be expressed by any ratio*: what follows is not a very good explanation of the (perfectly correct) claim that there is no ratio of 4 to 0. Here is another. If *n* is any positive number, it is easy to see that the ratio of 4 to *n* increases as *n* decreases, and that when *n* = 0 the ratio (if it existed) would have to exceed every finite number.

215ᵇ19 *(unless a line is made up of points)*: Aristotle will argue that this is impossible in Ch. 1 of Book VI (231ᵃ21–ᵇ18); he takes the thesis to imply that a point has non-zero magnitude.

215ᵇ23 *to B and to D*: these letters are to be understood as at 215ᵃ31–ᵇ12, i.e. *B* is water and *D* is air. Similarly *E* is the time *A* takes to traverse *D*.

216ᵃ26 *what void is in its own right*: the arguments from here to the end of the chapter are apparently designed to show that it is a mistake to think of *space* (or place) as existing in its own right, for one might take this view even while conceding that no space is ever empty. Since Aristotle is here using 'void' to mean 'space', he finds no contradiction in the idea of a void fully *occupied* by matter (216ᵇ8–9).

216ᵃ30 *in the direction in which it is its nature to be displaced*: this is not well thought out. If you submerge a cube in water, the water-level will rise. Similarly, if you then transfer it from the water to the air, the air-level (i.e. the water-level) will fall.

216ᵇ10 *why cannot any number of things coincide?*: this is obviously unfair. The fact that a solid body would coincide with a previously empty space has no tendency to show that two solid bodies could coincide with one another.

216ᵇ14 *why . . . conceive of place . . . apart from . . . volume*: it seems very easy to answer this question: a thing may change its place without changing its volume. Cf. 211ᵃ2–3.

216ᵇ26 *Xuthus*: nothing more is known of him.

216ᵇ33 *not separable void*: for the theory of 'inseparable void', see Introduction, p. xliii.

217ᵃ2 *in the same way that wineskins*: Aristotle is apparently thinking of the fact that an inflated wineskin will hold up, in water, whatever is attached to it. Similarly, the void in a thing might be supposed to move upwards itself, and thereby to raise the thing it is in.

217ᵃ4 *void of void*: this appears to be a play on words, since the correct thing to say would be 'full of void', but that sounds like a contradiction.

217ᵃ10 *because the speeds are incomparable*: that is, the upwards speed of pure void would have no ratio to the upwards speed of air or fire, just as the downwards speed of a heavy object *through* the void would have no ratio to its downwards speed through a genuine medium. From this we conclude that in each case there is no such speed, i.e. heavy objects do not move downwards through a void, and similarly, pure void does not move upwards. (The theory of an 'inseparable void', as I understand it, would deny the existence of any 'pure void'.)

217ᵃ21 *considerations . . . already established*: in Ch. 7 of Book I?

217ᵇ20 *whether . . . absolutely separate or . . . contained within rare things*: that is, whether containing bodies (as the space between worlds would), or contained in bodies (in small pockets). These are both kinds of 'separate' void, and are contrasted with 'inseparable' void, which is called in the next line 'potential' void.

217ᵇ27 *in what way there is . . . void*: Aristotle's view is, of course, that in the proper sense of the word there is *no* way in which there is void. But here he is presumably referring to the fact that there genuinely is a cause of rarity, and hence of upward movement, and some people might insist on calling this void (but 'inseparable').

218ᵃ3 *Moreover*: this paragraph simply restates, in more precise terms, the argument of the previous paragraph. It may be noted that Aristotle nowhere offers a response to this argument. (No doubt he could claim that time exists so long as change does—if we set aside 223ᵃ21–9—but the argument could easily be adapted to throw doubt on the existence of change.)

218ᵃ8 *time . . . does not . . . consist of nows*: this will be argued in Book VI, principally at 231ᵃ21–ᵇ18.

218ᵃ9 *whether it always stays the same or whether it is always different*: Aristotle usually uses his invented noun 'a now' simply to mean an instant, i.e. any instant, whether present or not. But this puzzle is specifically a puzzle about the *present* instant. Aristotle will answer (at 219ᵇ9–33) that in *one* sense this is always the same, but in another not. (His answer does not show how he would meet the problem stated at 218ª16–21, which asks *when* an instant ceases to exist. From elsewhere, i.e. *Metaphysics B*, 1002ª28–ᵇ11, one finds that his answer is that there is *no* time when this happens, for at every later instant it has already happened.)

218ª33 *Some say . . . others*: respectively Plato (*Timaeus* 39c–d) and, it is said, Pythagoras.

218ᵇ19 *For the moment let us assume*: Aristotle will distinguish between variation (*kinēsis*) and change (*metabolē*) in Chs. 1–2 of Book V, but for the most part he uses the two Greek words interchangeably, and when that is so, both words are here translated 'change'.

218ᵇ23 *as in the story*: there are different versions of this story, but all involve people sleeping for an inordinately long time.

219ª12 *It is because magnitude is continuous that change is too*: in Ch. 4 of Book VI Aristotle will argue that *every* change is continuous, but elsewhere he appears to recognize some exceptions (viii, 253ᵇ14–26). In any case, the change that he has primarily in mind here is change from one place to another, so that the 'magnitude' in question is distance.

219ª20 *what is before and after is a change*: in this passage Aristotle is apparently construing 'what is before and after' as point-like, i.e. as limits that bound a stretch, for he will say in the next paragraph that in the case of time what is before and after is a now. If that is right, then he is saying here that a stretch of change is bounded by instantaneous states *which are changes*. But it is an objection to this interpretation that in Book VI he will claim that there is no instantaneous state of change (234ª24–ᵇ9, 239ª10–22). Alternatively, his thought may be that all these instantaneous states together constitute whatever change is in question. But it is an objection to this interpretation that in Book VI he will claim that lines are not made up out of points (231ª24–ᵇ18), and the same reasoning would evidently show that changes are not made up out of instantaneous states.

219ᵇ6 *'number' is ambiguous*: for example, the number ten, considered in itself, is 'a number by which we number', whereas ten horses is also a number (i.e. a number of horses), but in this case a number in the sense of 'that which is numbered'. Aristotle's thought, then, is that the number ten is not a time, but ten days is a time.

219^{b}11 *what it is to be the one now is different*: here Aristotle apparently means that if we have two different changes x and y, which
end simultaneously, then what it is to be the first instant when x is
completed is not the same as what it is to be the first instant when
y is completed, though the one instant is the same instant as the
other. But in the next paragraph the 'nows' being considered are not
simultaneous. The sense in which they may nevertheless all be counted
as 'single and identical' is just that each of them is a *present* instant.

219^{b}26 *the now exists in so far as the before and after are numerable*:
as noted earlier (on 219^{a}20) Aristotle apparently construes what is
before and after in change as the instantaneous states of that change.
The correct thing to say, then, would be that a now exists in so far
as these states can be simultaneous, i.e. can fall under the same
instant. It *may* be that that is all that Aristotle means by saying that
they are 'numerable', and implying that they are 'numbered' by the
now (cf. 220^{a}21–4). If so, one can only protest that this is a misuse
of the notion of a number. If not, and Aristotle means something
else, then it is quite unclear what. For a now is not a quantity.
(Note, incidentally, that here it is the now which is always different
that Aristotle calls a 'number', not the now which is always the
same. See next note.)

220^{a}4 *as it were, a unit of number*: in this passage it *appears* to be the
now which is always the same that Aristotle compares to a unit of
number, since this is what corresponds to the moving object, which
exists all through the movement. This comparison is wholly obscure. *Perhaps* Aristotle means to suggest that this now 'measures
out' the time taken by the whole movement, namely by 'travelling
through' that whole time, and so it can be compared to a unit which
'measures out' a number of units by repeating itself so many times.

220^{a}7 *the change and the movement are unities*: in Ch. 4 of Book V
Aristotle sets out the conditions for a movement to be a unity. The
condition that the moving object be the same object throughout is
only one of those conditions.

220^{a}10 *by holding it together*: note that the Greek word for 'continuous' literally means 'held together'. Aristotle *may* be thinking of a
line as generated by a moving point, and held together in this way,
as a movement is generated by a moving body, and a time is generated by a 'moving now'. If so, one must complain that the case
of the line and the point is supposed to be one *not* involving any
movement. Alternatively he *may* be thinking that the point 'holds
together' a continuous line because the two halves of the line share
a single point as their common limit. This is what is demanded by

the definition of 'continuous' that is used at the start of Book VI.
But in that case there is no real parallel between the way that a line
is continuous and the way that a movement is a unity. Compare Ch.
13, 222ª10–20, which does not resolve the uncertainty.

220ª13 *there must be a pause*: this is argued in Book VIII, at 262ª12–
264ᵇ9. The implication of the next sentence is that one cannot
pause at a now, so a now cannot really be treated as two, and so
it does not really divide time. Nevertheless Aristotle still wishes to
say that it divides time *potentially*. Cf. 222ª10–14.

220ª14 *So time is a number*: this sentence explaining how time is and
is not a number makes no sense to me, and I suspect that something
has gone wrong with the text. The second part of the sentence, after
the semicolon, seems to be concerned with a way in which *the now*
is not a number, i.e. not in the way that the parts of a line may be
the number of the line (as when the line has ten parts, each 1 inch
long). *Perhaps* the whole sentence, as Aristotle first wrote it, referred not to time but to the now.

220ª22 *but in so far as it numbers*: the text here is insecure. With the
reading we adopt, the meaning *may* be that the *moving* now, which
has been compared to a unit of number at 220ª4, is a time (i.e. a
period of time). Or it *may* be that the now as a particular instant
is something temporal, not because it is itself a limit, but because
it is 'a number' in the sense of a universal, applying to all simultaneous limits. The continuation strongly supports the second suggestion. But it can be objected that Aristotle never says elsewhere that
the now *is* (a) time, since he very consistently uses 'a time' to mean
a period of time, and not an instant.

(The reading adopted has the authority of Philoponus, but almost
all other sources have the banal reading 'but in so far as it numbers
it is a number'.)

220ª30 *there is no smallest number*: Aristotle is here identifying a magnitude, such as length, with 'a particular kind of number' (ª27). His
usage is distinctly unexpected, since the Greeks did not standardly
recognize any numbers other than whole numbers (usually counting
two as the smallest, but occasionally one; cf. 206ᵇ31–2, 207ᵇ1–11).
Even if one takes into account his distinction between 'a number
without qualification' and 'a particular kind of number', still it is
puzzling that he should think of a length as a number. For we
certainly cannot suppose that he thought that every length could be
assigned a number, e.g. as a measure of the ratio between it and
some unit length. On the contrary, he was well aware that some
lengths will be incommensurable with whatever is chosen as the unit

length, which means that the ratio in question cannot be described by means of any numbers that the Greeks knew of.

220^{b}4 *the numbers by which we number*: the context clearly implies that time *is* such a number, in contradiction to 219^{b}5–7 above and 220^{b}8–9 below. Is this perhaps a slip of the pen?

220^{b}14 *the same time*: it is odd that Aristotle does not explicitly note that this use of the phrase is a different use from that discussed just above, where sameness implies simultaneity.

221^{b}13 *as I explained earlier*: 202^{a}3–5.

222^{a}10 *as I have said*: 220^{a}4–5.

222^{a}23 *the Flood*: standard Greek myth, like Hebrew myth, places the Flood in the past. Aristotle, who believed in the eternity of the universe, and in periods of scientific discovery followed by a loss of knowledge, perhaps supposes that the Flood recurs time and again. (But his *Meteorologica* i. 14 presents a more sophisticated view.)

222^{a}29 *all time will be finite*: Aristotle ignores the reason given for this suggestion, since it is clear that it has no force. He replies (i) that time is infinite because change is (^{a}29–30), and (ii) that in any case the nature of the now ensures that time cannot begin or end (^{a}33–^{b}7). His reasons for claiming that change is infinite are given in Ch. 1 of Book VIII, and one such reason is based upon the second argument given here. (As for the suggestion that 'the same time might recur again and again', Aristotle's brief reply here is somewhat enigmatic, but presumably he thought it impossible. Cf. 220^{b}6–8.)

222^{b}18 *Paron the Pythagorean*: nothing more is known of him.

222^{b}21 *earlier*: 221^{a}30–^{b}2.

222^{b}24 *without changing even in the slightest*: presumably Aristotle is thinking of death by old age, but it is a surprising statement.

223^{a}24 *number is . . . that which is numerable*: Aristotle here appears to deny the existence of a 'number by which we number' (219^{b}5–7), or a 'number without qualification' (220^{a}27), i.e. a number considered in itself, in abstraction from its concrete instances. We can offer two explanations: (*a*) that he has already said that time is not such an abstract number, so such numbers can be ignored here; (*b*) that an abstract number has no existence of its own, but exists only in its concrete instances; so if they do not exist, then neither does it.

223^{b}15 *as we said*: at 220^{b}22–4.

223^{b}21 *there is uniform movement*: in Ch. 7–8 of Book VIII Aristotle will argue at length that circular movement is the primary kind of change, and the only change that can be uniform. He simply

assumes that there is such a uniform change, namely the rotation of
the heavens. (And no one saw any reason to challenge this assumption until it was noticed that the rotation of the sun (i.e. the solar
day) was not uniform with respect to the rotation of the fixed stars
(i.e. the sidereal day).) Since, of course, Aristotle was not familiar
with modern clocks, it is hardly surprising that he does not notice
that for the measurement of time what is required is a regular
periodic change (e.g. the swings of a pendulum), and it does not
actually matter whether the movement within each cycle is a uniform movement.

BOOK V

224ᵃ21 *coincidental change*: as with causes, so also with changes, Aristotle holds that there is some one preferred way of describing the
agent of change, the object changed, and the initial and final states
of the change. The preferred description is said to describe 'a change
in its own right', and other descriptions are said to describe 'a
coincidental change'. Again, as with causes, Aristotle gives no cri-
teria for the preferred description, but we get some idea of what he
has in mind by attending to his examples.

224ᵇ5 *form . . . does not cause change*: presumably what Aristotle means
is that form is not an *agent* of change; the Greek word here translated 'cause change' could also be rendered 'set a change in motion'.
Even so, this claim is not too easy to reconcile with the claim of
Book II that form should be counted as efficient cause and formal
cause and (in particular) final cause of many natural changes
(198ᵃ24). See also note on 224ᵇ11.

224ᵇ10 *already*: in Chs. 1–2 of Book III.

224ᵇ11 *forms or affections or place*: probably 'form' is here to be understood as limited to 'substantial forms', i.e. to predicates in the
category of substance. Also 'affection' is used here for qualities, and
quantities are omitted from the list. (Similarly at ᵇ5 above, where
qualities are omitted from the list.) But at ᵇ25 below, 'form' is
apparently used to cover all four of the categories in which change
takes place, as in Ch. 7 of Book I. (Cf. 225ᵇ24.)

224ᵇ14 *affections*: as in English, the Greek word 'affection' can be used
both for the process of being affected by something and for the
product of that process.

224ᵇ18 *change to an object of thought*: suppose that I bleach the shirt so that it becomes white, and meanwhile you are thinking of the colour white. Of course it does not follow that the shirt comes to be thought of (by you), but it may be said to follow that the shirt comes to be something (namely white) that is being thought of (by you).

224ᵇ29 *opposites or intermediates, or in contradiction*: in Ch. 5 of Book I Aristotle argued that all change is between opposites or intermediates, but in Ch. 7 this was subsumed under the more general claim that all change is between a form and its privation. This included generation and destruction, for in such changes the *matter* was held to persist throughout, now with the form and now with the privation. But here Aristotle will not draw attention to anything that persists through generation and destruction (save possibly at 225ᵃ27), and he will say that this case, and this case only, is covered by 'contradiction' but not by 'opposites or intermediates'. (See also note on 225ᵇ3.)

225ᵃ1 *the word itself*: the word 'change' (*metabolē*) is a compound word, with the word 'after' (*meta*) as its first component.

225ᵃ10 *change from a non-entity to a non-entity is impossible*: what changes from weighing less than a pound to weighing more than a pound must also change from weighing neither 1 pound nor more to weighing neither 1 pound nor less, but no doubt Aristotle would dismiss the latter description as merely 'coincidental'.

225ᵃ20 *'Not being' is ambiguous*: in Greek 'is not' can be used to mean 'is not the case', or 'does not exist', or (in context) 'is not so-and-so, e.g. pale'. These are Aristotle's three examples. He claims that what 'is not' in either of the first two ways cannot vary, but his real concern here is just with what does not exist.

225ᵃ27 *it is coincidentally coming to be*: this is an obscure phrase. *Perhaps* Aristotle means, in accordance with the doctrine of Ch. 7 of Book I, that there is always a kind of coincidence involved when anything comes to be. For example, when a statue comes to be, that is because some bronze has become statue-shaped, and then this bronze *is* the statue, but only coincidentally. But one could wish that he had been more explicit.

225ᵃ29 *cannot be at rest either*: when the two standard words for change in Greek are distinguished as 'change' (*metabolē*) and 'variation' (*kinēsis*), it is *the latter* that contrasts with 'rest' (*ēremia*). So, on Aristotle's account, to be at rest is to be capable of varying but not actually varying. Hence he infers that since what does not exist is not capable of varying, it cannot be said to be at rest either (cf. 261^{b}11–12).

225^{b}3 *a privation*: of the examples that follow, it would seem that 'naked' and 'toothless' are examples of opposites that are privations, whereas 'dark' is an example of an opposite that is not a privation. Aristotle claims here that words for privations are positive. (The word for 'toothless' does not have a negative prefix or suffix in Greek.) This may be contrasted with his practice in Ch. 7 of Book I, where the privation is always expressed by a negative term, and with the way that he goes on to identify the privation with not being in Chs. 8 and 9 (191^{b}15–16, 192^{a}3–6).

225^{b}5 *The different kinds of variation*: this heading is inserted where the sense demands it. But the MSS mark the beginning of Ch. 2 as occurring two sentences later, at 225^{b}10.

225^{b}7 *action or affection*: note that time is omitted from this list. (See Introduction, p. xlvii.) Note also that the phrase 'action or affection' seems here to be treated as the name of a single category.

225^{b}31 *namely, becoming healthy*: the argument is that if a person undergoing the change from being healthy to being ill is to change from undergoing that change to undergoing another, then he must change to the opposite change, i.e. the change from being ill to being healthy. Aristotle apparently thinks that this is contradictory, for the first change will lead to his being ill *at the same time as* the second makes him healthy. But the truth seems to be that we have here a perfectly coherent case of one who is getting an illness, but this process is then reversed (e.g. by a dose of medicine), so he never does go down with the disease but ends up as healthy as at first.

225^{b}32 *a change from remembering to forgetting*: Aristotle writes as though he supposes that remembering and forgetting are themselves changes (rather than states). This is strange.

226^{a}6 *Thirdly*: the argument is this. Suppose that from t_0 to t_1 a thing x is coming to be a thing that comes to be, and that from t_1 to t_2 it is coming to be. The premiss is that a process of coming to be must be answered by a corresponding process of ceasing to be, so there must also be a period when x is ceasing to be a thing that comes to be. But when could this be? Not before t_1, for x does not become a thing that comes to be until t_1; not between t_1 and t_2, for at those times x *is* a thing coming to be, and so cannot also be ceasing to be such a thing; and finally not after t_2, for by then x *has* ceased to be a thing coming to be, and so cannot *be* ceasing to be such. (Compare the puzzle at 218^{a}16–21 on when an instant of time

ceases to be.) However, there does not seem to be anything wrong
with the reply that x is ceasing to be a thing that is coming to be
during the later part of the period from t_1 to t_2.

226^{a}10 *Fourthly*: by the doctrine of Ch. 7 of Book I, if anything comes
into being there must be some matter that is first in one state and
then in another, and the matter in its final state constitutes the thing
that comes to be. Aristotle asks what this matter could be, and what
its final state is, when what comes into being is itself a change or
a coming to be. (The plural 'they change to' at ^{a}13 shows that he
thinks that the matter would have to be different in the two cases
under discussion, i.e. the coming into being of (i) a change (i.e.
variation) and (ii) a coming into being. The final sentence of the
paragraph assumes that the matter, i.e. 'the underlying thing', would
itself have to be a change (i.e. variation) of some kind. It is not clear
how this assumption is justified.)

226^{a}25 *each of these categories admits opposition*: some qualities have
opposites (e.g. pale and dark) and some do not (e.g. round, square,
triangular), but any quality can be the end-point of a change. No
specific quantity (such as 3 feet long) has an opposite, but again any
such quantity can be the end-point of a change. (The 'opposition'
that Aristotle sees in this category is that of 'more and less'.) Similarly, no specific place is opposite to any other (but Aristotle is
probably thinking of the opposition between up and down).

226^{a}29 *an affective quality*: affective qualities are defined at *Categories*
9^{a}28–10^{a}10 as a special class of qualities (cf. note on 244^{b}5), but
here Aristotle presumably means the term to include *all* qualities
except those that are essential to the subject in question. (Loss of an
essential quality is destruction.)

226^{a}33 *despite the fact that*: the word here used for 'movement' is more
literally 'being carried', and is standardly used of the movements of
non-living things. It would not naturally be used of, e.g., walking or
running or jumping.

226^{b}22 *when they coincide in a single immediate place*: that is, when
they completely occupy exactly the same place. Aristotle will apply
this notion not to three-dimensional bodies but to limits, e.g. points
and surfaces, and two limits which touch do occupy exactly the
same place where they touch. (But one may observe that Aristotle's
account of place in Book IV does not allow such a limit to have a
place.)

226^{b}27 *change is continuous*: what follows is not the definition of continuity promised at the beginning of the chapter; that is coming later

at 227^{a}10, and it will explain what it is for one object to be continuous with another. Nor is it Aristotle's usual account of what makes a change continuous, for that does require that there be no gaps in the time (as e.g. in the next chapter; see note on 228^{a}20). Rather, this is an *ad hoc* meaning for 'continuous', adopted simply to clarify the idea of one thing being between others. The thought is that *x* is between *y* and *z* if a natural change from *y* to *z* would go through *x*.

227^{a}16 *what makes them continuous*: more literally, what holds them together, since the word 'continuous' literally means 'held together' in Greek. (Cf. note on 220^{a}10.) Aristotle is thinking that if the limits of *x* and *y* 'are identical' where they touch, then the whole which has *x* and *y* as parts will move as a piece. But (*a*) I should not have thought that nailing or gluing two objects together would count as making their limits 'identical', and (*b*) the top half and the bottom half of the water in the jug surely do have limits that 'are one', but they do not move as a piece.

227^{a}18 *contact . . . implies successiveness*: this is a mistake, on the existing definitions, for Aristotle has (very naturally) defined 'successive' in such a way that if *x* succeeds *y* then *x* must come *after y* in some suitable ordering. But contact does not require any ordering from before to after.

227^{a}29 *contact is a property of points*: Aristotle cannot mean that one point can touch another; he always (and correctly) denies this. I take it that he must mean, somewhat loosely, that points are involved in contact, for one thing will touch another *at* one or more points.

227^{b}19 *have we already decided*: the correct answer seems to be 'No', but perhaps Aristotle thinks that at 227^{b}7 the phrase 'in the same indivisible species' should be taken as covering 'what the change is in' as used here, i.e. the path of the change (straight or circular) and the manner of the change (by walking or by rolling).

228^{a}6 *the same alteration, but not a single alteration*: that is, the same in species but not in number.

228^{a}8 *one in substance*: Aristotle usually uses this phrase to mean 'one in definition' (and it is so translated in the other books of the *Physics*), but here it apparently means 'one in number'. Cf. 228^{b}13.

228^{a}12 *this much difference between them*: the text of the rest of this sentence is very uncertain; the reading adopted here is designed to give a reasonable argument, namely this. Suppose one says that my state of health this morning cannot be the same state as my state of health this afternoon, because they are not states of the same thing, on the ground that they are states of the body and the body *is* 'changing and in flux' (so it is 'in this way'—i.e. because of the flux—'numerically two'). Then it will follow by the same argument that my activity of walking this morning is equally not the same activity as my activity of walking this afternoon. On the other hand, if we insist that the body *is* still the same body (despite its flux), then we can maintain that there is only one state of health, existing both in the morning and in the afternoon, while still insisting that there are two distinct activities of walking, one in the morning and a different one in the afternoon. This is because the two activities are separated by a temporal gap, during which there is no walking.

It may be noted that Aristotle here equates activities with changes (such as alterations, ^{a}3–6). But elsewhere he distinguishes them, and counts walking as a change but *not* an activity. (The *locus classicus* is *Metaphysics* Θ6, 1048^{b}18–35.)

228^{a}20 *Every change is continuous*: elsewhere Aristotle defines a continuous thing as one that is divisible (only) into parts that are themselves further divisible, and hence as something that is infinitely divisible (*De Caelo* 268^{a}6–7; cf. *Physics* vi. 232^{b}24–5). That appears to be the definition he is relying on here, when he says: 'since every change is divisible [sc. into smaller *changes*]'. The definition, however, leaves out a point that he relies on in the opening argument of Book VI, and that he evidently regards as crucial, namely that the parts into which a continuous thing is divided must share their limits. I suspect that he would welcome a definition such as this. A continuous change occupies a continuous period of time, i.e. one with no gaps; moreover, at every instant during that period the changing object is in a state that is different from its states at suitably nearby instants (i.e. it is not at rest at any instant in the period); and further, its state at one instant of the period differs by as little as you please from its states at suitably nearby instants (i.e. it never 'leaps' instantaneously from one state to a different state).

228^{b}12 *generically, specifically, or in substance*: see note on 228^{a}8.

228^{b}30 *It follows*: because the difference in heaviness between one piece of earth and another is the difference in the speeds of their natural motion.

229^{a}4 *and therefore*: the argument is that a change which combines changes of two different species, one after the other, *cannot* proceed uniformly, but every single change *can* proceed uniformly (even if it does not always do so).

229^{a}22 *in a moment*: at ^{a}27 ff.

229^{b}10 *changes, but not variations*: at 225^{a}14–15 Aristotle had suggested that a change from not being pale to being pale could be viewed as a kind of coming to be, i.e. a coming to be of pallor, but presumably this can *also* be viewed as a variation in the underlying thing which is first not pale and later pale. Here he perhaps has in mind not something which turns pale but something which comes into existence as a pale thing (e.g. a growing mushroom).

229^{b}29 *two entities*: this apparently means 'two places'. Cf. 225^{a}3–7.

230^{a}10 *no such thing as rest*: recall that for Aristotle rest is opposed to variation, and variation does not include coming to be or ceasing to be. Cf. note on 225^{a}29.

230^{b}3 *even when it is not compressed?*: Ross's explanation is that corn will grow and ripen even when it is not packed down in the earth (but—e.g.—grown on blotting paper?). This strikes me as improbable, but I have no better explanation to offer.

230^{b}30 *it seems to retain what is being left behind*: Aristotle is thinking of an extended thing moving from its place, and noting that some part of the thing will still be in the place even after the motion has started (cf. 234^{b}10–17, 240^{b}20–31). He goes on to suggest (implausibly) that *this* part may be said to be at rest.

BOOK VI

231^{a}21 *our earlier definitions*: in Ch. 3 of Book V. But note that what Aristotle defined there was: '*x* is continuous with *y*', and not: '*x* is a continuum', which is what the argument needs here. See Introduction, §14.

231^{b}2 *contact is always between wholes*: this means that the whole of the one thing is in contact with the other, which can only be the case if the thing itself *is* a limit. So Aristotle here allows that one point may be said to touch another (e.g. a point on the surface of one sphere will touch a point on the surface of another, where the two spheres meet), but he observes that this means that the two points occupy exactly the same place; they do not together make up anything that can be divided into parts occupying different places.

231^{b}9 *always a line between points*: Aristotle leaves it to us to add: and on every line there is a point, so there is always a point between any two points. (Similarly for nows.)

231^{b}12 *between the parts of a continuum*: the suggestion is that a continuum might be made up of two sorts of parts, namely: (i) points,

273

and (ii) something else joining the points together. But Aristotle points out that this something else can only be the continuum itself (i.e. what joins two points is itself a line). (It may be noted that his argument here assumes that what is infinitely divisible will also satisfy the original criterion for being a continuum, namely of having parts that share limits. The same assumption underlies all the rest of the first two chapters.)

232ᵃ1 *a distance A which is not divisible into parts*: the argument is this. Suppose that the leading edge of X moves from one side of A to the other. Then there must be a time when that leading edge *is crossing A*, and at that time X will occupy part, but only part, of A. But this is impossible if A has no parts. Hence, if A has no parts, there can be no time when something *is moving* across A, but only times when it *has moved* across A.

232ᵃ23–233ᵇ32 *Further proofs that distance and time are continua*: the structure of this chapter is confusing. At 232ᵃ23–7 Aristotle claims that since every (spatial) magnitude is a continuum (i.e. infinitely divisible, cf. note on 231ᵇ12) the relation 'faster than' will have certain properties. Then at 232ᵃ27–ᵇ20 he gives a deduction of these properties, concentrating in particular on the claim that the faster object covers an equal distance in a shorter time. He apparently takes this to follow from the premiss that any distance is divisible, since this premiss is used at ᵇ2 to obtain the distance CF. (But he also offers a second argument at ᵇ14–20, which does not use this premiss.) Using this result reached about faster objects, and using also the further premiss that for any movement there is a faster movement, he then offers to prove that time, as well as distance, must be infinitely divisible (232ᵇ23–4). The proof that follows (232ᵇ20–233ᵃ12) does incidentally demonstrate the infinite divisibility of both time and distance. Similarly the next paragraph (233ᵃ13–21) promises to prove the divisibility of distance from the divisibility of time, but actually offers a proof that if either is divisible, so is the other. This leads to a digression on infinite divisibility and Zeno's argument (233ᵃ21–31), and a further digression on the idea that the time taken by a movement, or the distance covered, might be infinite in extent (233ᵃ31–ᵇ15). Finally Aristotle returns to his main topic at 223ᵇ15–31, offering another argument which shows equally that time is infinitely divisible, and that distance is too, this time using the premiss that the speed of one body may be to the speed of another in the ratio of 3 to 2.

The real premisses to these arguments, hardly acknowledged as such by Aristotle, are: (i) that some objects move faster than others (and in particular, for the last argument, that the ratio of their speeds may be as 3 to 2); and (ii) that *motion* is continuous. For if the latter is denied then all the arguments collapse.

232^{a}25 *it necessarily follows*: one might well take it to be obvious anyway that if x is moving faster than y, and if y moves a distance d in a time t, then (i) x moves a distance greater than d in the time t, and (ii) x moves the distance d in a time less than t. But Aristotle thinks that the premiss of divisibility is needed to ensure his third condition, namely (iii) that there is some distance d^+ greater than d, and some time t^- less than t, such that x moves the distance d^+ in the time t^-. He proceeds to deduce (iii) from (i), assuming the divisibility of distance, and then (ii) from (iii).

232^{b}9 *GH, say ... GI, say*: GH and GI may be taken to be the distances CF and CE of the previous paragraph. Similarly the times PQ and X introduced shortly may be identified with the times MO and MN mentioned earlier. It is not clear why Aristotle thought that a new set of letters would facilitate comprehension.

232^{b}19 *also*: that is, in addition to traversing a greater distance in an equal time, which Aristotle takes as the definition of 'faster'.

232^{b}26 *as we have demonstrated*: by the two proofs at ^{b}5–14 and ^{b}14–20.

233^{a}18 *if time is infinite in extent, distance will be too*: this claim is unexpected, since Aristotle's own position is that time is infinite in extent but distance is not. The explanation is that he is thinking here of the time taken by an object moving at a uniform speed in a straight line, as the following arguments show (^{a}31–^{b}15).

233^{b}15 *These arguments*: this refers to the arguments of Ch. 1 (in particular, 231^{b}21–232^{a}17), and the arguments in this chapter (in particular 233^{a}13–21) which *precede* the digression on infinity at 233^{a}21–^{b}15.

233^{b}33 *The now ... in the primary sense*: that is, construed strictly as the division between the past and the future. Aristotle realizes that the word 'now' is in practice used more loosely; see 222^{a}20–4.

234^{a}12 *there will be something of the past in the future*: the hypothesis is that the now is a stretch of time bounded at one end by the limit of the past, and at the other end by the limit of the future. Aristotle points out that any stretch of time is divisible, and he assumes that a division in this stretch will itself have what is past on one side of it and what is future on the other. This would entitle him to conclude that there will be a part of the past outside the supposed limit of the past with which we began; but he seems to be going too far when he infers that this part will be in the future.

234ᵃ18 *not always be the same now*: if one tries to posit an atomic now, during which nothing changes, Aristotle objects that it will be divisible into a past part and a future part, and that this division will not always be made in the same place; on the contrary, it will move from one end of the 'atomic now' to the other. But then something *is* changing during that supposed 'now', so it cannot after all be one and the same now all the time.

234ᵃ24 *nothing moves in the now*: the Greek words here translated are ambiguous. They may mean (i) 'during a now, nothing moves, i.e. gets from one place to another', and they may mean (ii) 'at a now, nothing is moving, i.e. is in motion'. Aristotle's first argument (ᵃ24–31) argues for sense (i); his second (ᵃ31–4) applies the point to rest, and apparently it slips into sense (ii); at any rate the third (ᵃ34–ᵇ5) is certainly an argument for (ii), and the same seems to be true of the fourth (ᵇ5–7).

234ᵇ4 *will simultaneously be at rest and in motion*: the argument is that the instant which divides a period of rest from a period of motion is *in* both periods. So if it is possible for a thing to be at rest at an instant, and to be moving at an instant, then at this dividing instant it must be in both states. But that is impossible.

234ᵇ17 *the first end-point*: the assumption that there always is a *first* end-point should be noted. The assumption must be false, if the change in question is genuinely continuous (as motion is), and this observation destroys the argument. Cf. 236ᵇ8–18, 237ᵃ28–34.

234ᵇ21 *two ways in which change is divisible*: the assumption that every change takes time, implicit in the previous paragraph, is here made explicit as the 'first way'. The result of the previous paragraph then yields a 'second way'. To support this latter claim, Aristotle argues at length (234ᵇ23–235ᵃ8) that the change of the whole is the sum of the changes of the parts. From the length of his argument, one would suppose that he must have regarded this thesis as controversial, but it is difficult to see why.

234ᵇ33 *we found*: in Ch. 4 of Book V.

235ᵃ15 *the time, the change, the changing, the changing object, and the respect*: we cannot tell what distinction Aristotle may have intended between 'the change' and 'the changing'. In what follows he argues for corresponding divisions of the time and the change at ᵃ18–25, and of the change and the changing at ᵃ25–34, and at ᵃ34 he remarks that the same argument could be applied also to the stretch covered by the change. But he does not explain how, when someone walks from Athens to Thebes, there will be *corresponding* divisions of the time taken and the person walking. (His final remark at ^{a}36, that the divisibility of any one of the things mentioned will imply the divisibility of all the others, is evidently an exaggeration.)

235^{b}1 *we have found*: there appears to be no place where it has already been argued that the infinite divisibility of the changing object is the foundation for that of other things. (But it has been said in Ch. 11 of Book IV that the properties of time depend on those of change, and the properties of change on those of distance.) It should be noted that 'infinite' in this passage means 'infinitely divisible'; the next two chapters do focus on this notion.

236^{a}17 *it would follow that the nows are consecutive*: Aristotle assumes that there is a last instant before the change begins, so if AD were a first instant at which the change has begun, these two instants would be next to one another. But that is impossible.

236^{a}18 *it is at rest at A*: Aristotle here allows himself to speak of something being at rest at an instant, contrary to 234^{a}31–^{b}9. This may be just a slip on his part, or it may be that he thinks that his opponent in this passage, i.e. the time-atomist, would have to permit rest at an instant.

236^{a}29 *we have proved*: at 234^{b}10–20.

236^{b}12 *something else which is indivisible*: Aristotle is assuming that AB and BC are equal and consecutive stretches of the distance, so that if one is indivisible then both are.

236^{b}24 *how we have been defining 'immediate'*: see 235^{b}33–4.

236^{b}33 *must have changed earlier*: earlier than what? Aristotle's intention must be that for any instant at which a thing is changing there is an earlier instant at which it has completed some part of that change. But it is quite difficult to express this point without using the idea of an instant at which a thing is changing, and Aristotle holds that that idea makes no sense (234^{a}24–^{b}9, 237^{a}14).

237^{a}24 *we have already shown*: at 235^{b}6–13.

237^{a}24 *the two nows*: which two nows? One of them is clearly the first instant at which our object is in state *B*; the other can only be taken to be the last instant at which it was in state *A*, and Aristotle must be assuming that there is such a last instant.

237^{a}32 *something without parts is consecutive*: in the present context this claim is unexplained. Aristotle is apparently repeating it from

236ᵇ12, where he had made the same point, but in a way which did explain the phrase.

237ᵇ2 *all we have to do is to take the time*: note that this assumes that every change does take time, as was argued at ᵃ20–5.

237ᵇ11 *in the case of things which are divisible and continuous*: Aristotle has argued at 234ᵇ10–20 that *every* changing thing is divisible, and hence continuous; but here he gives a slight hint that there may be exceptions when the change is a coming to be or ceasing to be. Some exceptions are indeed noted elsewhere (e.g. viii, 258ᵇ16–20; *De Caelo* 280ᵇ25–8; and frequently in the *Metaphysics*, e.g. 1002ᵃ28–ᵇ11, 1039ᵇ20–7, 1044ᵇ21–9, 1070ᵃ15–17).

238ᵃ9,11 *traversing the whole distance takes infinite time*: Aristotle takes it to follow from this that traversing any part of the distance less than the whole takes less than an infinite time, and hence takes a finite time. This is the basic mistake in his argument. (Let the whole journey be divided into infinitely many parts, as in Zeno's dichotomy (Introduction, §15). Suppose that I undertake the journey, but my speed constantly diminishes, so that I cover the first part in one day, the second part in two days, and in general the nth part in n days. Suppose also that I never give up. Then (i) I shall need an infinite amount of time to complete this finite journey, and (ii) for *any* way of dividing the journey into two parts I shall complete the earlier part in a finite time, but will need an infinite time to complete the later part.)

238ᵃ12 *it is impossible for infinity to consist of finite components*: it is true that infinity cannot consist of *finitely many* finite components, but it can consist of infinitely many finite components. (To illustrate by continuing the example of the previous note, the sum $1 + 2 + 3 + \ldots + n + \ldots$ is infinite, and it is made up of infinitely many components, each of which is finite.)

238ᵃ26 *as long as it is finite*: the basic fallacy is the same as before: Aristotle assumes that, in a time less than the whole time, the moving body must move a distance less than the whole distance, and therefore a finite distance.

238ᵇ13 *something infinite will not traverse something infinite*: presumably Aristotle is thinking of an example such as this. Take a straight line infinite in both directions, and divide it into a left part x and a right part y. Then 'something infinite will have traversed something infinite' if x traverses y, so that all of x ends up to the right of all of y.

238ᵇ23–239ᵇ4 *There is no last time of coming to rest, and no first time of being at rest*: this title does not use Aristotle's own terminology, for his claim is that in each case 'there is no immediate time', but I take it that the title gives his reason for claiming this. (See next note.) The argument about 'coming to rest/to a standstill' would apply also to coming to complete any change whatever, and it *may* be that Aristotle intended it thus generally. But the argument about being at rest relies on the premiss that nothing can be at rest at an instant, whereas Aristotle claims that there is always a first instant of being in the end-state of any change (235ᵇ6–236ᵃ7), so he could not apply his argument here to having completed any change.

The title does not cover the last paragraph of the chapter, which concerns being opposite to something for a time or at an instant. It appears that this is designed to lead up to the criticism of Zeno's paradox of the flying arrow, which opens the next chapter.

239ᵃ2 *there is no first stage*: I take it that what Aristotle means to say is that there is no first stage of changing and no *last* stage of coming to a standstill, for this is the respect in which the two are analogous.

239ᵃ5 *we have shown*: at 238ᵇ23–6.

239ᵃ7 *we have already seen*: at 238ᵇ31–6.

239ᵃ12 *because*: the arguments that follow recapitulate 234ᵃ24–ᵇ9.

239ᵇ13 *earlier*: at 233ᵃ21–31.

239ᵇ30 *a short while ago*: at ᵇ5–9.

239ᵇ33 *his fourth argument*: see Introduction, §15.

240ᵃ28 *without being entirely in either*: Aristotle's reply amounts to this. Any change has a preferred description (cf. note on 224ᵃ21) in which it is described, not as a change between contradictory states (being *A* and being not-*A*), but as a change between opposites (being entirely *A* and being entirely not-*A*) which leave room for intermediate states. This allows us to continue to maintain that every change takes time, without denying the law of excluded middle, that at any instant a thing is either *A* or not-*A*. (The explanation includes the case where 'being *A*' = 'existing', but is not confined to that case.)

240ᵇ8 *the next thing to prove*: a (fallacious) proof has already been given at 234ᵇ10–20. The present attempt at a proof is rather more careful, in so far as it is (supposed to be) restricted to things that change 'in their own right', i.e. not because they are carried along by something else which is changing. (But it will be observed that the argument actually given at ᵇ17–30 makes no use of this restriction, and is essentially the same as that given earlier.)

240^b20 *from AB to BC*: clearly Aristotle is mainly thinking of the movement of an allegedly indivisible atom, which first occupies a length *AB* and then moves so as to occupy the adjacent length *BC*. His argument is rather less convincing if *AB* and *BC* are taken as adjacent 'forms', e.g. colours; it carries no conviction at all if they are taken as genuinely contradictory states, such that at all times a thing must be in one or other of those states. (It may be noted that change of quantity—e.g. of size—is here omitted, perhaps because in this case one cannot speak of 'adjacent' states.)

240^b25 *we found*: this is apparently a reference back to the earlier proof at 234^b10–20.

240^b32 *only . . . if time consisted of nows*: this means: only if time is atomic, i.e. consists of a succession of atoms, each having duration, but such that nothing changes within an atom. It is true that if that were the nature of time then a thing could change only by being in one state during one time-atom, and in a different state during the next time-atom, without ever being partly in the one state and partly in the other. But it is not true conversely that if movement from place to place were atomic in this way then time would also have to be atomic. For it might be the case that movement was constrained to be atomic because *space* was atomic, whereas time was infinitely divisible, in the sense that within any period of time, however small, there was some state of affairs which obtained for part, but not all, of the period. The various versions of atomism are not in fact equivalent to one another, despite Aristotle's claim at 241^a4–6.

241^a2 *we have already proved*: at 231^b18–232^a22.

241^a7 *a point*: it is clear that this paragraph, and the next, treat a point as something with magnitude, i.e. in the same way as the previous paragraph treated 'things with no parts'. The argument collapses if one adopts Aristotle's own view that points have no magnitude.

241^a19 *we have already demonstrated*: in Chs. 1–2.

241^a26–^b20 *Can change be infinite?*: At ^a26–^b2 Aristotle clearly argues that coming to be and ceasing to be cannot be infinite, and nor can change of quality or change of quantity. In each case the argument is just that there must be both a starting-point and a finishing-point for each such change. You would think that the same argument should apply equally to movement from one place to another, even if the different places are not opposites. But Aristotle notes at ^b2–11 that there is the objection that a thing may *be moving* towards a certain place, and may be in this state for all time, if it never gets

there. (Again, you would think that the same objection would apply
to all kinds of change.) He replies, however, that one cannot count
as the goal of a movement a position which it will never reach, and
so apparently he concludes that every change is finite. But in the
final paragraph (^{b}11–20) he evidently wishes to introduce the possibility of an infinite change, namely the eternal rotation of the
heavens. This will be crucial to the argument of Book VIII. He does
not explain here how this eternal change could conform to his
axiom that every change is from a starting-point to a finishing-
point, but he does attempt an explanation in Book VIII (265ᵃ27–ᵇ8).

241ᵃ26 *as we have seen*: this has been a constant assumption. The
reference may be to 224ᵇ35–225ᵃ2 of Book V, or it may reach back
to Ch. 7 of Book I.

241ᵇ1 *the limit consisting in the removal of the thing's natural size*: an
upper limit to increase of size, set by nature, seems highly implausible to us. One might have expected a lower limit to decrease of size
to seem implausible even to Aristotle. But cf. 187ᵇ13–188ᵃ2.

BOOK VII

241ᵇ34–242ᵃ49 *Everything that changes must be changed by something*:
Aristotle apparently means: is changed by something *other than
itself*. This is explicit in version β, which has 'is changed by itself'
in place of 'is not changed by anything' at 241ᵇ43 and ᵇ45, and
which has 'something else' in place of 'something' at 242ᵃ37 and
ᵃ46 (twice). These occurrences are marked with a double dagger (‡).

242ᵃ36 *it must be that*: the principle is: if it is true that when *x* stops
changing so does *y*, then *y*'s change is caused by something (else).
This sounds plausible, because the thought seems to be that *y*'s
change is caused by *x*'s change, or perhaps by some larger change
of which *x*'s change is a part. But Aristotle will apply the principle
to the case when *x* is itself a part of *y*, and in this application the
principle is clearly quite unreasonable.

242ᵃ40 *every changing object is divisible*: this has been argued in Book
VI, first at 234ᵇ10–20 and then at greater length in Ch. 10.

242ᵃ66 *these limits guarantee that it is not infinite*: cf. 241ᵃ26–ᵇ20 at
the end of Book VI.

242ᵇ41 *we have discussed this issue earlier*: in Ch. 4 of Book V. (But
note that the present recapitulation omits the qualification introduced at 227ᵇ14–20.)

242^{b}49 *we are assuming what is possible*: this is a misleading description of the argument, which is this: each movement is caused by a
movement either equal to it or greater than it; hence each movement
is at least as great as the final movement *E*; hence the sum of all
these infinitely many movements must be infinite. We do not need
to make any assumption about whether all the movements are equal,
or whether they constantly increase, or whether some are equal and
some increase. For *all* of the possible assumptions lead to the same
result, i.e. that the sum of the movements is infinite. (The misleading
description is repeated at ^{b}66, and again at the end of the argument
at 243^{a}30.)

242^{b}60 *either in contact with or continuous with*: the definition of 'continuous' in Ch. 3 of Book V ensures that if *x* is continuous with *y*
then *x* must be in contact with *y*, and so there was no need for
Aristotle to distinguish two cases; contact covers both. No doubt
what he is thinking of is the distinction between one body being
pushed by another body, and one part of a body being pushed by
another part of the same body. (Cf. 211^{a}29–34.)

242^{b}65 *the movement will be infinite in any case*: notice that Aristotle
here distinguishes the 'greatness' of the movement from the 'greatness' (in size or weight?) of the moving body.

242^{b}71 *the conclusion is impossible*: Aristotle is apparently relying on
the results of Ch. 7 of Book VI, but illegitimately. Suppose that we
have an infinite number of bricks, laid end to end in a straight line,
with one end here and the others stretching off infinitely to the
right. Suppose also that each brick, pushed by the next, moves 9
inches to the left. This apparently illustrates the situation Aristotle
has in mind at the end of the previous paragraph, and which he says
at 242^{b}53–9 has not yet been proved impossible. So he adds in this
paragraph that all the bricks may be regarded as together forming
some one object, since each touches the next, and that this one
object then 'traverses an infinite magnitude in a finite time' (^{b}70).
But (*a*) it hardly seems fair to say that the infinite line of bricks has
'traversed an infinite magnitude', when all that it has done is shifted
9 inches to the left, and (*b*) even if the locution be allowed, still *this*
kind of 'infinite traversal' is quite different from those proved to be
impossible in Book VI.

243^{b}9 *except for . . . coming to be and ceasing to be*: there appears to
be no good reason for this exception. Aristotle holds (against the
atomists) that there are cases of coming to be which are not in any
way combinations (e.g. of atoms), but he admits that some things
(e.g. houses) come into existence by the combination of their materials

(e.g. bricks, cf. 190^b8). I see no reason to say that *these* combinations cannot be reduced to pushings and pullings, whereas others can.

243^b20 *being pulled or pushed or rotated*: one or other of these alternatives will apply when the carrier is *not* a living thing which moves itself.

244^a9 *the movement of the object doing the pulling is faster*: in an ordinary case (e.g. a horse pulling a cart), the speed of the object pulling is the same as the speed of the object pulled. So Aristotle must mean that the pulling object exercises greater *force* in the one direction than the pulled object exercises in the other, and these two forces are then thought of as *potential* speeds.

244^a12 *wood draws fire*: Aristotle is apparently thinking that in a fire a flame will be drawn towards wood that is not yet ignited. (If this is right, it is an example where an object causing another to move makes contact with it only *at the end* of the movement; one might reasonably compare magnetic attraction. The example contrasts with that of a projectile, as described at 243^a20–^b2, where the contact occurs only *at the beginning* of the movement. But projectiles are reconsidered in greater detail in Book VIII, at 266^b27–267^a20.)

244^b5 *affective qualities*: the phrase means 'qualities that arise in a thing because of the way it is affected (= acted on)', but it is clear from the discussion that Aristotle is using it as a synonym for 'perceptible qualities'. In ch. 8 of the *Categories* he distinguishes four kinds of quality, namely dispositions, capacities, affective qualities, and shapes. He will argue in the next chapter that changes in disposition (= state), and changes of shape, do not count as alterations. (He does not consider changes in capacity.) He argues here that changes in quality which do count as alterations, i.e. changes in an 'affective' quality, all involve contact between the agent of change and the object changed.

244^b5 *every body differs from every other in its perceptible qualities*: this is an unnecessarily strong version of Leibniz's principle of the identity of indiscernibles, i.e. the principle that no two objects share all their qualities. Evidently there are many exceptions to Aristotle's version.

244^b6 *there is an underlying quality of which they are affections*: this claim is obscure, but perhaps Aristotle means that there is a determinable quality (being coloured, having some temperature) of which particular affective qualities are determinates (being pale or dark, being hot or cold). Whatever he means, one cannot see how this claim is supposed to provide a reason for the preceding claim that alteration in a perceptible quality is itself caused by a perceptible quality.

245ᵃ4 *are obviously contiguous in all these cases*: Aristotle supports this claim in what follows by first explaining how it applies to perception, where the organ of perception undergoes some alteration, and next claiming that the same applies when the alteration is not a perception. His point is that some senses are straightforwardly contact-senses, e.g. taste (ᵃ9), and the others require a medium, e.g. light for the perception of colour (ᵃ6–7), and air for the perception of distant heat (ᵃ5–6) and of sounding or smelling objects (ᵃ7–8). The medium is in contact both with the sense-organ that is altered and with the distant object sensed.

245ᵃ14 *the agent of decrease*: what Aristotle says must imply that when the water-level in the pond falls, as the water is evaporated by the sun, the *agent* of this decrease is not the sun but the water.

245ᵇ3–248ᵃ9 *Only perceptible qualities can be altered or can alter other things*: this title for the chapter mirrors what Aristotle claims in the first sentence, but the two points that are actually argued in the chapter are (i) that a change of shape is not an alteration (but a generation), and (ii) that a change of state (or disposition) is not an alteration (but a completion, or its opposite). This second point applies to states of the body (e.g. health), to states of character (e.g. the moral virtues), and to states of the intellect.

245ᵇ10 *we do not . . . describe a statue as bronze*: in English the word 'bronze' functions as both noun and adjective. The point that Aristotle is relying on is that in Greek 'bronze' is a noun and 'bronzen' is the related adjective. So his claim is that the statue is bronzen, i.e. is made of bronze, but it is not the same thing as the bronze it is made of.

245ᵇ16 *we describe both the matter and the affection in the same terms*: I take this to be short for: we describe the matter (the bronze) as liquid, and we also describe the liquid thing as some bronze. Thus it is the same object that is referred to both as '(some) bronze' and as '(the) liquid (thing)', whereas it is *not* the same object that is referred to by the two expressions '(some) bronze' and '(a) statue'.

246ᵃ3 *the coming into being of these . . . is not alteration*: notice (i) that when the bronze acquires a new shape, this could be *both* an alteration of the bronze *and* a generation of the statue (though it is not either a generation of the bronze or an alteration of the statue), and (ii) that there is no reason to say that *every* change of shape must at the same time be the generation of some new substance (consider a lump of Plasticine).

246ᵃ10 *states . . . are not alterations*: taken literally, this is evident. But what Aristotle means is that *the acquisition* (or loss) of a state is not an alteration. That is not at all evident, and the arguments that follow are not very convincing.

246ᵇ3 *good states . . . are . . . relative to something*: Aristotle has argued in Ch. 2 of Book V that coming to stand in a relation to something is not (in itself) a change, but even if this be granted, his conclusion still will not follow. Perhaps when I come to be healthy that is because the various constituents of which I am composed have come to be related to one another in a certain way, and so (on his account) none of *them* need have changed. But it does not follow that *I* have not changed; on the contrary, it is obvious that I have. And elsewhere Aristotle standardly counts becoming healthy as an example of an alteration.

246ᵇ10 *which . . . bring the state into being or destroy it*: the thought is that a good state (such as health) allows me to do certain things (e.g. exercise), and is acquired precisely by practising the doing of those things, and lost by ceasing to do them. Cf. *Nicomachean Ethics* 1103ᵃ26–ᵇ25, 1104ᵃ27–ᵇ3, and Plato, *Republic* 444c–e.

247ᵇ3 *states of this part . . . are not generated*: Aristotle argues first (ᵇ4–9) that the change from having knowledge, but not having it actively before the mind, to having it before the mind, is not a generation or an alteration. This is what he means when he speaks of the potential knower becoming (actually) a knower, or (actually) using his knowledge. His reasons are: (i) that this change from potentiality to actuality is due to a relation to something else, i.e. to the perception of an object of the known type in question; and (ii) that it takes no time to come into existence (and in this way it resembles seeing something, or a contact), whereas he has argued in Ch. 6 of Book VI that every change takes time, including a coming to be (237ᵇ9–22). In the next paragraph (ᵇ9–18) he will turn to consider how one comes to be, in this sense, a potential knower.

247ᵇ13 *as I explained earlier*: in Ch. 2 of Book V, particularly 225ᵇ33–226ᵃ23, Aristotle argued that there is no process by which a change comes into existence. Here he is relying on the point that what holds for change will hold for rest as well. (But note that a part of Ch. 8 of Book VI, i.e. 239ᵃ10–22, was concerned to discuss the process of coming to rest.)

248ᵃ12 *the circumference . . . will be equal to . . . a straight line*: of course the circumference *is* equal to a straight line, and it is astonishing that Aristotle should have thought otherwise. It appears that he is misinterpreting the fact that in his day the problem of *constructing* a straight line equal to a given circumference was unsolved. (And it is in fact insoluble, as was shown in 1882 by F. Lindemann.)

248^{b}19 *'one' probably becomes ambiguous*: elsewhere Aristotle is firmly of the opinion that 'one' *is* ambiguous as at 185^{b}5–7 in Book I. The *locus classicus* is *Metaphysics* Δ6.

248^{b}24 *clarity or quantity*: in Greek as in English both water and speech may be said to be 'clear', but in quite different senses of the word. (In Greek, 'clear speech' is 'easy to hear' rather than 'easy to understand'.) In Greek (but not in English), both water and speech may be assigned a quantity, but again in quite different senses. For the distinction between long and short syllables, which is fundamental to all Greek poetry, is said to be a distinction in 'quantity'.

249^{a}18 *Or perhaps this is wrong*: cf. 227^{b}14–20.

249^{b}23 *if substance is number*: Aristotle frequently credits this obscure thesis to the Pythagoreans, and to Plato's 'unwritten doctrines'.

249^{b}29 *it is causing movement and has caused movement*: this applies to 'causing movement' what was said in Book VI of 'being moved' (236^{b}32–237^{a}17).

250^{a}22 *moved as it fell*: the assumption is that the noise made is proportional to the amount of air moved.

250^{a}24 *no fragment even exists, except potentially*: Aristotle holds that an undetached part of a whole does not actually exist, but will actually exist if it becomes detached. The *locus classicus* is *Metaphysics* Z13, 1039^{a}3–7.

250^{b}7 *which was what we found in the case of weight*: it seems that Aristotle's point is that a force may have to act for a certain minimum time before it can accomplish any alteration at all (cf. 253^{b}14–26). This has some similarity to 'what we found in the case of weight', namely that a force must be greater than some minimum quantity before it can bring about any movement at all. But it is rather loose to say that in each case we have 'the same' qualification.

BOOK VIII

250^{b}18 *those who say*: principally the atomists, and perhaps Anaximander as well.

251^{a}3 *unchanging in a cycle*: Empedocles, Fr.17. 9–13. It appears that Aristotle must be taking the last four words to mean: 'they are periodically changeless', for this quotation will support his point only if they are so understood. But Empedocles himself surely did not mean this, but rather 'they are changelessly periodic'. (The subject 'they' refers to the four elements.)

251ᵃ9 *Our view is*: Aristotle cites the definition of change given in Ch. 1 of Book III, but he uses it to justify the claim that, for any change, the object that undergoes the change must exist beforehand. This claim, especially when applied to changes that are comings to be, would be better supported by a reference to Ch. 7 of Book I.

251ᵃ24 *there must have been some cause of its rest*: while this claim is no doubt entirely reasonable (from Aristotle's point of view), it is not needed for the main argument.

251ᵃ30 *knowledge . . . of both opposites at once*: the paradigm is the doctor, whose knowledge of what will make one well is at the same time knowledge of what will make one ill, so it enables him to act in either direction. (The point is made by Plato at *Republic* i, 333e–334a. Aristotle's main discussion is in *Metaphysics* Θ2.)

251ᵇ12 *if time is . . . or is*: in Book IV Aristotle argued that time is not any kind of change (218ᵇ9–20), but that it is an 'aspect' of change (218ᵇ21–219ᵃ10), and in particular that it is the number of change (219ᵇ2–9). It is therefore somewhat surprising that he should here include an alternative that was ruled out in Book IV.

251ᵇ17 *Plato*: *Timaeus* 28b, 38b.

251ᵇ19 *Anyway*: the argument that follows occurs also in Book IV, at 222ᵃ28–ᵇ7.

251ᵇ28 *The same argument*: this refers to the argument at 251ᵃ17–ᵇ10, not to the more recent argument at 251ᵇ10–28.

251ᵇ31 *does not stop changing and being capable of change at the same time*: the previous argument claimed that before any change starts there must already exist something capable of undergoing it. The symmetrical claim should be: after any change ends there must still exist something capable of having undergone it. But what Aristotle says here cannot be taken in that sense, and it is of course quite possible that what remains after the change should no longer be capable of *undergoing* that change. (To use his own example, when wood is burnt to ashes, the matter in question ceases to be burnable at the same time as it ceases being burned.)

In any case, the thrust of the previous argument was that there must be an explanation of why a change begins when it does, and this explanation has to invoke a *previous* change. Perhaps one could also claim that there must be an explanation of why a change ends when it does, but one surely cannot add that this explanation would have to invoke a *subsequent* change. So on this score there is an objection to a first event which is not matched by a corresponding objection to a last event.

252ᵃ1 *A will still need to cease to exist*: why? Aristotle appears to be supposing that change could cease only if every object ceased to exist. But that is not so, as is shown by our own theory of the so-called 'heat-death' of the universe (i.e. a state when all energy is equalized over the whole universe, so that, by the second law of thermodynamics, no more change is possible).

252ᵃ10 *Anaxagoras . . . would probably share this view*: this is a strange statement, since Anaxagoras held that change had a definite beginning (cf. 203ᵃ28–33). We do not know of anything that he may have said about change having an end.

252ᵃ28 *he should say in what cases this happens*: Aristotle's criticism is this. Empedocles can fairly say that the function of love is to combine, and of hatred to separate, because we can see that this is the effect of the familiar love and hatred between people. If he wishes to add that first one predominates and then the other, then he should point out to us cases where we can also see that this happens (e.g. between people, or perhaps in some other context). For if there is to be a reason to suppose that love and hatred behave globally as Empedocles proposes, the reason could only be a generalization from cases which are already familiar.

252ᵃ34 *Democritus' explanation*: it is a fair point that some regularities may be deduced from others, and only the basic regularities, which are not so deducible, should be taken as principles. But we do not know enough of Democritus' position to be able to say whether this is a fair criticism of him.

252ᵇ27 *it might be true of the infinite too*: 'the infinite' here may mean 'the indefinite', and may refer to the undifferentiated mixture of all things which, according to Anaxagoras, remained at rest until intelligence started a rotation which began the process of separating out.

252ᵇ34 *a different note on different occasions*: the question that should be under consideration is whether, when the same string sounds the same note once more, this is or is not *the same change*. (Aristotle's own answer is that it is not, since he requires a single change to occupy a single and continuous period of time; see e.g. 228ᵃ19–ᵇ10.)

253ᵃ2 *later*: In Ch. 8.

253ᵃ6 *the point . . . is to raise the question*: the question is discussed at length in Ch. 3, and Aristotle treats it as setting the topic for the rest of the book. But the main example that seemed to raise the question is dealt with in the remainder of this chapter.

253ᵃ21 *later*: at 259ᵇ1–16 of Ch. 6.

253ᵃ32 *the idea that everything is at rest*: it may be noted that Aristotle does not refer back to his discussion of this idea in Chs. 2–3 of Book I, though the objection he makes here at ᵇ2–6 repeats that made earlier at 184ᵇ25–185ᵃ17. (A further objection is made later at 254ᵃ23–33.)

253ᵇ9 *some people*: Heraclitus and his followers.

253ᵇ14 *there is also something . . . between them*: Aristotle means that if a thing is first increasing and later decreasing, there must be a time between when it is doing neither. The point will be argued at length in Ch. 8.

253ᵇ18 *men hauling a ship*: cf. 250ᵃ16–19.

253ᵇ30 *the stone . . . does not become harder or softer*: this brief sentence makes a quite different point from that so far developed, namely that it is stupid to claim that a thing is always altering in respect of *every* quality that it has. (It may be said that the point is taken from Plato's *Theaetetus* 181c–183b.)

254ᵃ16 *the arguments used above*: it is not clear which arguments Aristotle is referring to. The view that everything is sometimes at rest and sometimes changing is not refuted until it is established that there is something (namely the heavens) which always moves and never rests, and that task is not completed until Ch. 9.

254ᵃ23 *we have already shown*: at 253ᵃ32–ᵇ6.

254ᵃ25 *some people*: the reference is apparently to Melissus, since he held that 'being is infinite', while Parmenides did not. (But here Aristotle's target is their shared claim that being is unchanging.)

254ᵇ7 *either coincidentally or in their own right*: Cf. v, 224ᵃ21–ᵇ35.

254ᵇ31 *this division explains how the creature . . . moves itself*: the division Aristotle intends is presumably the division into soul or mind (as mover) and body (as object moved), or it might be better to say not the mind itself but that part of the body that is the seat of the mind (i.e. the brain on our theory, but the heart on Aristotle's theory). But it should be noted that in this book he never offers any elucidation of which part of an animal counts as moving the rest of the animal.

255ᵃ34 *not actually putting it to use*: notice that on Aristotle's account one does not count as 'actually knowing' something unless (*a*) one knows it, and (*b*) one is thinking of it. Cf. note on 247ᵇ3.

255ᵇ4 *he will actualize his knowledge*: Aristotle is claiming that if I know something then I will be thinking of it 'as long as nothing stops me'. This is what he needs for his parallel with the natural movement of the elements, for in that case he says that as soon as some air comes into being it will move upwards unless something stops it.

255ᵇ30 *which enables them to be affected*: Aristotle is trying to reconcile his old claim that a natural object contains its own source of movement (ii. 1), and his more recent claim that the natural objects now being considered do not have one part which moves another (255ᵃ5–18). His solution is that this source does not set the thing in motion, but enables it to be moved by something else which removes an obstacle to its motion.

256ᵃ17 *it is impossible for there to be an infinite series*: Ch. 1 of Book VII attempts to give a proof of this claim, but here Aristotle seems to take it to be too obvious to need proof.

256ᵇ2 *we find something which changes itself*: this appears to be a slip. Aristotle is thinking of changes initiated by animals, and here he overlooks the fact that the movement of the heavenly bodies is caused, not by something which is changed by itself, but by something which is never changed at all.

256ᵇ3 *the same conclusion*: the first argument (256ᵃ8–21) aimed to show that a first agent of change, if it was itself changed, must be changed by itself. The second (256ᵃ21–ᵇ3) was presumably intended to reach the same conclusion, but it was incautiously stated. (See previous note.) The argument beginning here, and running to 257ᵃ14, is designed to show that it is not true that everything that changes is changed by something which is itself changing, so for the first time it introduces the idea that there is an unchanging cause of change. The argument proceeds by a dilemma: if the thesis were true, it would be true either coincidentally or in its own right; but the first alternative is (allegedly) refuted at ᵇ7–13, and the second at ᵇ27–ᵃ14. In between Aristotle digresses to point out the reasonableness of the hypothesis that there is an unchanging cause of change.

256ᵇ9 *a time when nothing . . . is changing*: the hypothesis is that it is not always the case that an agent of change *must* itself be changing, or in other words that some agents can cause change without themselves changing. Aristotle himself thinks that this hypothesis is true, for in his view God is an unchanging cause of change. It is therefore quite unclear why he should here suppose that the hypothesis would entail that there could be a time when nothing is changing, for he evidently thinks that that is false.

256^{b}19 *up to a certain point*: this apparently acknowledges that the thrower does not maintain contact with the thing thrown throughout its movement.

257^{a}7 *there is a finite number of kinds of change*: this claim is crucial to the argument (to ensure that each change has in its causal ancestry another change of exactly the same kind), and it surely needs some support.

257^{a}13 *it means that a teacher is learning*: this is a mistake. It means, rather, that somewhere in the causal ancestry of my teaching you how to do long division is another case of one person teaching another how to do long division.

257^{a}23 *the one suggestion . . . and the other*: the first is the suggestion that what can cause a certain change can itself undergo that same change, and the second the suggestion that it must be able to undergo some different change.

257^{a}26 *something which is at rest*: elsewhere Aristotle insists that only what is capable of changing can be said to be 'at rest' (e.g. 234^{a}32–3). His expression here must therefore be counted as somewhat careless.

257^{a}34 *earlier*: at 234^{b}10–20 of Book VI, further elaborated at 240^{b}8–241^{a}26.

257^{b}2 *cannot cause itself to change everywhere*: Aristotle means that it cannot be true that all of the object causes all of itself to change, so that the whole object both causes and undergoes the change. His theme from here to 258^{a}5 is that in a self-changer there must be two parts, one of which causes the other to change.

257^{b}6 *we have established*: at 251^{a}9–16.

257^{b}12 *where the agent . . . has to have the quality it imparts*: see note on 198^{a}27. It would seem that the present case does not fall under this general description, for the change which an animal causes in itself is movement (253^{a}7–21, 259^{b}6–7), and it is not plausible to suppose that the part of an animal which causes it to move is already itself moving with the movement it imparts.

257^{b}15 *there would be no first agent of change*: if x causes y to change, and y's change simultaneously causes x to change, and this change of x's is what is causing y to change, then (i) x is causing x to change, and y is causing y to change, and (ii) neither x nor y is a *first* cause of change, since each is dependent on the other.

257^{b}23 *In the third place*: this argument scarcely differs from that offered 'in the second place'.

258ᵃ11 *C will eventually be reached*: it may be noted that this is a new claim. Aristotle has already stated that when we trace the causes of a given change we shall not find an infinite regress (256ᵃ17–19); he now states that when we trace the effects of a given change we shall again find that the chain is finite.

258ᵃ20 *contact between both parts or from one to the other*: Aristotle apparently holds that, when the mind moves the body, the mind must in a sense 'touch' the body, but the body does not in any sense 'touch' the mind; see *De Generatione et Corruptione* 323ᵃ31.

258ᵃ28 *if A, the unchanging agent . . . is a continuum*: presumably Aristotle does not regard this as a serious possibility. He has prepared the way for rejecting it at ᵃ20–2 above, and when giving his solution at ᵃ32–ᵇ4 he will concentrate on the other problem, that the object of change will be a continuum, and hence divisible.

258ᵇ2 *it would cease to be what it now is*: that is, *AB* may be a primary self-changer, even though *B* is divisible, so long as the change in question could not proceed if *B* were actually divided.

258ᵇ15 *free from . . . coincidental change*: this contrasts with the unchanging cause of change within each animal, for that is subject to coincidental change (259ᵇ16–20).

258ᵇ17 *without undergoing any process of coming into existence*: see note on 237ᵇ11.

259ᵇ6 *only one kind of change*: that is, movement. Cf. 253ᵃ14–15.

259ᵇ20 *moves itself by means of leverage*: Aristotle apparently means that the mind must stay with the body if it is to keep moving it, as a lever must stay in contact with the weight it is moving. But the simile is certainly unexpected.

259ᵇ30 *those whose movement is complex*: the name 'planet' means 'wanderer', and the apparent motion of a planet through the fixed stars is irregular. Aristotle subscribes to an astronomical theory (due to Eudoxus) by which the motion of each planet is constructed from the regular movements of several different spherical shells, nested one inside the other, rotating with different speeds and on axes inclined to one another. The innermost shell carries the planet on its diameter; each outer shell carries the axis of the one next inside it. The apparent motions of the sun and the moon are also slightly irregular, though to a much smaller extent, and the same device was used to explain their anomalies too. Thus, on Aristotle's theory, every heavenly body except a fixed star has a complex movement: the spherical shell which carries it has its own rotation, and is also moved 'coincidentally' by the rotation of the next spherical shell, which carries its axis. (On ancient astronomy generally, see D. R. Dicks, *Early Greek Astronomy to Aristotle* (London: Thames and Hudson, 1970); O. Neugebauer, *The Exact Sciences in Antiquity*, 2nd edn. (New York: Harper Torchbooks, 1962).)

259^b33 *whatever . . . is immediately changed by it*: this refers to the rotation of the sphere of the fixed stars, and apparently to the rotation of every other heavenly sphere as well. (See Introduction, p. lxix.)

260^a5 *anything which is changed by something . . . changed directly by that which is unchanging*: the things 'changed directly by that which is unchanging' are the heavenly spheres (previous note). The things changed by them are the heavenly bodies that they carry, and in particular the sun. The sun 'causes different kinds of change' because it stands in different relations to the earth at different times. In particular, it creates the difference between the seasons of the year, which in turn affects the growth of all plants, the mating of (almost) all animals, and so on.

260^a20 *a fresh start*: in Chs. 7–9 Aristotle argues that rotation is the primary kind of change, and that it is the only kind of change that can be eternal (or, as he says, 'continuous'). This is 'a fresh start' in so far as the argument does not make use of any points already established; but it is a continuation of what we have had already in so far as it has just been claimed that the first agent of all change directly causes an eternal change (259^b32–260^a1), though what that change is has not been specified (except in these notes).

260^a27 *change of size*: this expression here stands in place of the more usual 'change of quantity', and it is one step towards Aristotle's concentration on the rather special case of the natural growth of a living thing. But in fact there are many other changes of quantity which are not covered by the subsequent remarks on growth (e.g. any kind of decrease).

260^b11 *condensation and rarefaction are combination and separation*: this is not Aristotle's own view, as given at 217^a20–^b11 of Book IV, so it is not clear why he feels entitled to assume it here.

260^b19 *the primacy of movement in the first sense*: it has been argued already (primarily in ^a26–^b7, but with two supplementary arguments at ^b7–15) that no other kind of change can exist without movement. So what needs to be established here is that movement can exist without other kinds of change existing. This is indeed stated at ^b26–8, where the example intended is evidently the movement of the heavenly bodies. But the intervening matter at ^b19–26 is irrelevant to 'primacy in the first sense', and in any case it relies upon premises that have not yet been established.

260^{b}24 *later*: in Chs. 8–9, particularly 264^{b}9–28, 265^{a}27–^{b}8.

261^{a}29 *both just now and earlier*: the point was recently assumed at 260^{b}23–4. The 'earlier assumption' referred to is perhaps the claim at 259^{b}32–260^{a}1 that since the first agent of change is eternal the change that it causes must also be eternal, for this presupposes the possibility of an eternal change.

261^{b}6 *the change . . . cannot be continuous*: the argument is this. If we imagine something that changes continually, say from white to black and back again, then if there is no pause between the one change and the other we shall have to say that at the instant which ends the one change and begins the other it is undergoing *both* changes, i.e. it is *both* changing from white to black *and* changing from black to white; but that is impossible. (The reply to this argument, at least *ad hominem*, is that according to Aristotle's own doctrine one cannot speak of an object changing *at an instant* at all. Cf. 237^{a}16.)

261^{b}27–265^{a}12 *Only circular movement can be continuous and eternal*: Since this chapter is rather long and digressive, it may help if I give a brief synopsis. The main point argued is that an eternal movement backwards and forwards on the same straight line does not count as a single continuous movement. Five arguments are given, namely: (i) 261^{b}31–262^{a}12, (ii) 262^{a}12–^{b}8, (iii) 264^{a}7–21, (iv) 264^{a}21–^{b}1, (v) 264^{b}1–6, with a coda at ^{b}6–9. (Of these, (i), (ii), and (v) are said to be 'particularly appropriate to the subject', whereas (iii) and (iv) are called 'more abstract' and 'more general'.) After these five arguments, at 264^{b}9–265^{a}2 the conclusion is drawn that movement in a circle is the only kind of change that can be continuous and eternal, and this brings the chapter to an end (after a brief corollary at 265^{a}2–10). Between arguments (ii) and (iii) there are several digressions, which elaborate the thought behind argument (ii). In detail: at 262^{b}8–263^{a}4 it is noted how this thought will resolve a sophistic paradox, but that this resolution still leaves the main point intact; at 263^{a}4–^{b}9 Aristotle gives a new solution to Zeno's chief paradox on motion; at 263^{b}9–264^{a}6 he attends to the question of how to characterize the instant of change from not being A to being A, and vice versa. (This last digression also contains within itself a further digression on time-atomism, at 263^{b}26–264^{a}4.) The digressions, particularly on Zeno, are of independent interest; but they also serve to reinforce the claim of argument (ii).

261^{b}30 *it follows*: on Aristotle's premises, motion in an ellipse must be compounded from motion in a circle and motion in a straight line.

He is about to argue that motion in a straight line cannot be continuous and eternal. But it evidently does *not* follow that the same applies to elliptical motion.

262ª1 *we have already defined*: in Ch. 4 of Book V.

262ᵇ2 *the two events must have happened at different points of time*: notice that Aristotle is assuming that if *A* has reached *B* and (later) has left *B*, then there must be a first instant when it has reached *B* *and* a first instant when it has left *B*. *Given* this assumption, his conclusion does indeed follow. (There is a similar mistake at 237ª17–25 of Book VI.)

262ᵇ11 *when A is at the point B*: at ª29–30 above, Aristotle has allowed that we can say that a moving object *A* is at a point *B* on its path at an instant, but not for any period of time. Either, then, what he writes here is a slip, and he had meant to say 'when *A* has reached *B*' (as at ᵇ15 and ᵇ18, below), or he is relying on this train of thought: if we say 'when *A* is at *B*, *D* is moving' we imply that *A* is at *B* for a period of time, because we cannot talk of *D moving* at an instant, but only for a period. (The first alternative seems to me more probable.) Cf. 239ª33–ᵇ4.

263ª11 *originally*: at 233ª21–31 of Book VI.

263ᵇ7 *a coincidental property*: if it is part of the definition of a line that it is a continuum, and if a continuum is defined as at 231ª21–ᵇ18 of Book VI, then this property is part of the definition, and so is *not* coincidental.

263ᵇ15 *a stretch of time ACB*: Aristotle means: a stretch *AB* divided by an instant *C* into two parts. He proceeds to call the two parts not *AC* and *CB* (as one might expect), but *A* and *B*.

263ᵇ31 *there must be some process . . . between A and B*: there seems to be no reason why the time-atomist should accept this claim. Of course it is true that if *A* and *B* are indivisible *instants* of time, and are different instants (since the changing object is not white at *A*, but is white at *B*), then they must be separated by a period of time. But the time-atomist's 'atoms' are periods and not instants, and they can perfectly well be consecutive.

264ª5 *the time . . . is no greater*: this says that the time *A with* its end-point *C* is no greater than the time *A without* its end-point *C*.

264ª7 *These arguments*: the digression of 262ᵇ8–264ª6 is now concluded, and Aristotle comes back to his claim that an object which moves backwards and forwards on a straight line is not moving with a single continuous movement. So the arguments he refers to here are those of 261ᵇ31–262ª12 and 262ª12–ᵇ8. He now offers some 'more abstract' arguments for the same conclusion, by which he seems to mean arguments that do not rely upon his account of what a continuum is.

264ᵃ18 *changing from a state which it is not in*: while something is moving from A to C it appears to be changing from a state (being at A) which it is not still in. But the argument Aristotle intends is perhaps this. It has been claimed that if the change is one and continuous, then at the same time as it is moving from A to C it is also moving from C to A, and hence it is changing from a state (being at C) which it is not *yet* in. This does strike one as absurd. (But note that we should get the same, apparently absurd, result by applying Aristotle's reasoning to a movement, in a single direction, from A via C to somewhere else.)

264ᵃ23 *we found that our list ... was exhaustive*: in Ch. 2 of Book V.

264ᵃ26 *not a change which is merely part of a whole*: clearly Aristotle needs to put in this qualification, for if the movement from C to A is merely part of a longer movement from somewhere else, via C, to A, then we cannot suppose that the object must have rested at C. But the qualification spoils his argument, for his opponent is indeed claiming that the movement from C to A is part of a longer movement, namely the movement from A to C and back again.

264ᵇ6 *it did all three at the same time*: the argument needs filling out, in this way. Suppose that at t_1 our object ceases to be white, i.e. becomes not white. Suppose also that at t_2 it has ceased to be not white, i.e. has become white again. Then there is no interval between t_1 and t_2, for any such interval would be a period during which the object remains in the state of being not white, i.e. it would be a period of rest. But that must mean that t_1 and t_2 are the same instant, since any two different instants are separated by a period.

The argument depends upon the fact that white and not-white are strict contradictories, allowing of no intermediate positions. But (*a*) movement from A to B and back again is not like this, and anyway (*b*) Aristotle himself has to reinterpret change between contradictories so that it is 'really' change between contraries (see note on 240ᵃ28), and that reinterpretation would destroy the argument here.

264ᵇ7 *how could ... whiteness and blackness share a limit?*: Aristotle must mean, not the states of being white and being black, but the processes of changing from being white to being black and vice versa. However, he does not here give any further reason for his claim that an end-point limiting the one could not also be a starting-point limiting the other.

264^{b}19 *never covers the same points twice*: presumably Aristotle means
that a circular movement *from A back to A* does not traverse any
point twice, whereas movement on a straight line from *A* back to
A must do so. But as he goes on he writes as though a circular
movement never traverses any point more than once even though it
is continued for ever, and this cannot be defended (cf. 207^{a}4–7).
The relevant distinction is that continued circular movement passes
the same points always in the same direction, whereas continued
oscillation on a line with two end-points must pass the same points
now in one direction and now in the other, as is stated at ^{b}26.

264^{b}23 *opposite movements at the same time*: apparently Aristotle is
citing the conclusion of his third argument, at 264^{a}7–18.

265^{a}3 *those natural scientists*: Heraclitus and his followers.

265^{a}14 *to repeat*: 261^{b}28–9.

265^{a}18 *infinite in this sense*: that is, infinitely long (rather than infinitely
divisible). The point was proved in Ch. 5 of Book III.

265^{a}22 *priority in nature, in definition, and in time*: in Ch. 7, movement
was argued to be prior to other kinds of change in three ways, i.e.
in form, in time, and because it could exist without other kinds of
change existing, but not vice versa. It may be noted that priority in
time is common to both lists, and that priority in nature here corresponds to priority in form earlier (cf. 261^{a}14), but the remaining
priorities do not correspond. One thing is 'prior in definition' to
another when the definition of the first is part of the definition of
the second.

Aristotle's reasoning is difficult to reconstruct. Presumably circular motion is prior to rectilinear motion in *time* because it is eternal,
but the important point here is that it has not begun to exist (not
that it will not cease to exist). But eternity seems not to imply
priority in nature (or form), or priority in definition, and here
Aristotle is apparently relying on the point that circular motion is
'complete' whereas rectilinear motion is either composite (and hence
posterior to its components) or 'incomplete'. But he does not explain what this alleged 'incompleteness' is, or why it has the stated
consequence.

265^{a}27 *to have concluded*: in Ch. 8.

265^{b}5 *there is no point to act as the limit of its movement*: Aristotle's
argument may seem to imply that *no* circular movement can *ever*
stop, which would evidently be going too far. But note that he does
in fact believe this to be true of *natural* circular motion, and he
would have a different explanation for why an unnatural circular

motion should stop, namely that it will stop when force is no longer being applied.

The greater difficulty for him at this point is that he has said over and over again that *all* changes have a starting-point and an end-point, and what the change is is determined (at least partly) by these. It may be debated whether he has an adequate response to this difficulty in his suggestion that circular motion does have a starting-point and an end-point, but these are points that are not on the circular path but at the middle of it.

265b12 *the way things move on a straight line*: that is, in their *natural* motion upwards or downwards.

265b17 *All those who have treated of change*: Aristotle proceeds to cite Empedocles (b19–22), Anaxagoras (b22–3), the atomists (b23–9), Anaximenes (b30–2), and Plato (b32–266a1). (Plato defines the mind as 'that which changes itself' at *Phaedrus* 245c–246a.)

265b26 *as it were change of place*: Aristotle will not allow that it is really change of place, for on his definition of 'place' (iv. 4) an atom in the void does not have a place.

266a1 *the word 'change'*: Aristotle standardly uses the Greek word in question (*kinēsis*) to cover all kinds of change, but in Chs. 1–2 of Book V he limits it to changes in which the same subject persists throughout. (In this context it is translated by 'variation'; see note on 218b19.) Here he suggests that the word applies primarily to movement, and that it will be understood in this way unless, by adding a qualification, the speaker makes it clear that he has some other kind of change (i.e. variation) in mind. This suggestion does not actually fit his own usage, but it is a fair comment on the common usage.

266a6 *We have argued*: this paragraph briefly summarizes the main results of Chs. 1, 5–6, and 7–9.

266a11 *the relevant premisses*: Aristotle proceeds to argue (i) that no finite body can cause change for an infinite time (a12–23), (ii) that no finite body can possess an infinite power (a23–b6), and (iii) that no infinite body can possess a finite power (b6–24). Since he believes that there cannot be an infinite body, proposition (iii) is irrelevant to his argument, and is added only for symmetry. He appears to regard (ii) as generalizing (i), perhaps because an infinite power would be needed *either* to move a finite body for an infinite time *or* to move an infinite body for a finite time, but again it is only the first of these that is relevant to the argument.

266a16 *let A be the agent ... B ... the object ... C ... an infinite ... time*: Aristotle appears to be thinking of a finite agent A moving a finite weight B over some given distance, and the hypothesis is that it takes an infinite time to do so. But (i) it is not clear why he should suppose that there must then be a fraction D of A that moves a corresponding fraction E of B over the same distance in some *part* of the time C; (ii) in any case it is a mistake to suppose that a (proper) part of an infinite time cannot itself be infinite (cf. note on 238ᵃ9 of Book VI); (iii) the case that ultimately concerns us is the case of a finite power causing a spherical shell to rotate eternally; it is odd to suppose that this would be because one part of the power causes one part of the shell to rotate, and another part causes another part to rotate.

266ᵃ25 *infinite power . . . possessed by a finite magnitude*: notice that the argument which follows does not use this premiss stating that it is a finite magnitude that has the infinite power. So, if the argument were valid, it would prove simply that there could not be an infinite power (contrary to Aristotle's claim, implied at 267ᵇ24–5, that the first cause of all change does have an infinite power).

266ᵃ33 *it takes a time AB for a finite power to achieve the same result*: the assumption that there is a finite power which will achieve the same result as the infinite power, but in a longer time, is evidently unwarranted—especially if the result achieved by the infinite power is that the heavens rotate eternally.

266ᵇ13 *I will never exhaust AB*: by hypothesis, AB is an infinite body with a finite power, and BC is a finite part of AB with a finite power. Then by continuing to double the *power* of BC one will exhaust the *power* of AB, since that is finite. This observation destroys the argument.

266ᵇ27 *a certain difficulty concerning movement*: on this discussion, see Introduction, p. lxv.

267ᵇ8 *the movement of the circumference*: that is, the movement of the sphere of the fixed stars, which completes one revolution every 24 hours. No other heavenly sphere has a rotation that is nearly as fast as this.

267ᵇ9 *something . . . moving to cause continuous movement*: Aristotle appears to be arguing that, if x moves y by being in motion itself, e.g. by pushing it or pulling it, then y's movement cannot be strictly continuous, even if it lasts only for a short time. But perhaps we should understand 'continuous' here to mean 'eternal and continuous', as in many preceding passages.

267ᵇ21 *we proved earlier*: Ch. 5 of Book III.

TEXTUAL NOTES

I have translated the text of W. D. Ross, as found in his edition of *Physics* (London: Oxford University Press, 1936), except at the following points, indicated in the text by obelisks (†).

184ᵇ21–2 Reading τὸ γένος ἕν, σχήματι δὲ ἢ εἴδει διαφερούσας, <ἢ γένει διαφερούσας,> ἢ καὶ ἐναντίας (Bostock).

186ᵇ5 Reading ἀλλ' ἄλλα ἐκείνῳ (Bostock).

186ᵇ11–12 Reading σημαίνει <καὶ> ὅπερ ὄν (Natorp).

189ᵃ18 Reading ἐξ ἄλλων with MSS EVS.

190ᵃ35 Retaining καὶ ποτέ with the MSS.

190ᵇ2 Retaining ἄλλα with the MSS.

191ᵃ10 Retaining ἡ ὕλη καί with the MSS.

193ᵇ17–18 Reading εἰς τί οὖν φύεται; οὐχὶ εἰς τὸ ἐξ οὗ ἀλλ' εἰς τὸ εἰς ὅ with the majority of the MSS.

194ᵃ33 Reading ἔτι καὶ ποιοῦσιν (Bostock).

196ᵇ35 Reading τοῦτο <οὐ> τοῦ κομίσασθαι (Bostock).

197ᵃ4–5 Retaining κομιζόμενος with the MSS.

201ᵇ7 Retaining αὕτη with some MSS (both here and at *Metaphysics* 1065ᵇ35).

205ᵃ7–8: Omitting δεῖ . . . αἰσθητόν (Hussey).

205ᵃ25: Retaining καθάπερ εἴρηται πρότερον with some MSS.

205ᵃ25–8 Transposing καὶ . . . κάτω to ᵃ19 (Bostock).

205ᵃ31–2: Retaining πεπεράνθαι ἀναγκαῖον with some MSS.

206ᵃ7: Reading ἑκάστου (Hussey).

207ᵃ11 Reading τὸ κυρίως [, οἷον τὸ] ὅλον, οὗ . . . (Hussey).

216ᵇ16 Reading παρά after a conjecture by Ross.

216ᵇ26 Omitting ἀέρα καὶ ὕδωρ (Hussey).

217ᵃ18–20 Transposing ἀεὶ γὰρ . . . εἰς εὐθύ to ᵃ16 after ποιεῖν (Waterfield).

219ᵇ19 Reading ἡ στιγμή (Owen).

220ᵃ22 Reading ᾗ δ' ἀριθμεῖ [ἀριθμός] (Philoponus, Simplicius).

221ᵃ17 Reading ἐν τῷ χρόνῳ (Torstrik).

222ᵇ16 Omitting μεταβολὴ δὲ πᾶσα φύσει ἐκστατικόν (Hussey).

222ᵇ23 Reading πάσχειν instead of πράττειν (Hussey).

223ᵇ2 Reading ὥσθ' ἑκατέρας after a conjecture by Ross.

225ᵇ12 Reading μεταβάλλοντος <μή> (Schwegler).

225ᵇ30 Reading τυχοῦσαν, δεῖ γὰρ κἀκείνην ἔκ τινος εἰς τι ἕτερον, ὥστε with MS H.

Textual Notes

228^a13 Reading διὰ τὸ οὕτως τῷ ἀριθμῷ <δυοῖν εἶναι> (Bostock).

228^b26 Retaining ποῦ with the MSS.

230^b2 Omitting ταχύ (Bostock).

234^a16 Reading [οὐ] τὸ καθ' αὑτό (Bostock).

236^a24 Reading εἰ δ' with the MSS.

236^b1 Reading εἰς ὅ . . . ἢ καθ' ὅ (Prantl).

239^a5 Omitting ἂν with some MSS.

240^a15–16 Retaining ἴσον . . . ὥς φησιν with the MSS.

244^a10 Retaining τοῦ ἕλκοντος with the MSS.

263^b21–2: Reading καὶ εἰ ἐγίγνετο λευκὸν καὶ εἰ ἐφθείρετο λευκόν (Simplicius).

267^b10 Punctuating with a comma after ἀλλὰ μή (Bostock).

The Oxford World's Classics Website

www.worldsclassics.co.uk

- Browse the full range of Oxford World's Classics online

- Sign up for our monthly e-alert to receive information on new titles

- Read extracts from the Introductions

- Listen to our editors and translators talk about the world's greatest literature with our Oxford World's Classics audio guides

- Join the conversation, follow us on Twitter at OWC_Oxford

- Teachers and lecturers can order inspection copies quickly and simply via our website

www.worldsclassics.co.uk

American Literature

British and Irish Literature

Children's Literature

Classics and Ancient Literature

Colonial Literature

Eastern Literature

European Literature

Gothic Literature

History

Medieval Literature

Oxford English Drama

Poetry

Philosophy

Politics

Religion

The Oxford Shakespeare

A complete list of Oxford World's Classics, including Authors in Context, Oxford English Drama, and the Oxford Shakespeare, is available in the UK from the Marketing Services Department, Oxford University Press, Great Clarendon Street, Oxford OX2 6DP, or visit the website at www.oup.com/uk/worldsclassics.

In the USA, visit www.oup.com/us/owc for a complete title list.

Oxford World's Classics are available from all good bookshops. In case of difficulty, customers in the UK should contact Oxford University Press Bookshop, 116 High Street, Oxford OX1 4BR.